Lecture Notes in Computer Science 7369

Commenced Publication in 1973
Founding and Former Series Editors:
Gerhard Goos, Juris Hartmanis, and Jan van Leeuwen

W0079262

Ferruh Özbudak
Francisco Rodríguez-Henríquez (Eds.)

Arithmetic
of Finite Fields

4th International Workshop, WAIFI 2012
Bochum, Germany, July 16-19, 2012
Proceedings

 Springer

Volume Editors

Ferruh Özbudak
Middle East Technical University
Institute of Applied Mathematics
Ankara, Turkey
E-mail: ozbudak@metu.edu.tr

Francisco Rodríguez-Henríquez
Centro de Investigación y de Estudios
Avanzados del Instituto Politécnico Nacional (CINVESTAV-IPN)
Departamento de Computación
Av. IPN No. 2508, Col. San Pedro Zacatenco, Mexico, D.F. 07360, Mexico
E-mail: francisco@cs.cinvestav.mx

ISSN 0302-9743   e-ISSN 1611-3349
ISBN 978-3-642-31661-6   e-ISBN 978-3-642-31662-3
DOI 10.1007/978-3-642-31662-3
Springer Heidelberg Dordrecht London New York

Library of Congress Control Number: 2012940985

CR Subject Classification (1998): I.1, G.2, E.3, K.6.5, D.4.6, F.2.1

LNCS Sublibrary: SL 1 – Theoretical Computer Science and General Issues

Typesetting: Camera-ready by author, data conversion by Scientific Publishing Services, Chennai, India

Printed on acid-free paper

Springer is part of Springer Science+Business Media (www.springer.com)

Preface

These are the proceedings of WAIFI 2012, the 4th International Workshop on the Arithmetic of Finite Fields, held in Bochum, Germany, during July 16–19, 2012. The three previous editions of this workshop were held in Madrid, Spain (WAIFI 2007), Siena, Italy (WAIFI 2008), and Istanbul, Turkey (WAIFI 2010). Since 2008, WAIFI has been held every even year, bringing together mathematicians, computer scientists, engineers and physicists who conduct research in different areas of finite field arithmetic. WAIFI 2012 was organized by the Ruhr-Universität Bochum, Germany, in cooperation with the International Association for Cryptologic Research (IACR). The General Chair of the conference was Christopher Wolf.

The program consisted of four invited talks and 13 contributed papers. The invited speakers were Shay Gueron (University of Hafia, Israel), Florian Hess (Universität Oldenburg, Germany), Alexander Pott (Universität Magdeburg, Germany), and Emmanuel Thome (INRIA, France). The papers supporting the four invited talks were also included in the proceedings. The contributed talks were selected from 29 submissions each of which was assigned to at least three committee members. Additionally, the Program Committee had a significant online discussion phase for several days.

We are very grateful to the Program Committee members and to the external reviewers for their dedication and profesionalism. Special thanks go out to Christopher Wolf, the General Chair, for his hard work in leading the overall organization and dealing with various local arrangements with meticulous care. We would also like to thank Jean-Jacques Quisquater and Çetin Kaya Koç, who helped us to negotiate the publication of WAIFI 2012 proceedings as a volume of *Lecture Notes in Computer Science*. We are also very grateful to José Luis Imaña for dilingently maintaining the workshop website. We heartily thank the members of the Steering Committee of the workshop series for their constant support and encouragement.

The submission and selection of papers were done using the EasyChair conference management system. Hence, thank you EasyChair! Finally, but most importantly, we deeply thank all the authors who submitted their papers to the workshop and the participants all over the world who chose to honor us with their attendance.

July 2012
Ferruh Özbudak
Francisco Rodríguez-Henríquez

WAIFI 2012

International Workshop on the Arithmetic of Finite Fields

Bochum, Germany
July 16–19, 2012

Organized by
Ruhr-Universität Bochum

In cooperation with
The International Association for Cryptologic Research (IACR)

Steering Committee

Claude Carlet   University of Paris 8, France
Jean-Pierre Deschamps   University Rovira i Virgili, Spain
José Luis Imaña   Complutense University of Madrid, Spain
Çetin Kaya Koç   University of California Santa Barbara, USA
Christof Paar   Ruhr-Universität Bochum, Germany
Jean-Jacques Quisquater   Université Catholique de Louvain, Belgium
Berk Sunar   Worcester Polytechnic Institute, USA
Gustavo Sutter   Autonomous University of Madrid, Spain

General Chair

Christopher Wolf   Ruhr-Universität Bochum, Germany

Program Chairs

Ferruh Özbudak   Middle East Technical University, Turkey
Francisco Rodríguez-Henríquez   CINVESTAV-IPN, México

Local Organizing Committee

Marina Efimenko   Ruhr-Universität Bochum, Germany
Sebastian Uellenbeck   Ruhr-Universität Bochum, Germany
Christian Walter   Ruhr-Universität Bochum, Germany

Program Committee

Jean-Claude Bajard	LIP6 CNRS/Université Pierre et Marie Curie, France
Stephane Ballet	Institut de Mathématiques de Luminy, France
Jean-Luc Beuchat	University of Tsukuba, Japan
Luca Breveglieri	Politecnico di Milano, Italy
Debrup Chakraborty	CINVESTAV-IPN, Mexico
Ricardo Dahab	University of Campinas, Brazil
Jérémie Detrey	INRIA, France
Haining Fan	Tsinghua University, China
Olav Geil	Aalborg University, Denmark
Guang Gong	University of Waterloo, Canada
Jorge Guajardo	Robert Bosch LLC, USA
Anwar Hasan	University of Waterloo, Canada
Tor Helleseth	University of Bergen, Norway
José L. Imaña	Complutense University of Madrid, Spain
Koray Karabina	University of Waterloo, Canada
Alexander Kholosha	University of Bergen, Norway
Tanja Lange	Technical University of Eindhoven, The Netherlands
Ivan Landjev	Bulgarian Academy of Sciences, Bulgaria
Julio López	University of Campinas, Brazil
Edgar Martínez-Moro	University of Valladollid, Spain
Gary Mullen	Pennsylvania State University, USA
Harald Niederreiter	Austrian Academy of Sciences, Austria
Arash Reyhani-Masoleh	University of Western Ontario, Canada
Erkay Savaş	Sabanci University, Turkey
Peter Schwabe	Academia Sinica, Taiwan
Igor Semaev	University of Bergen, Norway
Patrick Solé	Télécom ParisTech, France and AbdelAziz University, Saudi Arabia
Arne Winterhof	Austrian Academy of Sciences, Austria

External Reviewers

Diego Aranha
Jean-Philippe Aumasson
Selçuk Baktır
Razvan Barbulescu
Daniel Bernstein
Alessandro Barenghi
Qi Chai
Fernando Hernando
Hans Hüttel

Stéphane Louboutin
Cuauhtemoc Mancillas-López
Marc Mouffron
Mehran Mozaffari-Kermani
Christophe Negre
Matthew Parker
Gerardo Pelosi
Christiane Peters
Thomas Plantard

Damien Robert
Robert Rolland
Sumanta Sarkar

Reza Sohizadeh
Zilong Wang
Yang Yang

Sponsoring Institutions

Ruhr-Universität Bochum, Germany
Mercator Foundation, Essen, Germany

Table of Contents

Invited Talk 3

Finite Field Arithmetic

Equations and Functions

Invited Talk 4

Polynomial Factorization and Permutation Polynomial

Generalised Jacobians in Cryptography and Coding Theory

Florian Hess

Carl-von-Ossietzky Universität Oldenburg, Germany
http://www.staff.uni-oldenburg.de/florian.hess

Abstract. The use of generalised Jacobians in discrete logarithm based cryptosystems has so far been rather limited since they offer no advantage over traditional discrete logarithm based systems. In this paper we continue the search for possible applications in two directions.

Firstly, we investigate pairings on generalised Jacobians and show that these are insecure. Secondly, generalising and extending prior work, we show how the discrete logarithm problem in generalised Jacobians can be reduced to the minimal non zero weight word and maximum likelihood decoding problems in generalised algebraic geometric codes.

1 Introduction and Summary

The multiplicative group of finite fields and the point groups of Jacobian varieties of regular algebraic curves over finite fields, in particular point groups of elliptic curves, are the essential traditional building blocks of discrete logarithm based cryptography. Generalised Jacobian varieties of regular algebraic curves over finite fields can be thought of a combination of both of these traditional building blocks into one mathematical structure. Their use in cryptography has first been suggested in [6]. Soon thereafter, the lack in efficiency in comparison with multiplicative groups of finite fields and point groups of Jacobian varieties has been discussed in [10] and [7]. As a consequence, the use of generalised Jacobians in discrete logarithm based cryptosystems has so far been rather limited. In this paper we continue the search for possible interesting and useful applications of generalised Jacobians in two directions.

The first direction concerns pairings. Pairings have been a major topic in cryptography over the past decade. The only known generally suitable pairings are the Weil and Tate-Lichtenbaum pairings on Jacobians, and these pairings are bilinear. The construction of pairings in different mathematical contexts and the construction of multilinear pairings are important open research problems. Motivated by this we investigate a generalised form of the Tate-Lichtenbaum pairing on generalised Jacobians. As it turns out, these pairings are efficiently computable and also yield multilinear pairings. On the other hand we also have to show that the domain of these pairings suffers from a weak discrete logarithm problem, whence we do not get a useful application to cryptography.

F. Özbudak and F. Rodríguez-Henríquez (Eds.): WAIFI 2012, LNCS 7369, pp. 1–15, 2012.

The second direction concerns reductions of discrete logarithm problems to code based problems. Motivated by [3,5] we define some generalised algebraic-geometric codes and show how the discrete logarithm problem in generalised Jacobians can be reduced to the minimal non zero weight word problem and to the maximum likelihood decoding problem in these generalised algebraic-geometric codes. An essential element of this reduction is the construction of efficient sets of generators for generalised Jacobians. Following the methodology of [17] we prove a theorem on the existence (and efficient construction) of such efficient sets of generators. The discrete logarithm reductions of [3,5] then turn out to be essentially special cases of our general reduction.

The implication of the discrete logarithm reductions are not so clear at the moment. One point of view is that we obtain hardness results for the above mentioned code based computational problems. In this direction a further study of efficient generating sets and of special high genus curves could be of interest for the construction of codes over $\mathbb{F}_q$ with small q and large lower complexity bounds for the minimal non zero word or maximum likelihood decoding problems. Another point of view is that efficient algorithms for solving these code based computational problems might speed up algorithms for computing discrete logarithms. This latter point of view is for example taken in [1].

A different, but related way of combining the multiplicative group of finite fields and the point groups of Jacobian varieties of regular algebraic curves over finite fields into one mathematical structure is to consider Jacobians of singular algebraic curves over finite fields. This approach is taken in [18] where it is shown that one can achieve a compressed representation of finite fields elements from certain subgroups in this way. We do not consider Jacobians of singular curves in this paper though.

2   Preliminaries

Let C be an absolutely irreducible complete regular curve defined over the finite field $\mathbb{F}_q$ of characteristic p. The function field of C is denoted by $\kappa(C)$ and the genus of C by g.

We consider divisors as finite sums of places (closed points) of C. Let $\mathfrak{m}$ be an effective divisor of C. The group of divisors of C coprime to $\mathfrak{m}$ is denoted by $\mathcal{D}^{\mathfrak{m}}(C)$, and the subgroup of $\mathcal{D}^{\mathfrak{m}}(C)$ of principal divisors by $\mathcal{P}^{\mathfrak{m}}(C)$. Let $f \in \kappa(C)^{\times}$. Then by definition $f \equiv 1 \bmod \mathfrak{m}$ if $\mathrm{ord}_{\mathfrak{p}}(f - 1) \geq 1$ for all $\mathfrak{p} \in \mathrm{supp}(\mathfrak{m})$. The ray modulo $\mathfrak{m}$ is defined as

$$\mathcal{P}_{\mathfrak{m}}(C) = \{f \in \kappa(C)^{\times} \mid f \equiv 1 \bmod \mathfrak{m}\}.$$

The ray class group modulo $\mathfrak{m}$ is

$$\mathrm{Pic}_{\mathfrak{m}}(C) = \mathcal{D}^{\mathfrak{m}}(C)/\mathcal{P}_{\mathfrak{m}}(C)$$

and the degree zero ray class group modulo $\mathfrak{m}$ is

$$\mathrm{Pic}_{\mathfrak{m}}^0(C) = \{y \in \mathrm{Pic}_{\mathfrak{m}}(C) \mid \deg(y) = 0\}.$$

The generalised Jacobian of C modulo $\mathfrak{m}$ is a semi-abelian variety $\mathrm{Jac}_{\mathfrak{m}}(C)$ that represents the functor $k \mapsto \mathrm{Pic}^0_{\mathfrak{m}}(C \times k)$ from $\mathbb{F}_q$-algebras to finite abelian groups. The generalised Jacobian $\mathrm{Jac}_{\mathfrak{m}}(C)$ exists for any C and $\mathfrak{m}$, see [20]. In this paper we will not (directly) use the geometric structure of $\mathrm{Jac}_{\mathfrak{m}}(C)$, and thus will essentially only consider the groups $\mathrm{Pic}^0_{\mathfrak{m}}(C)$ and $\mathrm{Pic}_{\mathfrak{m}}(C)$. For $\mathfrak{m} = 0$ we have $\mathrm{Jac}_{\mathfrak{m}}(C) = \mathrm{Jac}(C)$, $\mathrm{Pic}^0_{\mathfrak{m}}(C) = \mathrm{Pic}^0(C)$ and $\mathrm{Pic}_{\mathfrak{m}}(C) = \mathrm{Pic}(C)$ as usual.

By the approximation theorem there is an exact sequence

$$1 \to \mathbb{F}_q^{\times} \to \prod_{\mathfrak{p} \in \mathrm{supp}(\mathfrak{m})} \left(\mathcal{O}_{C,\mathfrak{p}} / \mathfrak{p}^{\mathrm{ord}_{\mathfrak{p}}(\mathfrak{m})} \right)^{\times} \to \mathrm{Pic}^0_{\mathfrak{m}}(C) \to \mathrm{Pic}^0(C) \to 1. \qquad (1)$$

Moreover,

$$\left(\mathcal{O}_{C,\mathfrak{p}} / \mathfrak{p}^{\mathrm{ord}_{\mathfrak{p}}(\mathfrak{m})} \right)^{\times} \cong \kappa(\mathfrak{p})^{\times} \times (1 + \mathfrak{p})/(1 + \mathfrak{p}^{\mathrm{ord}_{\mathfrak{p}}(\mathfrak{m})}), \qquad (2)$$

where $\kappa(\mathfrak{p}) = \mathcal{O}_{C,\mathfrak{p}}/\mathfrak{p}$ is the residue class field of $\mathfrak{p}$ and $(1 + \mathfrak{p})/(1 + \mathfrak{p}^{\mathrm{ord}_{\mathfrak{p}}(\mathfrak{m})})$ is a p-group. Some details about these definitions and facts can be found in [15].

In discrete logarithm based cryptography one considers cyclic groups G of prime order ℓ that are embedded into suitable algebraic groups. The standard cases are

$$G \subseteq \mathbb{G}_m(\mathbb{F}_q) = \mathbb{F}_q^{\times}, \text{ and } G \subseteq \mathrm{Jac}(C)(\mathbb{F}_q) \text{ or } G \subseteq \mathrm{Pic}^0(C)$$

respectively. The use of

$$G \subseteq \mathrm{Jac}_{\mathfrak{m}}(C)(\mathbb{F}_q)$$

has first been suggested in [6]. Since the group laws and maps in (1) and (2) are effective and G has prime order, this case leads to an effectively computable isomorphism $\phi : G \to H$ with either

$$H \subseteq \mathrm{Jac}(C)(\mathbb{F}_q), \ H \subseteq \mathbb{G}_m(\kappa(\mathfrak{p})) \text{ or } H \subseteq (1 + \mathfrak{p})/(1 + \mathfrak{p}^{\mathrm{ord}_{\mathfrak{p}}(\mathfrak{m})})$$

for some $\mathfrak{p} \in \mathrm{supp}(\mathfrak{m})$. Since the discrete logarithm problem in $(1 + \mathfrak{p})/(1 + \mathfrak{p}^{\mathrm{ord}_{\mathfrak{p}}(\mathfrak{m})})$ is easy (see for example [15]) the use of $G \subseteq \mathrm{Jac}_{\mathfrak{m}}(C)(\mathbb{F}_q)$ offers no advantage over the standard cases in terms of efficiency and security. This argumentation is carried out in more detail in [10]. For a discussion from a slightly different point of view see [7].

On the other hand, given $G \subseteq \mathbb{G}_m(\kappa(\mathfrak{p}))$ there can exist efficiently computable isomorphisms $\phi : G \to H$ with $H \subseteq \mathrm{Jac}_{\mathfrak{m}}(C)(\mathbb{F}_q)$ or $H \subseteq \mathrm{Pic}^0_{\mathfrak{m}}(C)$ respectively. Such H are embedded in a context with richer or at least different structure than the context of G. In the following we investigate whether this structure can be used to obtain some new features relevant for cryptography or coding theory.

3 Pairings on Generalised Jacobians

Class field theory yields arithmetic duality pairings on generalised Jacobians. In this section we argue that these pairings indeed yield efficiently computable pairings different from the pairings that have been used in cryptography so far. These pairings can even be applied iteratively, resulting in multilinear pairings. As demonstrated in [2,19], non-degenerate multilinear pairings would be of great interest in cryptography, but no suitable such pairings have been found so far.

Unfortunately, we also have to argue that the domains of these pairings suffer from a weak discrete logarithm problem. The construction is thus not suitable for cryptography.

The construction of multilinear pairings is also discussed in [16] using a different approach. Contrary to our case, the obstacle in this work appears to be the efficient computability of the pairing.

3.1   A Generalised Tate-Lichtenbaum Pairing

Let n be a prime number such that $q \equiv 1 \bmod n$ and abbreviate $F = \mathbb{F}_q(C)$. Let

$$\mathrm{Sel}_{n,\mathfrak{m}}(C) = \{f \in F^\times \mid \mathrm{ord}_\mathfrak{p}(f) \equiv 0 \bmod n \text{ for all places } \mathfrak{p} \notin \mathrm{supp}(\mathfrak{m})\}$$

and denote the group of n-th root of unity of $\mathbb{F}_q^\times$ by μ_n. Class field theory yields the existence of a surjective pairing

$$\pi_\mathfrak{m} : \mathrm{Sel}_{n,\mathfrak{m}}(C) \times \mathrm{Pic}_\mathfrak{m}^0(C) \to \mu_n.$$

The left and right kernel of this pairing are $\mathbb{F}_q^\times \cdot (F^\times)^n$ and $n \cdot \mathrm{Pic}_\mathfrak{m}^0(C)$ respectively. Factoring out kernels thus yields a non-degenerate pairing

$$\pi_\mathfrak{m}' : \mathrm{Sel}_{n,\mathfrak{m}}(C) \,/\, \mathbb{F}_q^\times \cdot (F^\times)^n \times \mathrm{Pic}_\mathfrak{m}^0(C) \,/\, n \cdot \mathrm{Pic}_\mathfrak{m}^0(C) \to \mu_n.$$

A result of Hasse [12] on the algebraic representation of the Artin map gives an algebraic representation of the images under these pairings by means of function evaluation. The details of this are as follows.

Let $\mathfrak{d} = \sum_i \lambda_i \mathfrak{p}_i$ be a divisor of C where the $\mathfrak{p}_i$ are places. If $f \in F$ is a function with no pole at the place $\mathfrak{p}$ then $f + \mathfrak{p}$ denotes the image of f in the residue class field $\kappa(\mathfrak{p}) = \mathcal{O}_{C,\mathfrak{p}}/\mathfrak{p}$. Using the norm map $\mathrm{N}_{\kappa(\mathfrak{p}_i)/\mathbb{F}_q}$ of the extension $\kappa(\mathfrak{p}_i)/\mathbb{F}_q$ we can define an evaluation at divisors via

$$f(\mathfrak{d}) := \prod_i \mathrm{N}_{\kappa(\mathfrak{p}_i)/\mathbb{F}_q}(f + \mathfrak{p}_i)^{\lambda_i},$$

provided f has no zero at $\mathfrak{p}_i$ when $\lambda_i < 0$ and no pole when $\lambda_i > 0$. The evaluation map is multiplicative in f and additive in $\mathfrak{d}$.

Let $x = f \in \mathrm{Sel}_{n,\mathfrak{m}}(C)$ and $y = \mathfrak{d} + \mathcal{P}_\mathfrak{m}(C) \in \mathrm{Pic}_\mathfrak{m}^0(C)$. By the approximation theorem applied to f or $\mathfrak{d}$, f can be chosen modulo $(F^\times)^n$ or $\mathfrak{d}$ modulo $\mathcal{P}_\mathfrak{m}(C)$ such that $\mathrm{supp}(f) \cap \mathrm{supp}(\mathfrak{d}) = \emptyset$. Then

$$\pi_\mathfrak{m}(x,y) = f(\mathfrak{d})^{(q-1)/n}.$$

This definition also allows an efficient computation of $\pi_\mathfrak{m}(x,y)$. Using

$$\mathrm{Sel}_{n,0}(C) \cong \mathrm{Pic}^0(C)[n]$$

under $f \mapsto \frac{1}{n}\mathrm{div}(f) + \mathcal{P}_0(C)$ we see that π_0 is essentially the usual Tate-Lichtenbaum pairing as discussed in [9,14] and widely used in cryptography.

3.2   Variations – New Bilinear and Multilinear Pairings

The case $\mathrm{supp}(\mathfrak{m}) = \{\}$ essentially yields the usual and well known Tate-Lichtenbaum pairing. Next suppose $\mathrm{supp}(\mathfrak{m}) = \{\mathfrak{p}\}$. If $\deg(\mathfrak{p}) \not\equiv 0 \bmod n$ then the group $\prod_{\mathfrak{p} \in \mathrm{supp}(\mathfrak{m})} \left(\mathcal{O}_{C,\mathfrak{p}}/\mathfrak{p}^{\mathrm{ord}_{\mathfrak{p}}(\mathfrak{m})}\right)^{\times}$ does not contain elements of n-power order other than those coming from $\mathbb{F}_q^{\times}$, and by (1) we are again reduced to the case of a Tate-Lichtenbaum pairing. If on the other hand $\deg(\mathfrak{p}) \equiv 0 \bmod n$ then $\pi_{\mathfrak{m}}$ cannot be efficiently computed anymore for cryptographic sizes of n.

So suppose now $\#\mathrm{supp}(\mathfrak{m}) \geq 2$. Then there are various choices of monomorphisms

$$\chi : \mu_n \to \mu_n^{\mathrm{supp}(\mathfrak{m})} = \prod_{\mathfrak{p} \in \mathrm{supp}(\mathfrak{m})} \mu_n$$

that extend according to $\mu_n \subseteq \mathbb{F}_q^{\times} \subseteq \kappa(\mathfrak{p})^{\times}$, (1) and (2) to monomorphisms

$$\phi : \mu_n \to \mathrm{Pic}_{\mathfrak{m}}^0(C).$$

If χ or C are suitably chosen, for example such that $\#\mathrm{Pic}_{\mathfrak{m}}^0(C) \not\equiv 0 \bmod n$, then we can also have that ϕ extends further to a monomorphism

$$\phi' : \mu_n \to \mathrm{Pic}_{\mathfrak{m}}^0(C) \, / \, n \cdot \mathrm{Pic}_{\mathfrak{m}}^0(C).$$

We obtain a surjective pairing

$$\psi : \mathrm{Sel}_{n,\mathfrak{m}}(C) \times \mu_n \to \mu_n$$

given by $\psi(x, y) = \pi_{\mathfrak{m}}(x, \phi(y))$. The right kernel is zero and the left kernel is a subgroup U with $U \supseteq \mathbb{F}_q^{\times} \cdot (F^{\times})^n$. Factoring out U yields a non-degenerate pairing

$$\psi' : \mathrm{Sel}_{n,\mathfrak{m}}(C)/U \times \mu_n \to \mu_n.$$

This pairing has the remarkable feature that pairing values can serve as second arguments. The r-fold iterated combination gives rise to a non-degenerate multilinear pairing

$$\psi'_r : \mathrm{Sel}_{n,\mathfrak{m}}(C)/U \times \cdots \times \mathrm{Sel}_{n,\mathfrak{m}}(C)/U \times \mu_n \to \mu_n$$

in $r + 1$ arguments.

The pairing ψ'_r is efficiently computable if the arguments from $\mathrm{Sel}_{n,\mathfrak{m}}(C)/U$ are for example represented as elements from $F^{\times}$ in a compact representation as power product of elements of $F^{\times}$ of small degree (see for example [13]).

3.3   Weak Discrete Logarithm Problem in the Domain

Suppose that the elements of $\mathrm{Sel}_{n,\mathfrak{m}}(C)/U$ are given by representatives in the group $\mathrm{Sel}_{n,\mathfrak{m}}(C)$. We will now argue that $\mathrm{Sel}_{n,\mathfrak{m}}(C)/U$ then has weak discrete logarithm problem. This implies, unfortunately, that the pairings ψ and ψ' are not useful in cryptography. A consequence of [11] is that all homomorphisms

$\mathrm{Sel}_{n,\mathfrak{m}}(C)/U \to \mu_n$, defined by their image on a generator, would necessarily be efficiently computable as well. This already hints a weak discrete logarithm problem.

We introduce some convenient notation. If $\zeta = (\zeta_{\mathfrak{p}})_{\mathfrak{p} \in \mathrm{supp}(\mathfrak{m})} \in \mu_n^{\mathrm{supp}(\mathfrak{m})}$ and $x = (x_{\mathfrak{p}})_{\mathfrak{p} \in \mathrm{supp}(\mathfrak{m})} \in \mathbb{Z}^{\mathrm{supp}(\mathfrak{m})}$ then let

$$\zeta^x = \prod_{\mathfrak{p} \in \mathrm{supp}(\mathfrak{m})} \zeta_{\mathfrak{p}}^{x_{\mathfrak{p}}} \in \mu_n.$$

If $f \in F^{\times}$ then let $\mathrm{ord}(f) = (\mathrm{ord}_{\mathfrak{p}}(f) \deg(\mathfrak{p}))_{\mathfrak{p} \in \mathrm{supp}(\mathfrak{m})}$.

Theorem 1. *Let* $f \in \mathrm{Sel}_{n,\mathfrak{m}}(C)$ *and* $\zeta \in \mu_n$. *Then*

$$\psi(f, \zeta) = \chi(\zeta)^{\mathrm{ord}(f)}.$$

Proof. Let $\phi(\zeta) = \mathfrak{d} + \mathcal{P}_{\mathfrak{m}}(C)$. A closer look at the third map in (1) shows that $\mathfrak{d} = \mathrm{div}(g)$ for some $g \in F^{\times}$ with $g \equiv \chi_{\mathfrak{p}}(\zeta) \bmod \mathfrak{p}$ for all $\mathfrak{p} \in \mathrm{supp}(\mathfrak{m})$. Then

$$\psi(f, \zeta) = f(\mathfrak{d})^{(q-1)/n} = f(\mathrm{div}(g))^{(q-1)/n} = g(\mathrm{div}(f))^{(q-1)/n},$$

where the last equality holds by Weil reciprocity. Furthermore

$$g(\mathfrak{p}) = \mathrm{N}_{\kappa(\mathfrak{p})/\mathbb{F}_q}(\chi_{\mathfrak{p}}(\zeta)) = \chi_{\mathfrak{p}}(\zeta)^{\deg(\mathfrak{p})}$$

since $\chi_{\mathfrak{p}}(\zeta) \in \mathbb{F}_q$. Now $\mathrm{ord}_{\mathfrak{p}}(f) \equiv 0 \bmod n$ for all $\mathfrak{p} \notin \mathrm{supp}(\mathfrak{m})$, so we get

$$g(\mathrm{div}(f))^{(q-1)/n} = \prod_{\mathfrak{p} \in \mathrm{supp}(\mathfrak{m})} g(\mathfrak{p})^{\mathrm{ord}_{\mathfrak{p}}(f)} = \prod_{\mathfrak{p} \in \mathrm{supp}(\mathfrak{m})} \chi_{\mathfrak{p}}(\zeta)^{\mathrm{ord}_{\mathfrak{p}}(f) \deg(\mathfrak{p})}$$

$$= \chi(\zeta)^{\mathrm{ord}(f)}$$

as desired.

The ord map can be seen as a homomorphism

$$\mathrm{ord} : \mathrm{Sel}_{n,\mathfrak{m}}(C) \to \mathbb{Z}^{\mathrm{supp}(\mathfrak{m})}.$$

Let $G = \mathrm{ord}(\mathrm{Sel}_{n,\mathfrak{m}}(C))$ and $H = \mathrm{ord}(U)$. Since $\ker(\mathrm{ord}) \subseteq U$ by Theorem 1 it is clear that

$$\mathrm{Sel}_{n,\mathfrak{m}}(C)/U \cong G/H \tag{3}$$

via ord and that ψ is equivalent using ord to the pairing

$$\omega : G \times \mu_n \to \mu_n, \quad (x, \zeta) \mapsto \chi(\zeta)^x$$

with right kernel one and left kernel H. We can extend ω to the pairing

$$\omega_1 : \mathbb{Z}^{\mathrm{supp}(\mathfrak{m})} \times \mu_n \to \mu_n, \quad (x, \zeta) \mapsto \chi(\zeta)^x$$

with right kernel one and left kernel denoted by H_1. Then $H = G \cap H_1$ and

$$G/H \cong \mathbb{Z}^{\mathrm{supp}(\mathfrak{m})}/H_1 \tag{4}$$

via the inclusion $G \subseteq \mathbb{Z}^{\mathrm{supp}(\mathfrak{m})}$. Now H_1 can be easily computed from χ and the factor group $\mathbb{Z}^{\mathrm{supp}(\mathfrak{m})}/H_1$ has a weak discrete logarithm problem. Combining the effective isomorphisms of (3) and (4) shows that $\mathrm{Sel}_{n,\mathfrak{m}}(C)/U$ has weak discrete logarithm problem. We thus conclude that these pairings are indeed not useful in cryptography.

4 Lower Complexity Bounds for Codes

We continue the quest for a useful application of generalised Jacobians looking at coding theory. In [5], a reduction of the discrete logarithm problem in the multiplicative group of finite fields to the problem of maximum likelihood decoding in an associated Reed-Solomon code is given. In [3], a reduction of the discrete logarithm problem in elliptic curves to the problem of computing a non zero word of minimal weight in an associated elliptic code is given.

We now adapt and extend some of the ideas of [3,5] to the case of generalised Jacobians. As it turns out, the discrete logarithm reductions of [3,5] can both be seen as special cases of our general framework.

4.1 Generalised Algebraic Geometric Codes

Let S be a finite set of pairwise coprime effective divisors of C that are also coprime to $\mathfrak{m}$. Let $\mathfrak{a}$ and $\mathfrak{b}$ be divisors of C that are of the form $\mathfrak{a} = \sum_{\mathfrak{d} \in S} \lambda_{\mathfrak{d}} \mathfrak{d}$ and $\mathfrak{b} = \sum_{\mathfrak{d} \in S} \mu_{\mathfrak{d}} \mathfrak{d}$ with $\lambda_{\mathfrak{d}}, \mu_{\mathfrak{d}} \in \mathbb{Z}$ such that $\mu_{\mathfrak{d}} > -\lambda_{\mathfrak{d}}$ for all $\mathfrak{d} \in S$ and thus $\mathfrak{a} + \mathfrak{b} \geq 0$. Let $\operatorname{ord}_{\mathfrak{d}}(\mathfrak{e})$ denote the maximal integer r such that $r\mathfrak{d} \leq \mathfrak{e}$ for any divisor $\mathfrak{e}$. Then $\mathfrak{a} = \sum_{\mathfrak{d} \in S} \operatorname{ord}_{\mathfrak{d}}(\mathfrak{a})\mathfrak{d}$ and $\mathfrak{b} = \sum_{\mathfrak{d} \in S} \operatorname{ord}_{\mathfrak{d}}(\mathfrak{b})\mathfrak{d}$. Define

$$\mathcal{O}_{C,\mathfrak{d}} = \mathcal{O}_C(\operatorname{supp}(\mathfrak{d})) \quad \text{and} \quad M_{\mathfrak{d},\mathfrak{a},\mathfrak{b}} = (\mathcal{O}_C(\mathfrak{a})/\mathcal{O}_C(-\mathfrak{b}))(\operatorname{supp}(\mathfrak{d})).$$

Then $\mathcal{O}_{C,\mathfrak{d}} = \cap_{\mathfrak{p} \in \operatorname{supp}(\mathfrak{d})}\mathcal{O}_{C,\mathfrak{p}}$ and $M_{\mathfrak{d},\mathfrak{a},\mathfrak{b}}$ is the factorisation of the fractional ideal of $\mathcal{O}_{C,\mathfrak{d}}$ defined by $\mathfrak{a}$ by the fractional ideal of $\mathcal{O}_{C,\mathfrak{d}}$ defined by $-\mathfrak{b}$. This is an $\mathcal{O}_{C,\mathfrak{d}}$-module and finite $\mathbb{F}_q$-vector space of dimension $\operatorname{ord}_{\mathfrak{d}}(\mathfrak{a}+\mathfrak{b}) \deg(\mathfrak{d})$. For example, if $\mathfrak{d}$ is a place of degree one, this essentially means that we are looking at truncated Laurent series rings starting at the exponent $-\lambda_{\mathfrak{d}} = -\operatorname{ord}_{\mathfrak{d}}(\mathfrak{a})$ and ending at the exponent $\mu_{\mathfrak{d}} = \operatorname{ord}_{\mathfrak{d}}(\mathfrak{b})$. Furthermore, define $U = \cup_{\mathfrak{d} \in S}\operatorname{supp}(\mathfrak{d})$ and

$$\mathcal{O}_{C,S} = \mathcal{O}_C(U) = \cap_{\mathfrak{d} \in S}\mathcal{O}_{C,\mathfrak{d}} \quad \text{and}$$
$$M_{S,\mathfrak{a},\mathfrak{b}} = (\mathcal{O}_C(\mathfrak{a})/\mathcal{O}_C(-\mathfrak{b}))(U) = \prod_{\mathfrak{d} \in S} M_{\mathfrak{d},\mathfrak{a},\mathfrak{b}}.$$

Then $M_{S,\mathfrak{a},\mathfrak{b}}$ is an $\mathcal{O}_{C,S}$-module and finite $\mathbb{F}_q$-vector space of dimension $\deg(\mathfrak{a}+\mathfrak{b})$. The canonical epimorphism $\mathcal{O}_C(\mathfrak{a}) \to \mathcal{O}_C(\mathfrak{a})/\mathcal{O}_C(-\mathfrak{b})$ gives us the "evaluation" map

$$\operatorname{ev}_{S,\mathfrak{a},\mathfrak{b}} : \mathcal{O}_C(\mathfrak{a})(U) \to M_{S,\mathfrak{a},\mathfrak{b}}.$$

It is also the product of the restrictions to $\mathcal{O}_C(\mathfrak{a})(U)$ of the residue class epimorphisms $\mathcal{O}_C(\mathfrak{a})(\operatorname{supp}(\mathfrak{d})) \to M_{\mathfrak{d},\mathfrak{a},\mathfrak{b}}$.

We regard the $\mathfrak{d}$-components of $M_{S,\mathfrak{a},\mathfrak{b}}$ as symbols and the elements of $M_{S,\mathfrak{a},\mathfrak{b}}$ as words. Given $x_{\mathfrak{d}} \in \mathcal{O}_C(\mathfrak{a})(\operatorname{supp}(\mathfrak{d}))$ with $x_{\mathfrak{d}} \neq 0$ let $z_{\mathfrak{d}}(x_{\mathfrak{d}}) = r \deg(\mathfrak{d})$ with r the maximal integer such that $x_{\mathfrak{d}} \in \mathcal{O}_C(\mathfrak{a} - r\mathfrak{d})(\operatorname{supp}(\mathfrak{d}))$. For $x_{\mathfrak{d}} = 0$ we let $z_{\mathfrak{d}}(x_{\mathfrak{d}}) = \infty$. If $x_{\mathfrak{d}} \in M_{\mathfrak{d},\mathfrak{a},\mathfrak{b}}$ we define $z_{\mathfrak{d}}(x_{\mathfrak{d}})$ similarly with the additional maximal value $z_{\mathfrak{d}}(0) = \operatorname{ord}_{\mathfrak{d}}(\mathfrak{a} + \mathfrak{b}) \deg(\mathfrak{d})$. We say that $x_{\mathfrak{d}}$ has a zero of multiplicity $z_{\mathfrak{d}}(x_{\mathfrak{d}})$ at $\mathfrak{d}$. The weight of $x_{\mathfrak{d}} \in M_{\mathfrak{d},\mathfrak{a},\mathfrak{b}}$ is defined as $w_{\mathfrak{d}}(x_{\mathfrak{d}}) = \operatorname{ord}_{\mathfrak{d}}(\mathfrak{a} + \mathfrak{b}) \deg(\mathfrak{d}) - z_{\mathfrak{d}}(x_{\mathfrak{d}})$.

The number of zeros of $x \in \mathcal{O}_C(\mathfrak{a})(U)$ and $x \in M_{S,\mathfrak{a},\mathfrak{b}}$ respectively (counted with multiplicity) is then

$$z_S(x) = \sum_{\mathfrak{d} \in S} z_{\mathfrak{d}}(x_{\mathfrak{d}})$$

where the $x_{\mathfrak{d}}$ are the images of x in $\mathcal{O}_C(\mathfrak{a})(\operatorname{supp}(\mathfrak{d}))$ and $M_{\mathfrak{d},\mathfrak{a},\mathfrak{b}}$ respectively. The weight of $x \in \mathcal{O}_C(\mathfrak{a})(U)$ and $x \in M_{S,\mathfrak{a},\mathfrak{b}}$ respectively is

$$w_S(x) = \sum_{\mathfrak{d} \in S} w_{\mathfrak{d}}(x_{\mathfrak{d}})$$

where the $x_{\mathfrak{d}}$ are the images of x in $\mathcal{O}_C(\mathfrak{a})(\operatorname{supp}(\mathfrak{d}))$ and $M_{\mathfrak{d},\mathfrak{a},\mathfrak{b}}$ respectively. The weight defines a Hamming metric on $M_{S,\mathfrak{a},\mathfrak{b}}$ in a standard way. Obviously,

$$z_S(x) + w_S(x) = \deg(\mathfrak{a} + \mathfrak{b})$$

for $x \in M_{S,\mathfrak{a},\mathfrak{b}}$. For example $z_S(0) = \deg(\mathfrak{a} + \mathfrak{b})$.

If $\mathfrak{d}$ is any divisor coprime to $\mathfrak{m}$, we define

$$L_{\mathfrak{m}}(\mathfrak{d}) = \{ f \in F^\times \mid \operatorname{div}(f) \geq -\mathfrak{d} \text{ and } f \equiv c \bmod \mathfrak{m} \text{ for some } c \in \mathbb{F}_q \} \cup \{0\}.$$

This is a finite dimensional $\mathbb{F}_q$-vector space. For $\mathfrak{m} = 0$ we recover the standard spaces $L(\mathfrak{d})$. Finally, let $\mathfrak{e}$ be a divisor of C coprime to $\sum_{\mathfrak{d} \in S} \mathfrak{d}$ and to $\mathfrak{m}$ with $\deg(\mathfrak{e}) < \deg(\mathfrak{b})$ and define a generalised algebraic-geometric code

$$C(S, \mathfrak{a}, \mathfrak{b}, \mathfrak{e}, \mathfrak{m}) = \{ \operatorname{ev}_{S,\mathfrak{a},\mathfrak{b}}(f) \mid f \in L_{\mathfrak{m}}(\mathfrak{a} + \mathfrak{e}) \}.$$

If S is a set of places of degree one, $\mathfrak{a} = \mathfrak{m} = 0$ and $\mathfrak{b}$ has no multiplicities, then $C(S, \mathfrak{a}, \mathfrak{b}, \mathfrak{e}, \mathfrak{m})$ is a standard algebraic-geometric code. If $\mathfrak{a}$ and $\mathfrak{b}$ are sums of places of degree one and $\operatorname{ord}_{\mathfrak{d}}(\mathfrak{b}) = -\operatorname{ord}_{\mathfrak{d}}(\mathfrak{a}) + 1$ then $C(S, \mathfrak{a}, \mathfrak{b}, \mathfrak{e}, \mathfrak{m})$ is a code with Hamming distance w_S in the standard sense.

Proposition 1. *For any $f \in L_{\mathfrak{m}}(\mathfrak{a} + \mathfrak{e})$ with $f \neq 0$ it holds that*

$$z_S(\operatorname{ev}_{S,\mathfrak{a},\mathfrak{b}}(f)) \leq \deg(\mathfrak{a} + \mathfrak{e}) \quad \text{and} \quad w_S(\operatorname{ev}_{S,\mathfrak{a},\mathfrak{b}}(f)) \geq \deg(\mathfrak{b} - \mathfrak{e}).$$

In particular, $\operatorname{ev}_{S,\mathfrak{a},\mathfrak{b}}(f) \neq 0$.

Proof. Since $f \neq 0$ we can write $\operatorname{div}(f) = \mathfrak{e}_1 + \mathfrak{e}_2$ with $\operatorname{supp}(\mathfrak{e}_1) \subseteq U$ and $\operatorname{supp}(\mathfrak{e}_2) \cap U = \emptyset$. Then $\mathfrak{e}_1 \geq \sum_{\mathfrak{d} \in S} \operatorname{ord}_{\mathfrak{d}}(\operatorname{div}(f))\mathfrak{d}$ and $\mathfrak{e}_2 \geq -\mathfrak{e}$. We obtain

$$z_S(\operatorname{ev}_{S,\mathfrak{a},\mathfrak{b}}(f)) \leq \deg(\mathfrak{a}) + \sum_{\mathfrak{d} \in S} \operatorname{ord}_{\mathfrak{d}}(\operatorname{div}(f))\mathfrak{d}$$
$$\leq \deg(\mathfrak{a}) + \deg(\mathfrak{e}_1) = \deg(\mathfrak{a}) - \deg(\mathfrak{e}_2)$$
$$\leq \deg(\mathfrak{a}) + \deg(\mathfrak{e}) = \deg(\mathfrak{a} + \mathfrak{e}).$$

The other inequality follows from $z_S(\operatorname{ev}_{S,\mathfrak{a},\mathfrak{b}}(f)) + w_S(\operatorname{ev}_{S,\mathfrak{a},\mathfrak{b}}(f)) = \deg(\mathfrak{a} + \mathfrak{b})$.

4.2 Reduction of the Discrete Logarithm Problem to Code Based Computational Problems

Words of minimal non zero weight in the code $C(S, \mathfrak{a}, \mathfrak{b}, \mathfrak{c}, \mathfrak{m})$ correspond to relations in the group $\mathrm{Pic}_{\mathfrak{m}}(C)$ in the following way. If $\mathfrak{d} \in \mathcal{D}^{\mathfrak{m}}(C)$ then $[\mathfrak{d}]_{\mathfrak{m}}$ denotes the class of $\mathfrak{d}$ in $\mathrm{Pic}_{\mathfrak{m}}(C)$.

Theorem 2. *Words of $C(S, \mathfrak{a}, \mathfrak{b}, \mathfrak{c}, \mathfrak{m})$ of minimal non zero weight $\deg(\mathfrak{b} - \mathfrak{c})$ correspond bijectively to linear combinations*

$$[\mathfrak{c}]_{\mathfrak{m}} = \sum_{\mathfrak{d} \in S} \gamma_{\mathfrak{d}} \cdot [\mathfrak{d}]_{\mathfrak{m}}$$

with $\mathrm{ord}_{\mathfrak{d}}(\mathfrak{b}) \geq \gamma_{\mathfrak{d}} \geq -\mathrm{ord}_{\mathfrak{d}}(\mathfrak{a})$ for all $\mathfrak{d} \in S$. If $x \in C(S, \mathfrak{a}, \mathfrak{b}, \mathfrak{c}, \mathfrak{m})$ is such a word then the corresponding linear combinations satisfies

$$\gamma_{\mathfrak{d}} = \mathrm{ord}_{\mathfrak{d}}(\mathfrak{b}) - w_{\mathfrak{d}}(x_{\mathfrak{d}})/\deg(\mathfrak{d})$$

for all $\mathfrak{d} \in S$.

Proof. Let $[\mathfrak{c}]_{\mathfrak{m}} - \sum_{\mathfrak{d} \in S} \gamma_{\mathfrak{d}} \cdot [\mathfrak{d}]_{\mathfrak{m}} = 0$. Then there is $f \in F^{\times}$ with $f \equiv 1 \bmod \mathfrak{m}$, $\mathrm{div}(f) = -\mathfrak{c} + \sum_{\mathfrak{d} \in S} \mathrm{ord}_{\mathfrak{d}}(\mathrm{div}(f))\mathfrak{d}$ and $\gamma_{\mathfrak{d}} = \mathrm{ord}_{\mathfrak{d}}(\mathrm{div}(f))$. Since $\mathrm{ord}_{\mathfrak{d}}(\mathfrak{b}) \geq \mathrm{ord}_{\mathfrak{d}}(\mathrm{div}(f)) \geq -\mathrm{ord}_{\mathfrak{d}}(\mathfrak{a})$ by assumption and $\deg(\mathrm{div}(f)) = 0$ we have

$$z_S(\mathrm{ev}_{S,\mathfrak{a},\mathfrak{b}}(f)) = \sum_{\mathfrak{d} \in S} z_{\mathfrak{d}}(\mathrm{ev}_{S,\mathfrak{a},\mathfrak{b}}(f)_{\mathfrak{d}}) = \sum_{\mathfrak{d} \in S} (\mathrm{ord}_{\mathfrak{d}}(\mathfrak{a}) + \mathrm{ord}_{\mathfrak{d}}(\mathrm{div}(f))) \deg(\mathfrak{d})$$

$$= \sum_{\mathfrak{d} \in S} \mathrm{ord}_{\mathfrak{d}}(\mathfrak{a}) \deg(\mathfrak{d}) + \sum_{\mathfrak{d} \in S} \mathrm{ord}_{\mathfrak{d}}(\mathrm{div}(f)) \deg(\mathfrak{d})$$

$$= \deg(\mathfrak{a}) + \deg(\mathfrak{c}) = \deg(\mathfrak{a} + \mathfrak{c}).$$

Thus $w_S(\mathrm{ev}_{S,\mathfrak{a},\mathfrak{b}}(f)) = \deg(\mathfrak{a} + \mathfrak{b}) - z_S(\mathrm{ev}_{S,\mathfrak{a},\mathfrak{b}}(f)) = \deg(\mathfrak{b} - \mathfrak{c})$. This maps every linear combination to a word of minimal non zero weight in a well defined way. From

$$\gamma_{\mathfrak{d}} = \mathrm{ord}_{\mathfrak{d}}(\mathrm{div}(f)) = z_{\mathfrak{d}}(\mathrm{ev}_{S,\mathfrak{a},\mathfrak{b}}(f)_{\mathfrak{d}})/\deg(\mathfrak{d}) - \mathrm{ord}_{\mathfrak{d}}(\mathfrak{a})$$

$$= \mathrm{ord}_{\mathfrak{d}}(\mathfrak{b}) - w_{\mathfrak{d}}(\mathrm{ev}_{S,\mathfrak{a},\mathfrak{b}}(f)_{\mathfrak{d}})/\deg(\mathfrak{d})$$

we see that the map is injective and also the formula for $\gamma_{\mathfrak{d}}$ is proved.

To prove surjectivity, assume that $f \in L_{\mathfrak{m}}(\mathfrak{a} + \mathfrak{c})$ such that $w_S(\mathrm{ev}_{S,\mathfrak{a},\mathfrak{b}}(f)) = \deg(\mathfrak{b} - \mathfrak{c})$. Then $z_S(\mathrm{ev}_{S,\mathfrak{a},\mathfrak{b}}(f)) = \deg(\mathfrak{a} + \mathfrak{c})$. All inequalities in the proof of Proposition 1 are thus equalities. We obtain

$$\mathrm{div}(f) = \mathfrak{c}_1 + \mathfrak{c}_2 = \sum_{\mathfrak{d} \in S} \mathrm{ord}_{\mathfrak{d}}(\mathrm{div}(f))\mathfrak{d} - \mathfrak{c}.$$

Since $f \in L_{\mathfrak{m}}(\mathfrak{a}+\mathfrak{c})$ and f is coprime to $\mathfrak{m}$ there is $c \in \mathbb{F}_q$ such that $cf \equiv 1 \bmod \mathfrak{m}$. Thus f indeed yields $[\mathfrak{c}]_{\mathfrak{m}} - \sum_{\mathfrak{d} \in S} \gamma_{\mathfrak{d}} \cdot [\mathfrak{d}]_{\mathfrak{m}} = 0$ with $\gamma_{\mathfrak{d}} = \mathrm{ord}_{\mathfrak{d}}(\mathrm{div}(f))$. Finally, $\gamma_{\mathfrak{d}} \geq \mathrm{ord}_{\mathfrak{d}}(\mathfrak{a})$ holds by $f \in L_{\mathfrak{m}}(\mathfrak{a} + \mathfrak{c})$, and

$$z_S(\mathrm{ev}_{S,\mathfrak{a},\mathfrak{b}}(f)) = \sum_{\mathfrak{d} \in S} (\mathrm{ord}_{\mathfrak{d}}(\mathfrak{a}) + \mathrm{ord}_{\mathfrak{d}}(\mathrm{div}(f))) \deg(\mathfrak{d})$$

implies $\mathrm{ord}_{\mathfrak{d}}(\mathfrak{b}) \geq \gamma_{\mathfrak{d}}$ for all $\mathfrak{d} \in S$.

It is also possible to make a reduction of the minimal non zero weight word problem to the maximum likelihood decoding problem as in [3], under some additional assumptions.

Theorem 3. *Let* $\mathfrak{q} \in \mathrm{supp}(\mathfrak{e})$ *be a place of degree one and let*

$$\deg(\mathfrak{a} + \mathfrak{e}) \geq 2g + \deg(\mathfrak{m}).$$

There is $h \in L_{\mathfrak{m}}(\mathfrak{a} + \mathfrak{e})$ *such that* $\mathrm{ord}_{\mathfrak{q}}(h) = \mathrm{ord}_{\mathfrak{q}}(\mathfrak{e})$. *Assume that* $C(S, \mathfrak{a}, \mathfrak{b}, \mathfrak{e}, \mathfrak{m})$ *contains a word of minimal non zero weight* $\deg(\mathfrak{b} - \mathfrak{e})$. *Let* $x \in C(S, \mathfrak{a}, \mathfrak{b}, \mathfrak{e} - \mathfrak{q}, \mathfrak{m})$ *such that*

$$w_S(\mathrm{ev}_{S,\mathfrak{a},\mathfrak{b}}(h) - x)$$

is minimal. Then $\mathrm{ev}_{S,\mathfrak{a},\mathfrak{b}}(h) - x$ *is a word in* $C(S, \mathfrak{a}, \mathfrak{b}, \mathfrak{e}, \mathfrak{m})$ *of minimal non zero weight* $\deg(\mathfrak{b} - \mathfrak{e})$.

Proof. We have $L_{\mathfrak{m}}(\mathfrak{a}+\mathfrak{e}-\mathfrak{q}) \subseteq L_{\mathfrak{m}}(\mathfrak{a}+\mathfrak{e})$ and $C(S, \mathfrak{a}, \mathfrak{b}, \mathfrak{e}-\mathfrak{q}, \mathfrak{m}) \subseteq C(S, \mathfrak{a}, \mathfrak{b}, \mathfrak{e}, \mathfrak{m})$. If one of them exists, the minimal non zero distance of $C(S, \mathfrak{a}, \mathfrak{b}, \mathfrak{e} - \mathfrak{q}, \mathfrak{m})$ and $C(S, \mathfrak{a}, \mathfrak{b}, \mathfrak{e}, \mathfrak{m})$ is thus equal to the minimal non zero weight of $C(S, \mathfrak{a}, \mathfrak{b}, \mathfrak{e}, \mathfrak{m})$.

Consider the $\mathbb{F}_q$-linear map $\phi : L(\mathfrak{a} + \mathfrak{e}) \to \mathcal{O}_{C,\mathfrak{m}}/\mathfrak{m}$. Since $\mathfrak{a} + \mathfrak{e} - \mathfrak{m}$ is non special, the adele spaces satisfy $A_F = A_F(\mathfrak{a} + \mathfrak{e} - \mathfrak{m}) + F$, whence ϕ is surjective (like in the proof of the strong approximation theorem, see for example [21, p. 33]). As $\dim_{\mathbb{F}_q} \mathcal{O}_{C,\mathfrak{m}}/\mathfrak{m} = \deg(\mathfrak{m})$ we get

$$\dim_{\mathbb{F}_q} L_{\mathfrak{m}}(\mathfrak{a} + \mathfrak{e}) = \dim_{\mathbb{F}_q} L(\mathfrak{a} + \mathfrak{e}) - \deg(\mathfrak{m}) + 1.$$

Now $\deg(\mathfrak{a} + \mathfrak{e}) \geq 2g$ implies $L(\mathfrak{a} + \mathfrak{e} - \mathfrak{q}) \subsetneq L(\mathfrak{a} + \mathfrak{e})$, whence $L_{\mathfrak{m}}(\mathfrak{a} + \mathfrak{e} - \mathfrak{q}) \subsetneq L_{\mathfrak{m}}(\mathfrak{a} + \mathfrak{e})$ by the dimension formula, and this yields the existence of h.

By the above observation, $w_S(\mathrm{ev}_{S,\mathfrak{a},\mathfrak{b}}(h) - x) \geq \deg(\mathfrak{b} - \mathfrak{e})$. To prove the reverse inequality, let $f \in L_{\mathfrak{m}}(\mathfrak{a} + \mathfrak{e})$ such that that $w_S(\mathrm{ev}_{S,\mathfrak{a},\mathfrak{b}}(f)) = \deg(\mathfrak{b} - \mathfrak{e})$. This exists by assumption. Then $\mathrm{ord}_{\mathfrak{p}}(f) = \mathrm{ord}_{\mathfrak{p}}(\mathfrak{e})$ for all $\mathfrak{p} \in \mathrm{supp}(\mathfrak{e})$, including $\mathfrak{p} = \mathfrak{q}$. Since $\mathrm{ord}_{\mathfrak{q}}(h) = \mathrm{ord}_{\mathfrak{q}}(\mathfrak{e}) = \mathrm{ord}_{\mathfrak{q}}(f)$ and $\deg(\mathfrak{q}) = 1$ there is $\lambda \in \mathbb{F}_q^{\times}$ such that $h - \lambda f \in L_{\mathfrak{m}}(\mathfrak{a} + \mathfrak{e} - \mathfrak{q})$. Then $w_S(\mathrm{ev}_{S,\mathfrak{a},\mathfrak{b}}(h) - \mathrm{ev}_{S,\mathfrak{a},\mathfrak{b}}(h - \lambda f)) = \deg(\mathfrak{b} - \mathfrak{e})$ and thus $w_S(\mathrm{ev}_{S,\mathfrak{a},\mathfrak{b}}(h) - x) \leq \deg(\mathfrak{b} - \mathfrak{e})$, so we have

$$w_S(\mathrm{ev}_{S,\mathfrak{a},\mathfrak{b}}(h) - x) = \deg(\mathfrak{b} - \mathfrak{e}).$$

Clearly $\mathrm{ev}_{S,\mathfrak{a},\mathfrak{b}}(h) - x \in C(S, \mathfrak{a}, \mathfrak{b}, \mathfrak{e}, \mathfrak{m})$, and the theorem is proved.

We continue the discussion with ways of applying Theorem 2 and in particular working within $\mathrm{Pic}_{\mathfrak{m}}(C)$.

Theorem 4. *Every class of* $\mathrm{Pic}_{\mathfrak{m}}(C)$ *of degree greater or equal to* $2g + \deg(\mathfrak{m}) - 1$ *contains an effective divisor.*

Proof. Let $\mathfrak{d} \in \mathcal{D}^{\mathfrak{m}}(C)$ with $\deg(\mathfrak{d}) \geq 2g + \deg(\mathfrak{m}) - 1$. The divisor $\mathfrak{d} - \mathfrak{m}$ is not special, since $\deg(\mathfrak{d} - \mathfrak{m}) \geq 2g - 1$. The adele spaces then satisfy $A_F = A_F(\mathfrak{d} - \mathfrak{m}) + F$, whence there is $f \in L(\mathfrak{d})$ such that $f \equiv 1 \mod \mathfrak{m}$. The divisor $\mathfrak{a} = \mathfrak{d} + \mathrm{div}(f)$ is effective and defines the same class as $\mathfrak{d}$ in $\mathrm{Pic}_{\mathfrak{m}}(C)$.

Let $\mathfrak{c}$ be an arbitrary divisor of C. By replacing $\mathfrak{m}$ by $\mathfrak{m}$ plus the sum of the places in $\operatorname{supp}(\mathfrak{c})$ the theorem can also be applied to conclude that every class of $\operatorname{Pic}_{\mathfrak{m}}(C)$ contains an effective divisor coprime to $\mathfrak{c}$.

Corollary 1. *Let $\mathfrak{o}$ be a place of degree one of C. Every class of $\operatorname{Pic}_{\mathfrak{m}}^0(C)$ contains a divisor of the form*

$$\mathfrak{d} - \deg(\mathfrak{d})\mathfrak{o},$$

where $\mathfrak{d}$ is an effective divisor of degree less than or equal to $\leq 2g + \deg(\mathfrak{m}) - 1$. If $\mathfrak{d}$ has minimal degree then $\mathfrak{d}$ is uniquely determined.

Using these results we see that expressing any divisor $\mathfrak{c}$ as a linear combination of elements of any S in $\operatorname{Pic}_{\mathfrak{m}}(C)$, where the coefficients are bounded by $\mathfrak{a}$ and $\mathfrak{b}$, is equivalent to the minimal non zero weight word problem in $C(S, \mathfrak{a}, \mathfrak{b}, \mathfrak{c}, \mathfrak{m})$. Note that this reduction is polynomial time in the length of the input and the occurring degrees or multiplicities in $\mathfrak{a}$, $\mathfrak{b}$ and $\mathfrak{m}$, by well known algorithms for algebraic curves and function fields (see for example [13]). In order to get a meaningful reduction we need to exhibit supposedly hard instances of the problem of finding linear combinations or of the minimal non zero weight word problem that are related by sufficiently small S, $\mathfrak{a}$, $\mathfrak{b}$ and $\mathfrak{m}$.

Suppose S defines a generating set of $\operatorname{Pic}_{\mathfrak{m}}(C)$ with $\#S$ small such that every element of $\operatorname{Pic}_{\mathfrak{m}}^0(C)$ can be written as a linear combination of the $[\mathfrak{d}]_{\mathfrak{m}}$ with $\mathfrak{d} \in S$ with small coefficients bounded by $-\lambda$ from below and μ from above. Choose

$$\mathfrak{a} = \lambda \sum_{\mathfrak{d} \in S} \mathfrak{d} \quad \text{and} \quad \mathfrak{b} = \mu \sum_{\mathfrak{d} \in S} \mathfrak{d}.$$

Suppose $[\mathfrak{c}_1]_{\mathfrak{m}} = \lambda[\mathfrak{c}_2]_{\mathfrak{m}}$ in $\operatorname{Pic}_{\mathfrak{m}}^0(C)$. By the corollary, for any λ_1 and λ_2 there is an effective $\mathfrak{c}$ with $\deg(\mathfrak{c}) \leq 2g + \deg(\mathfrak{m}) - 1$ such that

$$[\mathfrak{c} - \deg(\mathfrak{c})\mathfrak{o}]_{\mathfrak{m}} = \lambda_1[\mathfrak{c}_1]_{\mathfrak{m}} - \lambda_2[\mathfrak{c}_2]_{\mathfrak{m}}.$$

The complexity of the computation of such an $\mathfrak{c}$ is polynomial in $\log(\lambda_1)$ and $\log(\lambda_2)$ and thus efficient if $\mathfrak{c}_1$ and $\mathfrak{c}_2$ satisfy a similar degree bound to start with (see [13]). Given an oracle for the minimal non zero weight word problem in the codes $C(S, \mathfrak{a}, \mathfrak{b}, \mathfrak{c}, \mathfrak{m})$ we can thus efficiently compute linear combinations

$$\lambda_1[\mathfrak{c}_1]_{\mathfrak{m}} - \lambda_2[\mathfrak{c}_2]_{\mathfrak{m}} = [\mathfrak{c} - \deg(\mathfrak{c})\mathfrak{o}]_{\mathfrak{m}} = \sum_{\mathfrak{d} \in S} \gamma_{\mathfrak{d}}[\mathfrak{d}]_{\mathfrak{m}}$$

for randomly chosen λ_1 and λ_2. After more than $\#S$ linear combinations we get linear dependencies of the right hand sides and can solve for λ in the usual way. The discrete logarithm problem in $\operatorname{Pic}_{\mathfrak{m}}^0(C)$ is thus reduced to the minimal non zero weight word problem in the codes $C(S, \mathfrak{a}, \mathfrak{b}, \mathfrak{c}, \mathfrak{m})$.

In a similar fashion, using random linear combinations of the $\mathfrak{d} \in S$ on the left hand side, we can also compute all relations between the $\mathfrak{d} \in S$ in $\operatorname{Pic}_{\mathfrak{m}}^0(C)$ and thus deduce the structure of $\operatorname{Pic}_{\mathfrak{m}}^0(C)$ as an abelian group. For both problems the best known algorithms have subexponential complexity or only exponential complexity on average. Depending on the point of view we can obtain hardness statements for the minimal non zero weight word problem in the codes $C(S, \mathfrak{a}, \mathfrak{b}, \mathfrak{e}, \mathfrak{m})$, or we get a possible way of improving said algorithms for the computation of discrete logarithms or for the computation of all relations.

4.3   Efficient Generating Sets of Generalised Jacobians

The crucial point to discuss now are suitable generating systems of $\mathrm{Pic}_\mathfrak{m}(C)$ in the above sense. Assume there exists a place $\mathfrak{o}$ of degree one of C. Let X be a multiset with $\mathrm{supp}(X) \subseteq \mathrm{Pic}^0_\mathfrak{m}(C)$ such that $-y \in X$ for all $y \in X$. The Cayley graph

$$\mathrm{Cay}(\mathrm{Pic}^0_\mathfrak{m}(C), X)$$

has the set of vertices $\mathrm{Pic}^0_\mathfrak{m}(C)$ and the multiset $\{(x, yx) \mid x \in \mathrm{Pic}^0_\mathfrak{m}(C), y \in X\}$ of edges. For every $x \in \mathrm{Pic}^0_\mathfrak{m}(C)$ there are exactly $\#X$ edges leaving x.

Theorem 5. *Let X_r be a multiset containing*

$$[(r/\deg(\mathfrak{p}))\mathfrak{p} - r\mathfrak{o}]_\mathfrak{m} \quad and \quad [r\mathfrak{o} - (r/\deg(\mathfrak{p}))\mathfrak{p}]_\mathfrak{m}$$

each with multiplicity $\deg(\mathfrak{p})$, where $\mathfrak{p}$ ranges over all places $\neq \mathfrak{o}$ with $\deg(\mathfrak{p})|r$ and $\mathfrak{p} \notin \mathrm{supp}(\mathfrak{m})$. Let $r, t \in \mathbb{Z}^{\geq 1}$ with

$$q^r - 2gq^{r/2} - \deg(\mathfrak{m}) > 0,$$
$$(q^r - 2gq^{r/2} - \deg(\mathfrak{m}))^t > 2(q^{1/2} + 1)^{2g}(q^{\deg(\mathfrak{m})} - 1)/(q-1)$$
$$\cdot ((2g + \deg(\mathfrak{m}))q^{r/2} + \deg(\mathfrak{m}))^t.$$

Let $x_0, x_1 \in \mathrm{Pic}^0_\mathfrak{m}(C)$. Then any random walk in $\mathrm{Cay}(\mathrm{Pic}^0_\mathfrak{m}(C), X_r)$ of length t starting in x_0 will end in x_1 with probability $p_{r,t}$ satisfying

$$\frac{1}{2}\#\mathrm{Pic}^0_\mathfrak{m}(C)^{-1} \leq p_{r,t} \leq \frac{3}{2}\#\mathrm{Pic}^0_\mathfrak{m}(C)^{-1}.$$

Proof. We will make use of the methodology in [17] that estimates rapid mixing properties of random walks of length t on the Cayley graph of $\mathrm{Pic}^0_\mathfrak{m}(C)$ with edges X_r. For this we need to estimate character sums over X_r.

Let $\chi : \mathrm{Pic}^0_\mathfrak{m}(C) \to \mathbb{C}^\times$ be a character. We extend χ to $\mathrm{Pic}_\mathfrak{m}(C)$ via $\chi([\mathfrak{o}]_\mathfrak{m}) = 1$. Then $\chi : \mathrm{Pic}_\mathfrak{m}(C) \to \mathbb{C}^\times$ is a character of finite order. Let $\mathfrak{f}(\chi)$ be its conductor. The theory of L-series gives

$$\sum_{\deg(\mathfrak{p})|r, \mathfrak{p} \nleq \mathfrak{f}(\chi)} \deg(\mathfrak{p}) \cdot \chi([\mathfrak{p}]_\mathfrak{m})^{r/\deg(\mathfrak{p})} = \rho_\chi(q^r + 1) - \sum_{i=1}^{2(g-1+\rho_\chi)+\deg(\mathfrak{f}(\chi))} \omega_i(\chi)^r,$$

where $\rho_\chi = 1$ if χ is the trivial character on $\mathrm{Pic}^0_\mathfrak{m}(C)$, $\rho_\chi = 0$ otherwise, and $\omega_i(\chi)$ are complex numbers of absolute value $q^{r/2}$. This gives

$$\lambda_\chi := \sum_{x \in X_r} \chi(x)$$
$$= \sum_{\deg(\mathfrak{p})|r, \mathfrak{p} \nleq \mathfrak{m}+\mathfrak{o}} \deg(\mathfrak{p}) \cdot \left(\chi\big((r/\deg(\mathfrak{p})) \cdot [\mathfrak{p}]_\mathfrak{m}\big) + \chi^{-1}\big((r/\deg(\mathfrak{p})) \cdot [\mathfrak{p}]_\mathfrak{m}\big) \right).$$

Furthermore,

$$\sum_{\deg(\mathfrak{p})|r,\mathfrak{p}\nleq\mathfrak{m}+\mathfrak{o}} \deg(\mathfrak{p}) \cdot \chi\big((r/\deg(\mathfrak{p}))\cdot[\mathfrak{p}]_{\mathfrak{m}}\big) = \sum_{\deg(\mathfrak{p})|r,\mathfrak{p}\nleq\mathfrak{m}+\mathfrak{o}} \deg(\mathfrak{p}) \cdot \chi([\mathfrak{p}]_{\mathfrak{m}})^{r/\deg(\mathfrak{p})}$$

$$= \sum_{\deg(\mathfrak{p})|r,\mathfrak{p}\nleq\mathfrak{f}(\chi)} \deg(\mathfrak{p}) \cdot \chi([\mathfrak{p}]_{\mathfrak{m}})^{r/\deg(\mathfrak{p})} - \sum_{\deg(\mathfrak{p})|r,\mathfrak{p}\nleq\mathfrak{f}(\chi),\mathfrak{p}\leq\mathfrak{m}+\mathfrak{o}} \deg(\mathfrak{p}) \cdot \chi([\mathfrak{p}]_{\mathfrak{m}})^{r/\deg(\mathfrak{p})}$$

$$= \rho_\chi(q^r+1) - \sum_{i=1}^{2(g-1+\rho_\chi)+\deg(\mathfrak{f}(\chi))} \omega_i(\chi)^r - \sum_{\deg(\mathfrak{p})|r,\mathfrak{p}\nleq\mathfrak{f}(\chi),\mathfrak{p}\leq\mathfrak{m}+\mathfrak{o}} \deg(\mathfrak{p}) \cdot \chi([\mathfrak{p}]_{\mathfrak{m}})^{r/\deg(\mathfrak{p})}.$$

We obtain

$$|\lambda_1/2 - (q^r+1)| \leq 2gq^{r/2} + \deg(\mathfrak{m}) + 1 \tag{5}$$

and

$$|\lambda_\chi/2| \leq (2g-2+\deg(\mathfrak{m}))q^{r/2} + \deg(\mathfrak{m}) + 1. \tag{6}$$

By the proof of [17, Lemma 2.1], the number of paths of length t starting and ending in any two fixed elements of $\mathrm{Pic}_{\mathfrak{m}}^0(C)$ is equal to

$$\frac{\lambda_1^t}{\#\mathrm{Pic}_{\mathfrak{m}}^0(C)} + w$$

with $|w| \leq \max_{\chi \neq 1}|\lambda_\chi|^t$. If r and t are chosen such that

$$\frac{\lambda_1^t}{\#\mathrm{Pic}_{\mathfrak{m}}^0(C)} > 2\max_{\chi\neq 1}|\lambda_\chi|^t \tag{7}$$

then the assertion of the theorem holds for S_r and t by [17, Lemma 2.1]. Observing

$$\#\mathrm{Pic}_{\mathfrak{m}}^0(C) \leq (q^{1/2}+1)^{2g}(q^{\deg(\mathfrak{m})}-1)/(q-1)$$

the inequality of the assertion follows by combining (5), (6) and (7).

Corollary 2. *Choose r and t as in the theorem and let S_r be the set containing the divisors $\mathfrak{o}$ and $(r/\deg(\mathfrak{p}))\mathfrak{p}$ for all places $\neq \mathfrak{o}$ with $\deg(\mathfrak{p})|r$ and $\mathfrak{p} \notin \mathrm{supp}(\mathfrak{m})$. Then S_r is a generating system of $\mathrm{Pic}_{\mathfrak{m}}(C)$ and every $[\mathfrak{c}]_{\mathfrak{m}}$ can be written as*

$$[\mathfrak{c}]_{\mathfrak{m}} = \sum_{\mathfrak{d}\in S_r} \lambda_{\mathfrak{d}}\mathfrak{d} \text{ with } \lambda_{\mathfrak{d}} \in \mathbb{Z} \text{ and } \sum_{\mathfrak{d}\in S_r}|\lambda_{\mathfrak{d}}| \leq 2t.$$

We remark that it is convenient but not really necessary to assume that there exists a place $\mathfrak{o}$ of degree one. We could also adapt the statements to the case that $\mathfrak{o}$ is only a divisor of degree one.

According to the corollary, S_r is a set of generators that can be used for an efficient reduction as described above. A drawback for the construction of the codes $C(S, \mathfrak{a}, \mathfrak{b}, \mathfrak{m}, \mathfrak{c})$ is that the $\lambda_{\mathfrak{d}}$ can also be negative, since we get non standard codes and Hamming weights. It is also possible to prove the existence of a small set of low degree prime generators such that $\lambda_{\mathfrak{d}} \in \{0,1\}$ for $\mathfrak{d} \neq \mathfrak{o}$, see the method of [4] in a special context. The bounds of this method appear to be weaker than the bounds in Theorem 5 though. Alternatively, using Theorem 5, we can probabilistically construct such a set of $\{0,1\}$-generators by the following theorem of [8].

Theorem 6. *Let G be an abelian group and $n = \#G$. Choose k elements $a_1,\ldots,a_k$ from G uniformly and independently at random. If $k \geq \log_2(n) + 2\log(\log(n))$, then*

$$G = \left\{ \sum_{i=1}^{k} \lambda_i a_i \mid \lambda_i \in \{0,1\} \right\}$$

with probability tending to 1 for $n \to \infty$.

Besides the additional probabilistic computation, the drawback here is that the a_i will be represented as $[\mathfrak{d} - \deg(\mathfrak{d})\mathfrak{o}]_{\mathfrak{m}}$ with $\mathfrak{d}$ effective by Corollary 1, where $\mathfrak{d}$ has in general degree of order $O(g + \deg(\mathfrak{m}))$. These degrees are much bigger than those of Theorem 5. The $\mathfrak{d}$ are not necessarily prime, but allowing for larger degrees, still of order $O(g + \deg(\mathfrak{m}))$, we could even achieve that the $\mathfrak{d}$ are places.

4.4   Examples

In [5], the discrete logarithm problem in $\mathbb{F}_{q^h} = \mathbb{F}_q[\alpha]$ is solved by decomposing $f(\alpha) \in \mathbb{F}_q[\alpha]$ as products of linear polynomials $\alpha + a$ in α with $a \in \mathbb{F}_q$ by an oracle to Reed-Solomon decoding. In our setting, we take $C = \mathbb{P}^1$, $\mathfrak{m}$ is defined as the place of degree h corresponding to the minimal polynomial of α over $\mathbb{F}_q$ and $\mathfrak{o}$ is the place "at infinity". By Corollary 1, the classes of $\mathrm{Pic}^0_{\mathfrak{m}}(C)$ can be represented as polynomials $f(\alpha)$ in α of degree $\leq h - 1$. In [5], h is chosen in comparison to q (as large as possible) such that the linear polynomials $\alpha + a$ in α with $a \in \mathbb{F}_q$ are a $\{0,1\}$-generating system. By Theorem 5 we obtain similar generating systems.

In [3], the discrete logarithm problem in an elliptic curve is cast as a minimal non zero weight word problem of an elliptic code. In our setting, we take C as elliptic curve, $\mathfrak{m} = 0$, $\mathfrak{a} = 0$, $\mathfrak{b}$ as divisor sum of point group multiples of a generator of the point group and $\mathfrak{e} = k\mathfrak{o}$, where $\mathfrak{o}$ is again a place "at infinity". Note that Theorem 5 does not apply in this context, since q is large. Estimates for incomplete character sums can be used to obtain a variant of Theorem 5 that does apply, but the resulting sizes of the generating systems are of order $q^{1/2}$ and hence completely unsatisfactory.

References

1. Augot, D., Morain, F.: Discrete logarithm computations over finite fields using Reed-Solomon codes (2012), http://hal.inria.fr/hal-00672050
2. Boneh, D., Silverberg, A.: Applications of multilinear forms to cryptography. In: Melles, C.G., et al. (eds.) Topics in Algebraic and Noncommutative Geometry; Proceedings in Memory of Ruth Michler, Luminy, France, Annapolis, MD, USA, July 20-22, October 25-28. American Mathematical Society (AMS), Providence (2001); Contemp. Math. 324, 71–90 (2003)

3. Cheng, Q.: Hard problems of algebraic geometry codes. IEEE Trans. Inform. Theory 54(1), 402–406 (2008)
4. Cheng, Q., Wan, D.: On the list and bounded distance decodability of Reed-Solomon codes. SIAM J. Comput. 37(1), 195–209 (2007)
5. Cheng, Q., Wan, D.: Complexity of Decoding Positive-Rate Reed-Solomon Codes. In: Aceto, L., Damgård, I., Goldberg, L.A., Halldórsson, M.M., Ingólfsdóttir, A., Walukiewicz, I. (eds.) ICALP 2008, Part I. LNCS, vol. 5125, pp. 283–293. Springer, Heidelberg (2008)
6. Déchène, I.: Arithmetic of Generalized Jacobians. In: Hess, F., Pauli, S., Pohst, M. (eds.) ANTS 2006. LNCS, vol. 4076, pp. 421–435. Springer, Heidelberg (2006)
7. Déchène, I.: On the security of generalized Jacobian cryptosystems. Adv. Math. Commun. 1(4), 413–426 (2007)
8. Erdős, P., Rényi, A.: Probabilistic methods in group theory. J. Analyse Math. 14, 127–138 (1965)
9. Frey, G., Rück, H.-G.: A remark concerning m-divisibility and the discrete logarithm in the divisor class group of curves. Math. Comp. 62, 865–874 (1994)
10. Galbraith, S.D., Smith, B.: Discrete Logarithms in Generalized Jacobians (2006), http://hal.inria.fr/inria-00537887
11. Galbraith, S.D., Hess, F., Vercauteren, F.: Aspects of pairing inversion. IEEE Trans. Inf. Theory 54(12), 5719–5728 (2008)
12. Hasse, H.: Theorie der relativ-zyklischen algebraischen Funktionenkörper, insbesondere bei endlichem Konstantenkörper. J. Reine Angew. Math. 172, 37–54 (1934)
13. Hess, F.: Computing Riemann-Roch spaces in algebraic function fields and related topics. J. Symbolic Comp. 33(4), 425–445 (2002)
14. Hess, F.: A note on the Tate pairing of curves over finite fields. Arch. Math. 82, 28–32 (2004)
15. Hess, F., Pauli, S., Pohst, M.E.: Computing the multiplicative group of residue class rings. Math. Comp. 72(243), 1531–1548 (2003) (electronic)
16. Huang, M.-D., Raskind, W.: A multilinear generalization of the Tate pairing. In: McGuire, G., et al. (eds.) Finite Fields. Theory and Applications. Proceedings of the 9th International Conference on Finite Fields and Applications, Dublin, Ireland, July 13-17, American Mathematical Society (AMS), Providence (2009); Contemporary Mathematics 518, 255–263 (2010)
17. Jao, D., Miller, S.D., Venkatesan, R.: Expander graphs based on GRH with an application to elliptic curve cryptography. J. Number Theory 129(6), 1491–1504 (2009)
18. Kohel, D.: Constructive and destructive facets of torus-based cryptography (2004), http://echidna.maths.usyd.edu.au/kohel/pub/torus.ps
19. Papamanthou, C., Tamassia, R., Triandopoulos, N.: Optimal Authenticated Data Structures with Multilinear Forms. In: Joye, M., Miyaji, A., Otsuka, A. (eds.) Pairing 2010. LNCS, vol. 6487, pp. 246–264. Springer, Heidelberg (2010)
20. Serre, J.-P.: Algebraic groups and class fields, Transl. of the French edn. Graduate Texts in Mathematics, vol. 117, ix, 207 p. Springer, New York (1988)
21. Stichtenoth, H.: Algebraic function fields and codes, 2nd edn. Graduate Texts in Mathematics, vol. 254, xiii, 355 p. Springer, Berlin (2009)

The Weight Distribution of a Family of Reducible Cyclic Codes[*]

Gerardo Vega[1] and Carlos A. Vázquez[2],[**]

[1] Dirección General de Cómputo y de Tecnologías de Información y Comunicación, Universidad Nacional Autónoma de México, 04510 México D.F., Mexico
gerardov@servidor.unam.mx
[2] Posgrado en Ciencia e Ingeniería de la Computación, Universidad Nacional Autónoma de México, 20059 México, D.F., Mexico
cvazquez@uxmcc2.iimas.unam.mx

Abstract. A remarkably general result which provides the evaluation of a family of exponential sums was presented by Marko J. Moisio (*Acta Arithmetica*, 93 (2000) 117-119). In this work, we use a particular instance of this general result in order to determine the value distribution of a particular exponential sum. Then, motivated by some new and fresh original ideas of Changli Ma, Liwei Zeng, Yang Liu, Dengguo Feng and Cunsheng Ding (*IEEE Trans. Inf. Theory*, 57-1 (2011) 397-402), we use this value distribution in order to obtain the weight distribution of a family of reducible cyclic codes. As we will see later, all the codes in this family are non-projective cyclic codes. Furthermore, they can be identified in a very easy way. In fact, as a by-product of this easy identification, we will be able to determine the exact number of cyclic codes in a family when length and dimension are given.

Keywords: Weight distribution, reducible cyclic codes and exponential sums.

1 Introduction

An important family of codes for error control in digital communications are the so-called cyclic codes. Therefore, finding the weight distribution of a q-ary cyclic code is not only a problem of theoretical interest, but also of practical importance. More specifically, the weight distribution of a code is important because it plays a significant role in determining the capabilities of error detection and correction of a given code. However, computing the weight distribution of a code, even on a computer, can be a formidable task. For cyclic codes this problem gains greater interest due mainly to their rich algebraic structure. Many authors have worked on the problem of determining the weight distribution of non-irreducible cyclic codes using different techniques (see for example [3], [11], [2], [9] and [7]). For a particular family of cyclic codes, it is quite common that

---

[*] Partially supported by PAPIIT-UNAM IN105611.
[**] Ph.D. student.

F. Özbudak and F. Rodríguez-Henríquez (Eds.): WAIFI 2012, LNCS 7369, pp. 16–28, 2012.
© Springer-Verlag Berlin Heidelberg 2012

one of these techniques consists in calculating the value distribution of a specific exponential sum. However, in most of the cases, the evaluation of an exponential sum is also a very hard problem. A remarkably general result that gives the evaluation of a family of exponential sums was presented in [8]. In this work, we determine the value distribution of a particular instance of this general result. Then, based on some new original ideas in [7], we take this value distribution in order to obtain the weight distribution of a family of reducible cyclic codes. As we will see later, all the codes in this family are non-projective cyclic codes. Furthermore, the codes in this family can be identified in a very easy way. In fact, as a by-product of this easy identification, we will be able to determine the exact number of cyclic codes in a family, when length and dimension are given.

This work is organized as follows: In Section 2 we establish some notation, recall some definitions and establish our main assumption that will be considered throughout this work. We also recall, for this section, some already known results. In particular, we present the evaluation of a specific exponential sum, which can be derived as an instance of a general result that was originally presented in [8]. Section 3 is devoted to presenting some general results. In Section 4 we use these results in order to obtain the weight distribution of a family of reducible and non-projective cyclic codes. We also present, in Section 4, an explicit formula for the number of cyclic codes that belong to one of these families, when length and dimension are fixed. Furthermore, some examples of this formula are included at the end of this section. Finally, Section 5 is devoted to conclusions.

2   Definitions, Notations, Preliminaries and Main Assumption

First of all, we set, for this section and for the rest of this work, the following:

Notation. By using p, t, q, k and Δ, we will denote five positive integers such that p is a prime number, $q = p^t$ and $\Delta = (q^k - 1)/(q - 1)$. From now on, γ will denote a fixed primitive element of $\mathbb{F}_{q^k}$. For any two integers i and j, we define $\mathcal{D}_i^{(j)} = \gamma^i \langle \gamma^j \rangle$, where $\langle \gamma^j \rangle$ denotes the subgroup of $\mathbb{F}_{q^k}^*$ generated by γ^j. For any integer a, the polynomial $h_a(x) \in \mathbb{F}_q[x]$ will denote the minimal polynomial of γ^{-a}. In addition, we will denote by "Tr", the absolute trace mapping from $\mathbb{F}_{q^k}$ to the prime field $\mathbb{F}_p$, and by "$\mathrm{Tr}_{\mathbb{F}_{q^k}/\mathbb{F}_q}$" the trace mapping from $\mathbb{F}_{q^k}$ to $\mathbb{F}_q$.

The following definitions are important for us:

A cyclic code is *irreducible* if its parity-check polynomial is irreducible (its polynomial representation is a minimal ideal).

An N-weight code is a code such that the cardinality of the set of nonzero weights is N.

A *projective code* is a linear code such that the minimum weight of its dual code is at least three (or, equivalently, if any two columns of its generator matrix are linearly independent).

For this work, we are particularly interested in non-irreducible cyclic codes, whose parity-check polynomials are factorizable in exactly two different irreducible factors, that is, we are interested in cyclic codes whose dual codes have two non conjugated zeros.

Now, we continue with this section by recalling the definition, and a basic property of the character or exponential sums (see, for example, [6]). In order to do this, let p, q, k and γ be as before, then the canonical additive character χ, of $\mathbb{F}_{q^k}$, is defined as

$$\chi(y) = \zeta_p^{\mathrm{Tr}(y)} , \qquad \text{for all } y \in \mathbb{F}_{q^k} ,$$

where $\zeta_p = \exp(\frac{2\pi\sqrt{-1}}{p})$. For the canonical additive character χ', of $\mathbb{F}_q$, the following orthogonal property will be useful for us:

$$\sum_{y \in \mathbb{F}_q} \chi'(y) = 0 . \tag{1}$$

The following is an easy preliminary result:

Lemma 1. *Let u and v be two positive integers such that u is odd. If 4 divides $u + 1$ then 4 divides $\frac{u^{2v}-1}{(u-1)}$.*

Proof. The result follows directly from the fact that $u + 1$ divides $\frac{u^{2v}-1}{(u-1)}$.   □

Now, we set, for this section and for the rest of this work, the following:

Main Assumption. From now on, we are going to suppose that q is an odd integer greater than 3 such that 4 divides $q + 1$. In addition, we will always assume that k is an even integer.

Remark 1. As a consequence of our main assumption, observe that $\frac{q-1}{2}$ is an odd integer. Furthermore, due to the previous lemma, observe that 4 divides Δ.

With our main assumption in mind, let χ be the canonical additive character of $\mathbb{F}_{q^k}$, and i be any integer. Now, owing to Remark 1, and due to $(q^k - 1) = \Delta(q-1)$, we have that $4|(q^k-1)$, therefore, $\sum_{x \in \mathbb{F}_{q^k}} \chi(\gamma^i x^4) = 1+4\sum_{z \in \mathcal{D}_i^{(4)}} \chi(z)$. Considering this fact, and since $4|(q+1)$, then the following result is a particular instance of a general result that was proved in [8] (see Theorem 1):

Theorem 1. *With our notation and main assumption, let i and w be two integers in such a way that w is given by*

$$w = \begin{cases} 0 & \text{if } (2 \mid \frac{k}{2}) \quad \text{or} \quad (2 \nmid \frac{k}{2} \text{ and } 2 \mid \frac{q+1}{4}) , \\ 2 & \text{otherwise} \end{cases} .$$

Also let η_0 and η_1 be the two integers given by

$$\eta_0 = \frac{(-1)^{\frac{k}{2}-1}3q^{\frac{k}{2}} - 1}{4},$$

$$\eta_1 = \frac{(-1)^{\frac{k}{2}}q^{\frac{k}{2}} - 1}{4}.$$

Then

$$\sum_{z \in \mathcal{D}_i^{(4)}} \chi(z) = \begin{cases} \eta_0 & \text{if } i \equiv w \pmod{4}, \\ \\ \eta_1 & \text{otherwise} \end{cases} \quad . \tag{2}$$

As we will see later, the previous theorem will be fundamental in determining the weight distribution of a family of reducible cyclic codes.

3   Some General Results

Lemma 2. *Let q, k and Δ be as before. Considering our main assumption, we have that Δ does not divides $2(q^s - 1)$ for any integer s such that $1 \le s < k$.*

Proof. On the contrary, suppose that $\Delta | 2(q^s - 1)$ for some integer s such that $1 \le s < k$. Thus, there must exist a positive integer l such that $2(q^s - 1) = l(q^{k-1} + q^{k-2} + \cdots + q + 1)$. Clearly, if $s < k - 1$ or $l > 1$ then $2(q^s - 1) < l(q^{k-1} + q^{k-2} + \cdots + q + 1)$. On the other hand, observe that equality $2(q^{k-1} - 1) = (q^{k-1} + q^{k-2} + \cdots + q + 1)$ is impossible if $q > 3$.   □

Lemma 3. *Let q, k and Δ be as before. Considering our main assumption, we also take λ to be a divisor of $q - 1$, and define $n = \lambda\Delta$. Let a_2 be an integer such that $a_2n \equiv 0 \pmod{q^k - 1}$. If $\gcd(\frac{\Delta}{2}, a_2) = 2$, then $\gcd(q^k - 1, a_2) = 2^d \frac{q-1}{\lambda'}$ for some integers d and λ', such that $d = 0, 1$ or 2, and λ' is the divisor of $q - 1$ satisfying $\gcd(q - 1, a_2) = \frac{q-1}{\lambda'}$. In addition, if we have that $d = 2$ in the previous assertion, then $\gcd(q^k - 1, a_2 + \frac{q^k - 1}{2}) = 2^{d'} \frac{q-1}{\lambda'}$, for some integer d' such that $d' = 0$ or 1.*

Proof. Since $a_2n \equiv 0 \pmod{q^k - 1}$ then observe that $a_2 = \frac{q-1}{\lambda}u$ for some integer u. Now, since we are supposing $\gcd(\frac{\Delta}{2}, a_2) = 2$, we have $\gcd(q^k - 1, a_2) = \gcd(2(q-1)\frac{\Delta}{2}, a_2) = \gcd(4(q-1), a_2) = \gcd(4(q-1), \frac{q-1}{\lambda}u) = 2^d \frac{q-1}{\lambda'}$, for some integers d and λ', such that $d = 0, 1$ or 2, and λ' is the divisor of $q - 1$ satisfying $\gcd(q - 1, a_2) = \frac{q-1}{\lambda'}$ (recall that $\frac{q-1}{2}$ is an odd integer, and observe that $\lambda'|\lambda$).

For the second assertion, we must first observe that $\gcd(\frac{\Delta}{2}, a_2 + \frac{q^k - 1}{2}) = \gcd(\frac{\Delta}{2}, a_2 + \frac{\Delta}{2}(q - 1)) = \gcd(\frac{\Delta}{2}, a_2) = 2$, and also that $\gcd(q - 1, a_2 + \frac{q^k - 1}{2}) = \gcd(q - 1, a_2 + \frac{\Delta}{2}(q - 1)) = \gcd(q - 1, a_2) = \frac{q-1}{\lambda'}$. Thus, for the first assertion of this lemma, we know that there must exist an integer d' in such a way that $\gcd(q^k - 1, a_2 + \frac{q^k - 1}{2}) = 2^{d'} \frac{q-1}{\lambda'}$ and $d' = 0, 1$ or 2. Now, if $d = 2$, then $2|a_2$, implying that $\frac{q-1}{\lambda'} = \gcd(q - 1, a_2)$ is an even integer. Therefore 8 divides $2^d \frac{q-1}{\lambda'}$, which in turn implies that $8|a_2$. Thus, by supposing $d' = d = 2$, we obtain that

$\gcd(q^k - 1, a_2 + \frac{q^k-1}{2}) = \gcd(\Delta(q-1), a_2 + \frac{\Delta}{2}(q-1)) = 2^d \frac{q-1}{\lambda'}$, and this will imply that $4|\frac{\Delta}{2}$. In consequence, we have $\gcd(\frac{\Delta}{2}, a_2) \geq 4$, a contradiction! Therefore $\gcd(q^k - 1, a_2 + \frac{q^k-1}{2}) = 2^{d'} \frac{q-1}{\lambda'}$, for some integer d' such that $d' = 0$ or 1. □

As a direct consequence of the previous lemma we have the following:

Corollary 1. *Consider the same notation and hypotheses as in previous lemma, then* $\gcd(q^k - 1, a_2) = 2^d \frac{q-1}{\lambda'}$ *or* $\gcd(q^k - 1, a_2 + \frac{q^k-1}{2}) = 2^d \frac{q-1}{\lambda'}$*, for some integers d and λ' such that $d = 0$ or 1, and λ' is the divisor of $q - 1$ satisfying* $\gcd(q - 1, a_2) = \frac{q-1}{\lambda'}$.

Lemma 4. *Let q, k, Δ and γ be as before. Considering our main assumption, we also take λ to be a divisor of $q - 1$, and define $n = \lambda\Delta$. Let a_2 be an integer such that $a_2 n \equiv 0 \pmod{q^k - 1}$. Suppose that $\gcd(q^k - 1, a_2) = 2^d \frac{q-1}{\lambda'}$ for some integers d and λ', such that $d = 0$ or 1, and λ' is the divisor of $q - 1$ satisfying $\gcd(q - 1, a_2) = \frac{q-1}{\lambda'}$. If $n' = \lambda' \frac{\Delta}{2^d}$ then*

$$\{(\gamma^{a_2 m}, (-1)^m) \mid 0 \leq m < n\} = \frac{2^d \lambda}{\lambda'} * \{(\gamma^{2^d \frac{q-1}{\lambda'} m}, (-1)^m) \mid 0 \leq m < n'\},$$

where $\frac{2^d \lambda}{\lambda'} * \{(\gamma^{2^d \frac{q-1}{\lambda'} m}, (-1)^m) \mid 0 \leq m < n'\}$ is the multiset in which each element of $\{(\gamma^{2^d \frac{q-1}{\lambda'} m}, (-1)^m) \mid 0 \leq m < n'\}$ appears with multiplicity $\frac{2^d \lambda}{\lambda'}$.

Proof. Since $0 \leq d \leq 1$ and $\gcd(q^k - 1, a_2) = 2^d \frac{q-1}{\lambda'}$, then n' is the least positive integer satisfying $2^d \frac{q-1}{\lambda'} n' \equiv 0 \pmod{q^k - 1}$. In addition, there must exist integers i and j such that $i a_2 + j(q^k - 1) = 2^d \frac{q-1}{\lambda'}$. Observe that i is necessarily an odd integer, because if not, then $2^{d+1} \frac{q-1}{\lambda'}|(i a_2)$, and since $2^{d+1} \frac{q-1}{\lambda'}|(q^k - 1)$ (recall that $0 \leq d \leq 1$ and $q^k - 1 = \Delta(q - 1)$), then we can conclude that $2^{d+1} \frac{q-1}{\lambda'}|2^d \frac{q-1}{\lambda'}$, and clearly this last condition is impossible. Thus, since $\langle\gamma^{a_2}\rangle = \langle\gamma^{2^d \frac{q-1}{\lambda'}}\rangle$, $|\langle\gamma^{a_2}\rangle| = n'$, and since i is an odd integer, then

$$\{(\gamma^{a_2 m}, (-1)^m) \mid 0 \leq m < n'\} = \{(\gamma^{2^d \frac{q-1}{\lambda'} m}, (-1)^m) \mid 0 \leq m < n'\}.$$

The result now follows from the fact that $\frac{n}{n'} = \frac{2^d \lambda}{\lambda'}$. □

The following is a modified version of Lemma 3 in [7].

Lemma 5. *With our notation and main assumption, let λ' and d be integers such that λ' is a divisor of $q-1$ and d is equal to zero or one. If $\gcd(\frac{\Delta}{2}, \frac{2^d(q-1)}{\lambda'}) = 2$, then, for any integer i, we have*

$$\{xy \mid x \in \mathcal{D}_i^{(\frac{2^{d+1}(q-1)}{\lambda'})} \text{ and } y \in \mathbb{F}_q^*\} = \frac{2\lambda'}{2^d} * \mathcal{D}_i^{(4)},$$

where $\frac{2\lambda'}{2^d} * \mathcal{D}_i^{(4)}$ is the multiset in which each element of $\mathcal{D}_i^{(4)}$ appears with multiplicity $\frac{2\lambda'}{2^d}$.

Proof. Since $\mathcal{D}_i^{(j)} = \gamma^i \mathcal{D}_0^{(j)}$, for all integers i and j, then, without loss of generality, we can simply suppose that $i = 0$. By hypothesis, we can see that $\frac{2^{d-1}(q-1)}{\lambda'}$ is a positive integer, and also that $\gcd(\Delta, \frac{2^{d+1}(q-1)}{\lambda'}) = 4$. Now, since $4|\Delta$ and $d = 0$ or 1, then observe that $(\frac{2^{d+1}(q-1)}{\lambda'})|(q^k - 1)$. Thus, for each $x \in \mathcal{D}_0^{(\frac{2^{d+1}(q-1)}{\lambda'})}$ and $y \in \mathbb{F}_q^*$, there exist unique integers l_1 and l_2, with $0 \le l_1 < \lambda'\frac{\Delta}{2^{d+1}}$ and $0 \le l_2 < (q - 1)$, such that

$$xy = \gamma^{\frac{2^{d+1}(q-1)}{\lambda'}l_1 + \Delta l_2},$$
$$= (\gamma^4)^{\frac{2^{d-1}(q-1)}{\lambda'}l_1 + \frac{\Delta}{4}l_2}.$$

Therefore $xy \in \mathcal{D}_0^{(4)}$. Now, trivially, we have $\frac{\Delta}{4}|\lambda'\frac{\Delta}{2^{d+1}}$ and $\frac{2^{d-1}(q-1)}{\lambda'}|(q - 1)$. Thus, since $\gcd(\frac{\Delta}{4}, \frac{2^{d-1}(q-1)}{\lambda'}) = 1$, and since $|\langle\gamma^4\rangle| = \frac{q^k-1}{4}$, we conclude that each element xy will appear with multiplicity $\lambda'\frac{\Delta}{2^{d+1}}(q - 1)/(\frac{q^k-1}{4}) = \frac{2\lambda'}{2^d}$. □

The following result will be important in order to determine the weight distribution of the class of non-irreducible cyclic codes that we are interested in.

Lemma 6. *For integers i and j, with $0 \le i, j \le 3$, let*

$$\mathcal{E}_{(i,j)} = \{(\alpha, \alpha) \mid \alpha \in \mathcal{D}_i^{(4)}\} \cup \{(\alpha, -\alpha) \mid \alpha \in \mathcal{D}_j^{(4)}\},$$

and

$$\mathcal{G}_{(i,j)} = \{(\alpha, \beta) \in \mathbb{F}_{q^k} \times \mathbb{F}_{q^k} \mid (\beta + \alpha) \in \mathcal{D}_i^{(4)} \text{ and } (\beta - \alpha) \in \mathcal{D}_j^{(4)}\}.$$

Also, let w be as in Theorem 1, and for any even integer r define the following six sets:

$$\mathcal{S}_1 = \{ (0,0)\}$$

$$\mathcal{S}_2 = \begin{cases} \mathcal{E}_{(w,w+2)} & \text{if } r/2 \text{ is odd} \\ \mathcal{E}_{(w,w)} & \text{if } r/2 \text{ is even} \end{cases},$$

$$\mathcal{S}_3 = \begin{cases} \mathcal{E}_{(w+2,w)} \quad \cup \mathcal{E}_{(1,1)} \cup \mathcal{E}_{(3,3)} & \text{if } r/2 \text{ is odd} \\ \mathcal{E}_{(w+2,w+2)} \cup \mathcal{E}_{(1,1)} \cup \mathcal{E}_{(3,3)} & \text{if } r/2 \text{ is even} \end{cases},$$

$$\mathcal{S}_4 = \begin{cases} ((\cup_{i=0}^3 \mathcal{G}_{(i,w+2)}) \cup (\cup_{j=0}^3 \mathcal{G}_{(w,j)})) \setminus \mathcal{G}_{(w,w+2)} & \text{if } r/2 \text{ is odd} \\ ((\cup_{i=0}^3 \mathcal{G}_{(i,w)}) \quad \cup (\cup_{j=0}^3 \mathcal{G}_{(w,j)})) \setminus \mathcal{G}_{(w,w)} & \text{if } r/2 \text{ is even} \end{cases},$$

$$\mathcal{S}_5 = \begin{cases} \mathcal{G}_{(w,w+2)} & \text{if } r/2 \text{ is odd} \\ \mathcal{G}_{(w,w)} & \text{if } r/2 \text{ is even} \end{cases},$$

$$\mathcal{S}_6 = \quad (\cup_{i=0}^3 \cup_{j=0}^3 \mathcal{G}_{(i,j)}) \setminus (\mathcal{S}_4 \cup \mathcal{S}_5),$$

where the subscript $w + 2$ is taken modulo 4 (thus, if $w = 2$ then, for example, $\mathcal{E}_{(w,w+2)} = \mathcal{E}_{(2,0)}$). Then, the sets $\mathcal{S}_l$, $l = 1, 2, 3, 4, 5, 6$, are pairwise disjoint, and $\mathbb{F}_{q^k} \times \mathbb{F}_{q^k} = \cup_{l=1}^6 \mathcal{S}_l$. Their cardinalities are: $|\mathcal{S}_1| = 1$, $|\mathcal{S}_2| = \frac{q^k-1}{2}$, $|\mathcal{S}_3| = \frac{3(q^k-1)}{2}$, $|\mathcal{S}_4| = \frac{6(q^k-1)^2}{16}$, $|\mathcal{S}_5| = \frac{(q^k-1)^2}{16}$ and $|\mathcal{S}_6| = \frac{9(q^k-1)^2}{16}$. Furthermore, let η_0 and η_1 be as in Theorem 1, thus if $\alpha, \beta \in \mathbb{F}_{q^k}$ then

$$\sum_{m=0}^{1} \sum_{z \in \mathcal{D}_{rm}^{(4)}} \chi(z(\beta + (-1)^m \alpha)) = \begin{cases} \frac{q^k-1}{2} & \text{if } (\alpha, \beta) \in \mathcal{S}_1 \\ \frac{q^k-1}{4} + \eta_0 & \text{if } (\alpha, \beta) \in \mathcal{S}_2 \\ \frac{q^k-1}{4} + \eta_1 & \text{if } (\alpha, \beta) \in \mathcal{S}_3 \\ \eta_0 + \eta_1 & \text{if } (\alpha, \beta) \in \mathcal{S}_4 \\ 2\eta_0 & \text{if } (\alpha, \beta) \in \mathcal{S}_5 \\ 2\eta_1 & \text{if } (\alpha, \beta) \in \mathcal{S}_6 \end{cases}.$$

Remark 2. Since $4|\Delta$, then $\mathbb{F}_q^* \subset \mathcal{D}_0^{(4)}$. Therefore, for any integer i, observe that $\alpha \in \mathcal{D}_i^{(4)}$ if and only if $2\alpha \in \mathcal{D}_i^{(4)}$. In addition, if i and l are integers and if $\rho \in \mathcal{D}_l^{(4)}$, then observe that

$$\sum_{z \in \mathcal{D}_i^{(4)}} \chi(z\rho) = \sum_{z \in \mathcal{D}_{i+l}^{(4)}} \chi(z).$$

Proof. Since $-1 \neq 1$, on the finite field $\mathbb{F}_{q^k}$, then the first assertion comes from a direct inspection of sets $\mathcal{S}_l$, $l = 1, 2, 3, 4, 5, 6$.

Trivially, $|\mathcal{S}_1| = 1$. On the other hand, since $4|(q^k - 1)$, then observe that $|\mathcal{D}_i^{(4)}| = \frac{q^k-1}{4}$, for all $0 \le i \le 3$, and this implies that $|\mathcal{E}_{(i,j)}| = 2|\mathcal{D}_0^{(4)}| = \frac{q^k-1}{2}$, for $0 \le i, j \le 3$. Therefore, $|\mathcal{S}_2| = \frac{q^k-1}{2}$ and $|\mathcal{S}_3| = \frac{3(q^k-1)}{2}$. Now, let

$$M = \begin{pmatrix} 1 & 1 \\ 1 & -1 \end{pmatrix}.$$

Clearly, matrix M is invertible over $\mathbb{F}_{q^k}$. Thus, the linear transformation $T_M : \mathbb{F}_{q^k}^2 \to \mathbb{F}_{q^k}^2$, given by the rule

$$T_M(\alpha, \beta) = \begin{pmatrix} \beta + \alpha \\ \beta - \alpha \end{pmatrix},$$

is a bijection among the vectors in $\mathbb{F}_{q^k}^2$. Thus, if $(\delta_i, \delta_j) \in \mathcal{D}_i^{(4)} \times \mathcal{D}_j^{(4)}$ then, there must exist an unique vector $(\alpha, \beta) \in \mathbb{F}_{q^k}^2$ such that $\beta + \alpha = \delta_i$ and $\beta - \alpha = \delta_j$. Therefore, $|\mathcal{G}_{(i,j)}| = |\mathcal{D}_i^{(4)} \times \mathcal{D}_j^{(4)}| = \frac{q^k-1}{4} \cdot \frac{q^k-1}{4} = \frac{(q^k-1)^2}{16}$, for $0 \le i, j \le 3$. Thus $|\mathcal{S}_4| = \frac{6(q^k-1)^2}{16}$, $|\mathcal{S}_5| = \frac{(q^k-1)^2}{16}$ and $|\mathcal{S}_6| = \frac{9(q^k-1)^2}{16}$.

Observe that

$$\sum_{m=0}^{1} \sum_{z \in \mathcal{D}_{rm}^{(4)}} \chi(z(\beta + (-1)^m \alpha)) = \begin{cases} \sum_{z \in \mathcal{D}_0^{(4)}} \chi(z(\beta + \alpha)) + \sum_{z \in \mathcal{D}_2^{(4)}} \chi(z(\beta - \alpha)), & \text{if } r/2 \text{ is odd} \\ \\ \sum_{z \in \mathcal{D}_0^{(4)}} \chi(z(\beta + \alpha)) + \sum_{z \in \mathcal{D}_0^{(4)}} \chi(z(\beta - \alpha)), & \text{if } r/2 \text{ is even} \end{cases},$$

and that $\sum_{z \in \mathcal{D}_i^{(4)}} \chi(0) = |\mathcal{D}_0^{(4)}| = \frac{q^k-1}{4}$, for $0 \le i \le 3$. On the other hand, recall that w can only take values 0 or 2. Therefore, owing to Remark 2, we can see now that the last assertion follows from (2) and from the definition of the sets $\mathcal{S}_l$, $l = 1, 2, 3, 4, 5, 6$. □

Table 1. Value distribution of $\displaystyle\sum_{m=0}^{1}\sum_{z\in\mathcal{D}_{rm}^{(4)}}\chi(z(\beta+(-1)^m\alpha))$, where r is any even integer

Value	Frequency
$\frac{q^k-1}{2}$	1
$\frac{q^k-1}{4}+\eta_0$	$\frac{q^k-1}{2}$
$\frac{q^k-1}{4}+\eta_1$	$\frac{3(q^k-1)}{2}$
$\eta_0+\eta_1$	$\frac{6(q^k-1)^2}{16}$
$2\eta_0$	$\frac{(q^k-1)^2}{16}$
$2\eta_1$	$\frac{9(q^k-1)^2}{16}$

Remark 3. As a direct consequence of the previous lemma observe that, regardless of the even integer r, the value distribution for the exponential sum $\sum_{m=0}^{1}\sum_{z\in\mathcal{D}_{rm}^{(4)}}\chi(z(\beta+(-1)^m\alpha))$ is given by Table 1.

As we will see in the next section, the value distribution of the previous exponential sum, will determine the weight distribution of a family of non-irreducible cyclic codes.

4   The Weight Distribution of a Family of Non-irreducible Cyclic Codes

We begin with this section recalling the following already known identity:

Let $\mathcal{C}$ be an N-weight linear code, over $\mathbb{F}_q$, of length n and dimension $2k$. Suppose that $w_1, w_2, \cdots, w_N$ are the nonzero weights of $\mathcal{C}$. For $1 \leq i \leq N$, let A_i be the number of words of weight w_i in $\mathcal{C}$ and let B_j be the number of words of weight j in $\mathcal{C}^\perp$ (the dual code of $\mathcal{C}$). Then, the third identity of Pless (see [4, p. 259] for the general result), for $\mathcal{C}$, is

$$\sum_{i=1}^{N} w_i^2 A_i = [n(q-1)(n(q-1)+1) - B_1(q+2(n-1)(q-1)) + 2B_2]q^{2k-2}. \quad (3)$$

In the context of the previous identity, observe that a linear code is projective if and only if B_1 and B_2 are zero in (3).

By keeping in mind this identity, we are now able to obtain the weight distribution of a family of non-irreducible cyclic codes.

Theorem 2. *Let q, k and Δ be as before. Considering our main assumption, we also take λ to be a divisor of $q-1$ and define $n = \lambda\Delta$. Let a_1 and a_2 be two integers such that $a_2 n \equiv 0 \pmod{q^k-1}$ and $a_2 q^j - a_1 \equiv \frac{q^k-1}{2} \pmod{q^k-1}$, for some integer j with $1 \leq q^j < q^k$. Let $\mathcal{C}_{(a_1,a_2)}$ be the cyclic code with parity-check*

Table 2. Weight distribution of $\mathcal{C}_{(a_1,a_2)}$

Weight	Frequency
0	1
$\frac{\lambda}{2}(q^{k-1} - 3(-q)^{(k-2)/2})$	$\frac{q^k-1}{2}$
$\frac{\lambda}{2}(q^{k-1} + (-q)^{(k-2)/2})$	$\frac{3(q^k-1)}{2}$
$\lambda(q^{k-1} - (-q)^{(k-2)/2})$	$\frac{6(q^k-1)^2}{16}$
$\lambda(q^{k-1} - 3(-q)^{(k-2)/2})$	$\frac{(q^k-1)^2}{16}$
$\lambda(q^{k-1} + (-q)^{(k-2)/2})$	$\frac{9(q^k-1)^2}{16}$

polynomial $h_{a_1}(x)h_{a_2}(x)$. If $\gcd(\frac{\Delta}{2}, a_2) = 2$, then the following three assertions are true:

(A) $\deg(h_{a_1}(x)) = \deg(h_{a_2}(x)) = k$ *and* $h_{a_1}(x) \neq h_{a_2}(x)$.

(B) $\mathcal{C}_{(a_1,a_2)}$ *is an* $[n, 2k]$ *cyclic code with the weight distribution given in Table 2.*

(C) *If B_1 and B_2 are, respectively, the number of words of weight 1 and 2 in the dual code of $\mathcal{C}_{(a_1,a_2)}$, then $B_1 = 0$ and $B_2 = \frac{n(q-1)(2\lambda-1)}{2}$. Therefore $\mathcal{C}_{(a_1,a_2)}$ is a non-projective cyclic code.*

Proof. First observe that, for any integer $\jmath$, we have that γ^{-a_2} and $\gamma^{-a_2 q^{\jmath}}$ are conjugates. Thus, without loss of generality, we can simply suppose that $a_1 = a_2 + \frac{q^k-1}{2}$.

Part (A): Let s be the smallest positive integer such that $a_2 q^s \equiv a_2 \pmod{n}$. Then $\lambda\Delta | a_2(q^s - 1)$. But $\gcd(\Delta, 2a_2) = 4$; thus, condition $\lambda\Delta | (2a_2 \frac{q^s-1}{2})$ implies that $\Delta | 2(q^s - 1)$. However, thanks to Lemma 2, we know that this last condition is impossible if $s < k$, therefore, $\deg(h_{a_2}(x)) = k$. On the other hand, $\gcd(\frac{\Delta}{2}, a_1) = \gcd(\frac{\Delta}{2}, a_2 + \frac{q^k-1}{2}) = \gcd(\frac{\Delta}{2}, a_2 + \frac{\Delta}{2}(q-1)) = \gcd(\frac{\Delta}{2}, a_2) = 2$, thus we can similarly conclude that $\deg(h_{a_1}(x)) = k$.

Now, suppose that $h_{a_1}(x) = h_{a_2}(x)$. Then, there must exist an integer $0 \leq s < k$ such that $a_2 q^s \equiv a_1 \pmod{q^k - 1}$. But $a_1 = a_2 + \frac{q^k-1}{2}$, thus the last congruence implies $a_2(q^s - 1) \equiv \frac{q^k-1}{2} \pmod{q^k - 1}$, which in turn implies that $a_2(q^s - 1) \equiv 0 \pmod{\frac{q^k-1}{2}}$. In consequence, $(q^k - 1) | 2a_2(q^s - 1)$, and therefore $\Delta | 2a_2((q^s - 1)/(q-1))$. Now since $\gcd(\Delta, 2a_2) = 4$, we have $\Delta | 4((q^s - 1)/(q-1))$, and therefore $(q^k - 1) | 4(q^s - 1)$. But such condition is impossible if $s < k$ and $q \geq 5$, thus, $h_{a_1}(x) \neq h_{a_2}(x)$.

Part (B): Since $a_1 n = a_2 n + (q^k - 1)\lambda\frac{\Delta}{2} \equiv 0 \pmod{q^k - 1}$, then the cyclic code, $\mathcal{C}_{(a_1,a_2)}$, has length n and its dimension is $2k$, due to Part (A).

Now, for each $\alpha, \beta \in \mathbb{F}_{q^k}$, we define $c(n, a_1, a_2, \alpha, \beta)$ as the vector of length n over $\mathbb{F}_q$, which is given by:

$$\left(\mathrm{Tr}_{\mathbb{F}_{q^k}/\mathbb{F}_q}(\alpha(\gamma^{a_1})^0 + \beta(\gamma^{a_2})^0), \cdots, \mathrm{Tr}_{\mathbb{F}_{q^k}/\mathbb{F}_q}(\alpha(\gamma^{a_1})^{n-1} + \beta(\gamma^{a_2})^{n-1})\right).$$

Thanks to Delsarte's Theorem (see, for example, [1]), it is well known that

$$\mathcal{C}_{(a_1,a_2)} = \{c(n, a_1, a_2, \alpha, \beta) \mid \alpha, \beta \in \mathbb{F}_{q^k}\} \, .$$

Thus the Hamming weight of any codeword $c(n, a_1, a_2, \alpha, \beta) \in \mathcal{C}_{(a_1,a_2)}$ is equal to $n - Z(\alpha, \beta)$, where

$$Z(\alpha, \beta) = \sharp\{m \mid 0 \le m < n, \text{ and } \mathrm{Tr}_{\mathbb{F}_{q^k}/\mathbb{F}_q}(\alpha\gamma^{a_1 m} + \beta\gamma^{a_2 m}) = 0\} \, .$$

Now, if χ' is the canonical additive character of $\mathbb{F}_q$, then, by the orthogonal property in (1), we know that for each $c \in \mathbb{F}_q$ we have

$$\sum_{y \in \mathbb{F}_q} \chi'(yc) = \begin{cases} q & \text{if } c = 0 \\ 0 & \text{if } c \ne 0 \end{cases} ,$$

thus

$$Z(\alpha, \beta) = \frac{1}{q} \sum_{m=0}^{n-1} \sum_{y \in \mathbb{F}_q} \chi'(\mathrm{Tr}_{\mathbb{F}_{q^k}/\mathbb{F}_q}(y(\alpha\gamma^{a_1 m} + \beta\gamma^{a_2 m}))) \, .$$

If χ denotes the canonical additive character of $\mathbb{F}_{q^k}$, then χ' and χ are related by $\chi'(\mathrm{Tr}_{\mathbb{F}_{q^k}/\mathbb{F}_q}(\varepsilon)) = \chi(\varepsilon)$ for all $\varepsilon \in \mathbb{F}_{q^k}$. Therefore, we have

$$Z(\alpha, \beta) = \frac{n}{q} + \frac{1}{q} \sum_{m=0}^{n-1} \sum_{y \in \mathbb{F}_q^*} \chi(y(\alpha\gamma^{a_1 m} + \beta\gamma^{a_2 m}))$$

$$= \frac{n}{q} + \frac{1}{q} \sum_{m=0}^{n-1} \sum_{y \in \mathbb{F}_q^*} \chi(\gamma^{a_2 m} y((-1)^m \alpha + \beta)),$$

where the last equality arises because $a_1 = a_2 + \frac{q^k-1}{2}$ and $\gamma^{\frac{q^k-1}{2}} = -1$. Now, since $\gcd(\frac{\Delta}{2}, a_2) = 2$ and $\mathcal{C}_{(a_1,a_2)} = \mathcal{C}_{(a_2,a_1)}$, then, thanks to Corollary 1, we can assume without loss of generality that $\gcd(q^k - 1, a_2) = 2^d \frac{q-1}{\lambda'}$ for some integers d and λ' such that $d = 0$ or 1, and λ' is the divisor of $q - 1$ satisfying $\gcd(q - 1, a_2) = \frac{q-1}{\lambda'}$. But these conditions are the hypotheses in Lemma 4, thus

$$Z(\alpha, \beta) = \frac{n}{q} + \frac{2^d \lambda}{q\lambda'} \sum_{m=0}^{n'-1} \sum_{y \in \mathbb{F}_q^*} \chi(\gamma^{2^d \frac{q-1}{\lambda'} m} y((-1)^m \alpha + \beta)) \, ,$$

where $n' = \lambda' \frac{\Delta}{2^d}$. But we know that $d = 0$ or 1, then $|\mathcal{D}_0^{(\frac{2^d(q-1)}{\lambda'})}| = n' = \lambda' \frac{\Delta}{2^d}$ is an even integer, thus observe that

$$\{\gamma^{2^d \frac{q-1}{\lambda'} m} \mid 0 \le m < n'\} = \mathcal{D}_0^{(\frac{2^d(q-1)}{\lambda'})} = \mathcal{D}_0^{(\frac{2^{d+1}(q-1)}{\lambda'})} \cup \mathcal{D}_{2^d \frac{q-1}{\lambda'}}^{(\frac{2^{d+1}(q-1)}{\lambda'})} \, .$$

Therefore,

$$Z(\alpha, \beta) = \frac{n}{q} + \frac{2^d \lambda}{q \lambda'} \sum_{m=0}^{1} \sum_{x \in \mathcal{D}^{(\frac{2^{d+1}(q-1)}{\lambda'})}_{2^d \frac{q-1}{\lambda'} m}} \sum_{y \in \mathbb{F}_q^*} \chi(xy((-1)^m \alpha + \beta)) . \tag{4}$$

Since $\gcd(\frac{\Delta}{2}, a_2) = 2$, then $2^d \frac{q-1}{\lambda'} = \gcd(q^k - 1, a_2)$ is an even integer such that $2^d \frac{q-1}{\lambda'} | a_2$, therefore $2 \le \gcd(\frac{\Delta}{2}, 2^d \frac{q-1}{\lambda'}) \le \gcd(\frac{\Delta}{2}, a_2) = 2$. Thus, the conclusion is that $\gcd(\frac{\Delta}{2}, 2^d \frac{q-1}{\lambda'}) = 2$. Therefore, after applying Lemma 5 to (4), we obtain

$$Z(\alpha, \beta) = \frac{n}{q} + \frac{2\lambda}{q} \sum_{m=0}^{1} \sum_{z \in \mathcal{D}^{(4)}_{2^d \frac{q-1}{\lambda'} m}} \chi(z(\beta + (-1)^m \alpha)) .$$

Since the Hamming weight of any codeword $c(n, a_1, a_2, \alpha, \beta) \in \mathcal{C}_{(a_1, a_2)}$ is equal to $n - Z(\alpha, \beta)$, and since $2^d \frac{q-1}{\lambda'}$ is an even integer, then the result now follows Remark 3.

Part (C): It is well known that there are no 1-weight words in the dual of any cyclic code (see, for example, [12]), therefore $B_1 = 0$. Since $q > 3$ and $B_1 = 0$, the assertion about B_2, follows directly from Table 2 and (3). Finally, since $n(q-1) \ne 0$ and $\lambda \ne \frac{1}{2}$, then $B_2 \ne 0$. Therefore $\mathcal{C}_{(a_1, a_2)}$ is a non-projective cyclic code. □

The following is a direct application of the previous theorem.

Example 1. If $q = 7$, $k = 2$, $\lambda = 1$ and $a_2 = 6$, then $\Delta = 8$ and $\gcd(\frac{\Delta}{2}, a_2) = 2$. In addition, observe that $\frac{\lambda}{2}(q^{k-1} + (-q)^{(k-2)/2}) = \lambda(q^{k-1} - 3(-q)^{(k-2)/2}) = 4$. Therefore, $\mathcal{C}_{(30,6)}$ is a 4-weight non-projective cyclic code, over $\mathbb{F}_7$, of length 8, dimension 4 and weight enumerator polynomial $A(z) = 1 + 24z^2 + 216z^4 + 864z^6 + 1296z^8$.

Observe that, for a divisor λ of $q - 1$ and for any integer a_2, Theorem 2 basically states that if a_2 satisfies the condition $\gcd(\frac{\Delta}{2}, a_2) = 2$, then, for any integer $\jmath$, the code $\mathcal{C}_{(a_2 q^{\jmath} \pm \frac{q^k - 1}{2}, a_2)}$ could be described by means of the three assertions in such theorem. Thanks to this easy-to-check condition we can now give the exact number of different cyclic codes $\mathcal{C}_{(a_1, a_2)}$ that satisfy the hypotheses in Theorem 2, but before this, we want to point out the following:

Remark 4. If ϕ denotes the Euler ϕ-function (see, for example, [5, p. 20]) and if ζ and $\frac{\Delta}{4}$ are two positive integers, then the number of integers between 1 and $\zeta \frac{\Delta}{4}$, relatively prime to $\frac{\Delta}{4}$, is $\zeta \phi(\frac{\Delta}{4})$.

Theorem 3. *Let q, k and Δ be as before. Considering our main assumption, we also take λ to be a divisor of $q - 1$ and define $n = \lambda \Delta$. For any integers a_1 and a_2, let $\mathcal{C}_{(a_1, a_2)}$ be the cyclic code with parity-check polynomial $h_{a_1}(x) h_{a_2}(x)$. Let $\mathcal{N}_{(q,k,\lambda)}$ be the number of cyclic codes, $\mathcal{C}_{(a_1, a_2)}$ of length n and dimension $2k$ that satisfy conditions in Theorem 2. Then*

$$\mathcal{N}_{(q,k,\lambda)} = \begin{cases} 0 & \text{if } \gcd(\frac{\Delta}{2}, \frac{q-1}{\lambda}) > 2 \\ \\ \frac{2\lambda\phi(\frac{\Delta}{4})}{\gcd(\lambda,2)k} & \text{otherwise} \end{cases}.$$

Proof. A cyclic code $\mathcal{C}_{(a_1,a_2)}$ belongs to the family of codes described by Theorem 2, if $a_2 n \equiv 0 \pmod{q^k - 1}$, $\gcd(\frac{\Delta}{2}, a_2) = 2$ and $a_1 = a_2 q^{\jmath} \pm \frac{q^k - 1}{2}$, for some integer $\jmath$ with $1 \le q^{\jmath} < q^k$. Then $a_2 = \frac{q-1}{\lambda} u$, for some integer u. If we suppose that $\gcd(\frac{\Delta}{2}, \frac{q-1}{\lambda}) > 2$ then, clearly, $\gcd(\frac{\Delta}{2}, a_2) > 2$. Therefore $\mathcal{N}_{(q,k,\lambda)} = 0$, if $\gcd(\frac{\Delta}{2}, \frac{q-1}{\lambda}) > 2$. Thus we will suppose that $\gcd(\frac{\Delta}{2}, \frac{q-1}{\lambda}) \le 2$. Now, since each one of the minimal polynomials $h_{a_1}(x)$ and $h_{a_2}(x)$ has exactly k different conjugate roots, and since we have $\gcd(\frac{\Delta}{2}, a_1) = \gcd(\frac{\Delta}{2}, a_2 q^{\jmath} \pm \frac{q^k - 1}{2}) = \gcd(\frac{\Delta}{2}, a_2 q^{\jmath} \pm \frac{\Delta}{2}(q - 1)) = \gcd(\frac{\Delta}{2}, a_2 q^{\jmath}) = \gcd(\frac{\Delta}{2}, a_2)$ for any integer $\jmath$, then

$$\begin{aligned} \mathcal{N}_{(q,k,\lambda)} &= \frac{\sharp\{a_2 \mid a_2 n \equiv 0 \,(\mathrm{mod}\ q^k - 1), \ \gcd(\frac{\Delta}{2}, a_2) = 2 \text{ and } 0 \le a_2 < (q^k - 1)\}}{2k} \\ &= \frac{\sharp\{u \mid \gcd(\frac{\Delta}{2}, \frac{q-1}{\lambda} u) = 2 \text{ and } 0 \le u < n\}}{2k}. \end{aligned}$$

But $\frac{q-1}{2}$ is an odd integer, and since we are supposing $\gcd(\frac{\Delta}{2}, \frac{q-1}{\lambda}) \le 2$, thus

$$\mathcal{N}_{(q,k,\lambda)} = \frac{\sharp\{u \mid \gcd(\frac{\Delta}{4}, u) = 1 \text{ and } 0 \le u < \frac{n}{\gcd(\lambda,2)}\}}{2k}.$$

Clearly $\frac{n}{\gcd(\lambda,2)} = \frac{4\lambda}{\gcd(\lambda,2)} \frac{\Delta}{4}$, therefore the result now follows directly from Remark 4. $\square$

The following are direct applications of Theorem 3.

Example 2. If $q = 7$, $k = 2$ and $\lambda = 3$, then $\Delta = 8$ and $\mathcal{N}_{(7,2,3)} = 3$. In fact, if $q = 7$, $k = 2$ and $\lambda = 3$, then the family of cyclic codes $\mathcal{C}_{(a_1,a_2)}$ described by Theorem 2 are $\mathcal{C}_{(2,26)}$, $\mathcal{C}_{(6,30)}$ and $\mathcal{C}_{(10,34)}$. These three codes are 4-weight non-projective cyclic codes of length 24, dimension 4 and weight enumerator polynomial $A(z) = 1 + 24z^6 + 216z^{12} + 864z^{18} + 1296z^{24}$.

Example 3. If $q = 11$, $k = 2$ and $\lambda = 10$, then $\Delta = 12$ and $\mathcal{N}_{(11,2,10)} = 10$. These ten codes are 5-weight non-projective cyclic codes of length 120, dimension 4 and weight enumerator polynomial $A(z) = 1 + 60z^{40} + 180z^{60} + 900z^{80} + 5400z^{100} + 8100z^{120}$.

5 Conclusion

In this work we presented the evaluation of a specific exponential sum, which can be derived directly as a particular instance of a general result that was originally presented in [8]. We continued by determining the value distribution of this specific exponential sum, and we used it in order to present the weight distribution of a family of reducible cyclic codes. As we have shown here, all the codes in this family are non-projective cyclic codes. Furthermore, we show that the codes in this family can be identified in a quite easy way, and as a by-product of this easy identification we were able to determine the exact number of cyclic codes in a family, when length and dimension are given.

Finally, it is interesting to realize that the family of codes studied in this work is completely different to the family of codes studied in [10]. However, the techniques that were used in order to study both families are quite similar. Thus, it is perhaps possible to develop a more general theory that includes these two families of codes.

Acknowledgement. The authors would like to thank the anonymous referees for their valuable time in reviewing the paper.

References

1. Delsarte, P.: On subfield subcodes of Reed-Solomon codes. IEEE Trans. Inform. Theory 21(5), 575–576 (1975)
2. Feng, K., Luo, J.: Weight distribution of some reducible cyclic codes. Finite Fields and Their Appl. 14(2), 390–409 (2008)
3. Helleseth, T.: Some two-weight codes with composite parity-check polynomials. IEEE Trans. Inform. Theory 22, 631–632 (1976)
4. Huffman, W.C., Pless, V.: Fundamental of Error-Correcting Codes. Cambridge University Press, Cambridge (2003)
5. Ireland, K., Rosen, M.: A Classical Introduction to Modern Number Theory. Springer, New York (1990)
6. Lidl, R., Niederreiter, H.: Finite Fields. Cambridge Univ. Press, Cambridge (1983)
7. Ma, C., Zeng, L., Liu, Y., Feng, D., Ding, C.: The Weight Enumerator of a Class of Cyclic Codes. IEEE Trans. Inf. Theory 57(1), 397–402 (2011)
8. Moisio, M.: A note on evaluations of some exponential sums. Acta Arith. 93, 117–119 (2000)
9. Vega, G.: Two-weight cyclic codes constructed as the direct sum of two one-weight cyclic codes. Finite Fields Appl. 14(3), 785–797 (2008)
10. Vega, G.: The Weight Distribution of an Extended Class of Reducible Cyclic Codes. IEEE Trans. Inform. Theory (in press, 2012), doi:10.1109/TIT.2012.2193376
11. Vega, G., Wolfmann, J.: New classes of 2-weight cyclic codes. Des. Codes Crypt. 42, 327–334 (2007)
12. Wolfmann, J.: Are 2-Weight Projective Cyclic Codes Irreducible? IEEE Trans. Inform. Theory 51(2), 733–737 (2005)

A New Method for Constructing Small-Bias Spaces from Hermitian Codes

Olav Geil[1], Stefano Martin[1], and Ryutaroh Matsumoto[1,2]

[1] Department of Mathematical Sciences, Aalborg University, Denmark
{olav,stefano}@math.aau.dk
[2] Department of Communications and Integrated Systems,
Tokyo Institute of Technology, Japan
ryutaroh@rmatsumoto.org

Abstract. We propose a new method for constructing small-bias spaces through a combination of Hermitian codes. For a class of parameters our multisets are much faster to construct than what can be achieved by use of the traditional algebraic geometric code construction. So, if speed is important, our construction is competitive with all other known constructions in that region. And if speed is not a matter of interest the small-bias spaces of the present paper still perform better than the ones related to norm-trace codes reported in [12].

Keywords: Small-bias space, balanced code, Gröbner basis, Hermitian code.

1 Introduction

Let $\boldsymbol{X} = (X_1, \dots, X_k)$ be a random vector that takes on values in $\mathbb{F}_2^k$. As shown by Vazirani [17] the variables $X_1, \dots, X_k$ are independent and uniformly distributed if and only if

$$\mathrm{Prob}\left(\sum_{i \in T} X_i = 0\right) = \mathrm{Prob}\left(\sum_{i \in T} X_i = 1\right) = \frac{1}{2} \qquad (1)$$

holds for every non-empty set of indexes $T \subseteq \{1, \dots, k\}$. In particular, if (1) is to hold for a space $\mathcal{X} \subseteq \mathbb{F}_2^k$ then necessarily $\mathcal{X}$ must be equal to $\mathbb{F}_2^k$. There is a need for much smaller spaces $\mathcal{X} \subseteq \mathbb{F}_2^k$ with statistical properties close to that of (1). In the following by a space we will mean a multiset $\mathcal{X}$ with elements from $\mathbb{F}_2^k$ (this we write $\mathcal{X} \subseteq \mathbb{F}_2^k$). The multiset $\mathcal{X}$ is made into a probability space by adjoining to each element $\boldsymbol{x} \in \mathcal{X}$ the probability $p(\boldsymbol{x}) = i(\boldsymbol{x})/|\mathcal{X}|$ where $i(\boldsymbol{x})$ denotes the number of times $\boldsymbol{x}$ appears in $\mathcal{X}$. As a measure for describing how close a given space $\mathcal{X}$ is to the above situation with respect to randomization, Naor and Naor [15], and Alon et. al. [1] introduced the concept of ϵ-biasness [15, Def. 3]. (See also [14]).

F. Özbudak and F. Rodríguez-Henríquez (Eds.): WAIFI 2012, LNCS 7369, pp. 29–44, 2012.
© Springer-Verlag Berlin Heidelberg 2012

Definition 1. *A multiset $\mathcal{X} \subseteq \mathbb{F}_2^k$ is called an ϵ-bias space if*

$$\frac{1}{|\mathcal{X}|} \left| \sum_{x \in \mathcal{X}} (-1)^{\sum_{i \in T} x_i} \right| \leq \epsilon \tag{2}$$

holds for every non-empty index set $T \subseteq \{1, \ldots, k\}$.

Clearly, the ϵ in Definition 1 can be taken to be a number between 0 and 1. Good randomization properties are achieved when ϵ is close to 0 as (2) becomes (1) when $\epsilon = 0$. Multisets with ϵ small are called small-bias spaces. Citing [15, Abstract] they are used to construct almost k-wise independent random variables. From [15, Abstract] we have the following list of applications:

- Derandomization of algorithms.
- Reducing the number of random bits required by certain randomized algorithms, e.g., verification of matrix multiplication.
- Exhaustive testing of combinatorial circuits.
- Communication complexity: Two parties can verify equality of strings with high probability exchanging only a logarithmic number of bits.
- Hash functions.

Further examples can be found in [15, Sec. 10].
Rather than saying that a multiset is an ϵ-bias space we will often just say that it is ϵ-biased. Another name for ϵ-bias space is ϵ-bias set [2, Def. 1] and [12, Def. 1.1]. This notion may be a little misleading as the item under consideration is actually a multiset.
One way of constructing small-bias spaces is through the use of error-correcting codes.

Definition 2. *A binary $[n, k]$ code is said to be ϵ-balanced if every non-zero code word $\mathbf{c}$ satisfies*

$$\frac{1 - \epsilon}{2} \leq \frac{w_H(\mathbf{c})}{n} \leq \frac{1 + \epsilon}{2}.$$

Here $[n, k]$ means that the code is linear, of dimension k and length n. Further, w_H denotes the Hamming weight.

There is a simple direct translation [1] between the concepts described in Definition 1 and Definition 2:

Theorem 1. *Let G be a generator matrix for an ϵ-balanced binary $[n, k]$ code. The columns of G constitute an ϵ-bias space $\mathcal{X} \subseteq \mathbb{F}_2^k$ of size n. Similarly, using the elements of an ϵ-bias space $\mathcal{X}$ as columns of a generator matrix an ϵ-balanced code is derived.*

The following example illustrates the above theorem. It also shows why it is important in Definition 1 to work with multisets rather than sets.

Example 1. Consider the matrix

$$G = \begin{bmatrix} 0 & 1 & 0 & 1 & 0 & 1 & 0 & 1 & 0 & 0 & 0 & 0 \\ 0 & 0 & 1 & 1 & 0 & 0 & 1 & 1 & 0 & 0 & 0 & 0 \\ 1 & 1 & 1 & 1 & 1 & 1 & 1 & 1 & 0 & 0 & 0 & 0 \end{bmatrix}.$$

The code having G as a generator matrix is ϵ-balanced with $\epsilon = 1/3$ and indeed the multiset made from the columns of G is $\epsilon = 1/3$ biased. Treating the columns as a set (rather than a multiset) we derive

$$\mathcal{X}' = \{(0,0,1),(1,0,1),(0,1,1),(1,1,1),(0,0,0)\}.$$

The smallest value of ϵ for which $\mathcal{X}'$ is ϵ-biased is $\epsilon = 3/5$.

A standard construction from [1] tells us how to make small-balanced codes (meaning ϵ-biased codes with ϵ small):

Theorem 2. *Let $q = 2^s$ for some integer s and consider a q-ary $[N, K, D]$ code C. Let C_s be the (binary) $[2^s, s]_2$ Walsh-Hadamard code, $s \geq 1$. The concatenated code derived by using C as outer code and C_s as inner code is an $\epsilon = (N - D)/N$-balanced binary code of length $n = N2^s$ and dimension $k = Ks$.*

Proof. The result relies on the fact that every non-zero codeword of C_s contains exactly as many 0s as 1s.

The literature contains various examples of small-bias spaces that cannot all be compared to each other. We refer to [2, Sec. 1] for more details. In the following we will concentrate on important families of multisets for which comparison can be made. We remind the reader of how bigO notation works when given functions of multiple variables. In our situation we have real valued positive functions $f_i(x, y), i = 1, 2$ where x can take on any value in $\mathbb{Z}^+$ but for every fixed choice of x the variable y can only take on values in an interval $I(x) \subseteq \mathbb{R}^+$. By $f_1(x, y) = \mathcal{O}(f_2(x, y))$ we mean that a witness (C, κ) exists such that for all x with $\kappa < x$ and all $y \in I(x)$ it holds that $f_1(x, y) \leq Cf_2(x, y)$. We are interested in upper bounding the size of $\mathcal{X}$ which will be done in terms of bigO estimates as above. At the same time we are interested in lower bounding the length of the words in the multiset $\mathcal{X}$. Such estimates are described using bigOmega notation. We remind the reader that by definition $f(x) = \Omega(g(x))$ if and only if $g(x) = \mathcal{O}(f(x))$. As we are only interested in bigOmega estimates the meaning of k changes accordingly. In the following list of results note that a family of ϵ-bias spaces is considered to behave well if when given ϵ and k the size of $\mathcal{X}$ is small.

- Using Reed-Solomon codes as outer codes in Theorem 2 one achieves [1,2] for all possible choices of ϵ and k

$$\mathcal{X} \subseteq \mathbb{F}_2^{\Omega(k)}, \quad |\mathcal{X}| = \mathcal{O}\left(\frac{k^2}{\epsilon^2 \log^2(k/\epsilon)}\right).$$

This is called the RS-bound.

– Let $P_1, \ldots, P_{\mathcal{N}-1}, Q$ be rational places of an algebraic function field over $\mathbb{F}_q$ and denote by g the genus. Assume $\mathcal{N} = (\sqrt{q}-1)g$. That is, we assume that the function field attains the Drinfeld-Vladut bound. Using codes $C_{\mathcal{L}}(U = P_1 + \cdots + P_{\mathcal{N}-1}, mQ)$ with $g < m$ as outer codes one gets for all ϵ and k (see Section 2 for a discussion)

$$\mathcal{X} \subseteq \mathbb{F}_2^{\Omega(k)}, \quad |\mathcal{X}| = \mathcal{O}\left(\frac{k}{\epsilon^3 \log(1/\epsilon)}\right).$$

This result which is in the folklore is known as the AG-bound.

– Using Hermitian codes with $m < g$ as outer codes one achieves [2] for $\epsilon \geq k^{-\frac{1}{2}}$

$$\mathcal{X} \subseteq \mathbb{F}_2^{\Omega(k)}, \quad |\mathcal{X}| = \mathcal{O}\left(\left(\frac{k}{\epsilon^2 \log(1/\epsilon)}\right)^{\frac{5}{4}}\right). \tag{3}$$

This we call the BT-bound after the authors of [2], Ben-Aroya and Ta-Shma.

– Using in larger generality Norm-Trace codes of low dimension as outer codes one achieves [12] for $l = 4, 5, \ldots$ and $\epsilon \geq k^{-\frac{1}{\sqrt{l}}}$ (see Section 5)

$$\mathcal{X} \subseteq \mathbb{F}_2^{\Omega(k)}, \quad |\mathcal{X}| = \mathcal{O}\left(\left(\frac{k}{\epsilon^{l-\sqrt{l}} \log(1/\epsilon)}\right)^{\frac{l+1}{l}}\right).$$

Here, $l = 4$ corresponds to the Hermitian case described in [2].

– The Gilbert-Varshamov bound also applies to the small-bias spaces (as usual in a non-constructive way). It is derived by plugging into the Gilbert-Varshamov bound for binary codes $d = n/2$ and to make a Taylor approximation on the resulting formula. The construction uses Theorem 1 directly. It guarantees for all ϵ and k the existence of multisets with

$$\mathcal{X} \subseteq \mathbb{F}_2^{\Omega(k)}, \quad |\mathcal{X}| = \mathcal{O}\left(\frac{k}{\epsilon^2}\right).$$

– The linear programming bound tells us that we cannot hope to produce ϵ-bias spaces with

$$\mathcal{X} \subseteq \mathbb{F}_2^{\Omega(k)}, \quad |\mathcal{X}| = \mathcal{O}\left(\frac{k}{\epsilon^2 \log(1/\epsilon)}\right).$$

One way of comparing the above results is to choose $\epsilon = k^{-\alpha}, \alpha \in \mathbb{R}^+$ and then to take the logarithm with base k. The bigO notation suggests that we then let k go to infinity. The origin of this point of view is [2, Sec. 1]. When making the above operation we must be careful to specify which choices of α are allowed. We remind the reader of the little-o notation. Given functions $f_i(x) : \mathbb{Z}^+ \to \mathbb{R}^+, i = 1, 2$ by $f_1(x) = o(f_2(x))$ we mean that for every choice of $c \in \mathbb{R}^+$ there exists a $\kappa(c) \in \mathbb{Z}^+$ such that when $\kappa(c) < x$ then necessarily $f_1(x) \leq cf_2(x)$. The above list of results translates to (note that when given α we want $\log_k(|\mathcal{X}|)$ to approach a low value):

- RS-bound: The family of concatenated codes from Theorem 2 with Reed-Solomon codes as outer codes gives

$$\log_k(|\mathcal{X}|) = 2 + 2\alpha + o(1)$$

for all choices of $\alpha \in \mathbb{R}^+$.

- AG-bound: The family of concatenated codes from Theorem 2 with algebraic geometric codes as outer codes and $g < m$ gives

$$\log_k(|\mathcal{X}|) = 1 + 3\alpha + o(1)$$

for all choices of $\alpha \in \mathbb{R}^+$.

- BT-bound: The family of concatenated codes from Theorem 2 with Hermitian codes as outer codes and $m < g$ gives

$$\log_k(|\mathcal{X}|) = \frac{5}{4} + \frac{5}{2}\alpha + o(1)$$

for all choices of $\alpha \in]1/2, \infty[$.

- The family of concatenated codes from norm-trace codes of low dimension gives

$$\log_k(|\mathcal{X}|) = \frac{l+1}{l}(1 + \alpha(l - \sqrt{l})) + o(1)$$

for $l = 4, 5, \ldots$, and for all $\alpha \in [1/\sqrt{l}, \infty[$ (see Section 5).

- The Gilbert-Varshamov bound and the Linear Programming bound in combination tell us that we can achieve

$$\log_k(|\mathcal{X}|) = 1 + 2\alpha + o(1)$$

for all choices of $\alpha \in \mathbb{R}^+$ but no better than this.

In the present paper we shall introduce a new family of small-bias spaces using a combination of Hermitian codes as outer code. This family gives

$$\log_k(|\mathcal{X}|) = \frac{4}{3} + \frac{8}{3}\alpha + o(1)$$

for all choices of $\alpha \in \mathbb{R}^+$. We allow $2g < m$ and it is therefore surprising that for $\alpha \in]1, \infty[$ the achievements are better than those of the Hermitian codes with $g < m$. Our small-bias spaces perform better than the ones derived from norm-trace codes for all $l \geq 5$ (see Section 5 for the proof). For $\alpha < 1$ they behave better than what can be achieved using Reed-Solomon codes as outer code. For $\alpha < 1$ admittedly the new ϵ-bias spaces perform worse than the spaces coming from the AG construction. This, however, is only part of the picture. It turns out that to construct the spaces with $\alpha < 1/2$ from the AG construction requires quite a number of operations. In contrast, our construction is considerable faster. We shall revert to this issue in Section 4. Before dealing with the new construction we will investigate how to ensure $\epsilon = k^{-\alpha}$ in the case of the AG bound. It turns out that for $\alpha < 1/2$ the situation is rather complicated. We include the description here, as to our best knowledge, the details cannot be found in the literature.

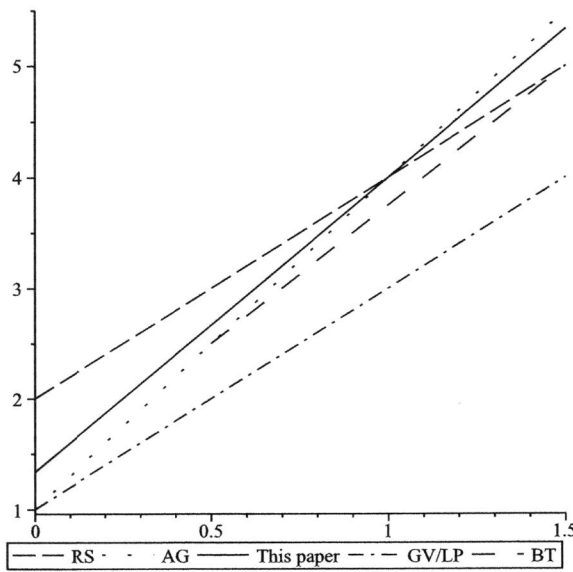

Fig. 1. Comparison of various constructions: First axis is α, second axis corresponds to $\log_k(|\mathcal{X}|)$ when $k \to \infty$.

2   The AG-Bound

Let q be a power of 2 and consider an algebraic function field over $\mathbb{F}_{q^2}$ of genus g with at least $\mathcal{N} = (q-1)g$ rational places. That is, the function field attains the Drinfeld-Vladut bound. As noted in the introduction Theorem 2 equipped with a one-point algebraic geometric code from the above function field produces ϵ-bias spaces $\mathcal{X} \subseteq \mathbb{F}_2^{\Omega(k)}$ with

$$|\mathcal{X}| = \mathcal{O}\left(\frac{k}{\epsilon^3 \log_2(\frac{1}{\epsilon})}\right). \tag{4}$$

In the following we investigate how to achieve corresponding values ϵ and k under the requirement $\epsilon = k^{-\alpha}$, $\alpha > 0$, and $k \to \infty$. Observe, that in this situation for any fixed α we have $\epsilon \to 0$. For completeness we start by proving (4) in this setting.

Consider rational places $P_1, \ldots, P_{\mathcal{N}-1}, Q$ and let $U = P_1 + \cdots + P_{\mathcal{N}-1}$ and $G = (ag)Q$ with $a \geq 1$. The code $C_{\mathcal{L}}(U, G)$ has parameters $N = (q-1)g - 1$, $K \geq \deg G - g = (a-1)g$, and $D \geq N - \deg G = ((q-1) - a)g - 1$. As we are interested in asymptotics we shall assume $N = (q-1)g$ and $D \geq ((q-1) - a)g$. From Theorem 2 we get ϵ-bias spaces with $\epsilon = a/(q-1)$, $\mathcal{X} \subseteq \mathbb{F}_2^{\Omega(k)}$. Here, $k = 2\log_2(q)(a-1)g$ and we have $|\mathcal{X}| = q^2 N = (q^3 - q^2)g$. As a is bounded below by 1 and $\epsilon \to 0$ we need $q \to \infty$ when $k \to \infty$. So the task basically boils down to establishing a sequence of function fields over increasingly large fields and a corresponding function $a(q)$ such that

$$|\mathcal{X}| = \mathcal{O}\left(\frac{2\log_2(q)(a-1)g}{\left(\frac{a}{q-1}\right)^3\log_2\left(\frac{q-1}{a}\right)}\right). \tag{5}$$

Note that the argument on the right side is a function in the single variable q as by construction now g is a function of q. We have

$$\frac{2\log_2(q)(a-1)g}{\left(\frac{a}{q-1}\right)^3\log_2\left(\frac{q-1}{a}\right)} \geq \frac{1}{2}\frac{\log_2(q)(a-1)}{a^3(\log_2(q-1)-\log_2(a))}|\mathcal{X}|$$

as $(q-1)^3 \geq \frac{1}{4}(q^3-q^2)$ holds for $q \geq 2$. In conclusion (5) holds if $a(q) = \mathcal{O}(1)$. We first assume that the sequence of function fields are the Hermitians which are function fields with $g = q(q-1)/2$. Here, actually the number of rational places is $2qg+q^2+1$ but we shall only use $(q-1)g$ of them. Let $a = 1+q^{-c}$ where $0 \leq c < 2$. Clearly, $a(q) = \mathcal{O}(1)$ as requested. We have $k = 2\log_2(q)q^{-c}g = q^{2-c}q^\beta$ where $\beta(q) \to 0$ for $q \to \infty$. Hence, asymptotically $\epsilon = k^{-\alpha}$ with $\alpha = 1/(2-c)$. In other words the situation is clear for $\alpha \in [\frac{1}{2}, \infty[$.

To achieve $\alpha \in]0, \frac{1}{2}[$ is more difficult. The problem is to keep $a(q) = \mathcal{O}(1)$ at the same time as having $\epsilon = k^{-\alpha}$. For this purpose we consider families of towers of function fields over $\mathbb{F}_{q^2}$ attaining the Drinfeld-Vladut bound [5]. We will need one tower for each value of q. Note that in such a tower for arbitrary $v \geq 2$ we can find a function field with $g \geq q^v$. Say $g = q^{v+d(q)}$, where $d(q) \geq 0$ holds. Let $a(q) = 1+q^{-d(q)}$ then clearly $a(q) = \mathcal{O}(1)$ holds. We have $k = 2\log_2(q)(a-1)g = q^{v+\beta}$ where $\beta(q) \to 0$ for $q \to \infty$. Also $\epsilon = q^{-1+\gamma}$ where $\gamma(q) \to 0$ for $q \to \infty$. Hence, $k^{-\alpha} = \epsilon$ asymptotically means $v\alpha = 1 \Rightarrow \alpha = 1/v$. As we only assumed $v \geq 2$ we have established that all $\alpha \in]0, \frac{1}{2}[$ can be attained.

For our purpose the best candidate for a family of good towers of function fields is the second construction by Garcia and Stichtenoth [5]. In [16] it was shown how to construct $C_{\mathcal{L}}(U,G)$ codes from this tower using

$$\mathcal{O}\left((N\log_q(N))^3\right) \tag{6}$$

operations over $\mathbb{F}_{q^2}$. Although we might only need codes of small dimension the method as stated requests us to find bases for all one-point codes. As shall be demonstrated in Section 4 the small-bias spaces of the present paper can be constructed much faster than what (6) guarantees for the AG construction.

3   The New Small-Bias Spaces

In the present paper we propose a new choice of outer codes in the construction of Theorem 2. As already mentioned this results in small-bias spaces with good properties. The new choice of outer codes is derived by combining two Hermitian codes as described below. The easiest way to explain the combination is by using the language of affine variety codes [4] and we therefore start our investigations with a presentation of Hermitian codes as such.

Definition 3. *Given a monomial ordering $\prec$ and an ideal $I \subseteq \mathbb{F}[X_1, \ldots, X_m]$ (here $\mathbb{F}$ is any field) the footprint is*

$$\Delta_\prec(I) := \{X_1^{\alpha_1} \cdots X_m^{\alpha_m} \mid X_1^{\alpha_1} \cdots X_m^{\alpha_m} \text{ is not a leading monomial}$$
$$\text{of any polynomial in } I\}.$$

We have the following two useful results [3, Pro. 4 and Pro. 8, Sec. 5.3].

Theorem 3. *The set $\{M + I \mid M \in \Delta_\prec(I)\}$ is a basis for $\mathbb{F}[X_1, \ldots, X_m]/I$ as a vector space over $\mathbb{F}$.*

As a corollary one gets the following result often referred to as the footprint bound [7,9].

Theorem 4. *Assume I is zero-dimensional (meaning that $\Delta_\prec(I)$ is finite). The variety $\mathbb{V}_{\mathbb{F}}(I)$ satisfies $|\mathbb{V}_{\mathbb{F}}(I)| \leq |\Delta_\prec(I)|$.*

Consider the Hermitian polynomial $X^{q+1} - Y^q - Y$ and the corresponding ideal

$$I = \langle X^{q+1} - Y^q - Y \rangle \subseteq \mathbb{F}_{q^2}[X, Y].$$

Define a monomial function w by $w(X) = q$ and $w(Y) = (q + 1)$ and consider the weighted degree monomial ordering $\prec_w$ given by $X^{\alpha_1} Y^{\beta_1} \prec_w X^{\alpha_2} Y^{\beta_2}$ if one of the following two conditions holds:

1. $w(X^{\alpha_1} Y^{\beta_1}) < w(X^{\alpha_2} Y^{\beta_2})$.
2. $w(X^{\alpha_1} Y^{\beta_1}) = w(X^{\alpha_2} Y^{\beta_2})$ but $\beta_1 < \beta_2$.

Observe for later use that no two different monomials in

$$\Delta_{\prec_w}(I) = \{X^i Y^j \mid 0 \leq i \text{ and } 0 \leq j < q\}$$

are of the same weight implying that $w : \Delta_{\prec_w}(I) \to \langle q, q+1 \rangle$ is a bijection. Observe also that the Hermitian polynomial $X^{q+1} - Y^q - Y$ contains exactly two monomials of highest weight. The implication of this is that

$$w(\mathrm{lm}(F(X, Y))) = w(\mathrm{lm}(F(X, Y) \text{ rem } \{X^{q+1} - Y^q - Y\}))$$

holds for any polynomial $F(X, Y)$ that possesses exactly one monomial of highest weight in its support.
Consider next the ideal

$$I_{q^2} := \langle X^{q^2} - X, Y^{q^2} - Y \rangle + I.$$

The variety $\mathbb{V}_{\mathbb{F}_{q^2}}(I) = \mathbb{V}_{\mathbb{F}_{q^2}}(I_{q^2})$ consists of $n = q^3$ different points $\{P_1, \ldots P_n\}$. The set $\{X^{q^2} - X, X^{q+1} - Y^q - Y\}$ constitutes a Gröbner basis for I_{q^2} with respect to $\prec_w$ and therefore

$$\Delta_{\prec_w}(I_{q^2}) = \{X^i Y^j \mid 0 \leq i < q^2, 0 \leq j < q\}$$

holds. It now follows from Theorem 3 that

$$\{X^i Y^j + I_{q^2} \mid 0 \leq i < q^2, 0 \leq j < q\}$$

is a basis for $\mathbb{F}_{q^2}[X,Y]/I_{q^2}$ as a vector space over $\mathbb{F}_{q^2}$. The code construction relies on the bijective evaluation map ev : $\mathbb{F}_{q^2}[X,Y]/I_{q^2} \to \mathbb{F}_{q^2}^n$ given by $\mathrm{ev}(F(X,Y)+I_{q^2}) = (F(P_1), \ldots, F(P_n))$. Theorem 4 tells us that we can estimate the Hamming weight of a word $\boldsymbol{c} = \mathrm{ev}(F(X,Y) + I_{q^2})$ by

$$w_H(\boldsymbol{c}) \geq n - |\Delta_{\prec_w}(\langle F(X,Y) \rangle) + I_{q^2})|.$$

Without loss of generality we can assume $\mathrm{Supp}(F) \subseteq \Delta_{\prec_w}(I_{q^2})$. From the discussion prior to the definition of I_{q^2} we conclude that no two different monomials in $F(X,Y)$ are of the same weight. As a consequence

$$w(\mathrm{lm}(X^\alpha Y^\beta F(X,Y))) = w(\mathrm{lm}(X^\alpha Y^\beta F(X,Y) \text{ rem } \{X^{q+1} - Y^q - Y\})$$

holds for all $X^\alpha Y^\beta$. Write $\Lambda = w(\Delta_{\prec_w}(I)) = \langle q, q+1 \rangle$, $\Lambda^* = w(\Delta_{\prec_w}(I_{q^2})) \subseteq \Lambda$ and $\lambda = w(\mathrm{lm}(F)) \in \Lambda^*$. We have

$$|\Delta_{\prec_w}(\langle F(X,Y) \rangle) + I_{q^2})| \leq |(\Lambda^* - (\lambda + \Lambda))| \leq |(\Lambda \backslash (\lambda + \Lambda)| = \lambda,$$

where the last equality comes from [10, Lem. 5.15]. Hence, $w_H(\boldsymbol{c}) \geq n - \lambda$ holds. Observe that

$$\Lambda^* = \{\lambda_1, \ldots, \lambda_g\} \cup \{2g, \ldots, n-1\} \cup \{\lambda_{n-g+1}, \ldots, \lambda_n\}, \tag{7}$$

where $\lambda_i \leq g-1+i$ for $i = 1, \ldots, g$. This is a general result for Weierstrass semigroups and not particular for the Hermitian function field. Having described the Hermitian codes as affine variety codes we are now ready to introduce the combination of codes on which our construction of small-bias spaces rely. Consider the ideal

$$I_{q^2}^{(2)} := \langle X_1^{q+1} - Y_1^q - Y_1, X_2^{q+1} - Y_2^q - Y_2, X_1^{q^2} - X_1, Y_1^{q^2} - Y_1, X_2^{q^2} - X_2, Y_2^{q^2} - Y_2 \rangle$$

and the corresponding variety

$$\mathbb{V}_{\mathbb{F}_{q^2}}(I_{q^2}^{(2)}) = \mathbb{V}_{\mathbb{F}_{q^2}}(I_{q^2}) \times \mathbb{V}_{\mathbb{F}_{q^2}}(I_{q^2}) = \{Q_1, \ldots, Q_{q^6}\}.$$

Define a monomial function $w^{(2)}$ given by $w^{(2)}(X_1) = (q,0)$, $w^{(2)}(Y_1) = (q+1,0)$, $w^{(2)}(X_2) = (0,q)$, and finally $w^{(2)}(Y_2) = (0,q+1)$. Let $\prec_{\mathbb{N}_0^2}$ be any monomial ordering on $\mathbb{N}_0^2$ and define $\prec_{w^{(2)}}$ by

$$X_1^{\alpha_1^{(1)}} Y_1^{\beta_1^{(1)}} X_2^{\alpha_2^{(1)}} Y_2^{\beta_2^{(1)}} \prec_w^{(2)} X_1^{\alpha_1^{(2)}} Y_1^{\beta_1^{(2)}} X_2^{\alpha_2^{(2)}} Y_2^{\beta_2^{(2)}}$$

if one of the following two conditions holds:

1. $w^{(2)}(X_1^{\alpha_1^{(1)}} Y_1^{\beta_1^{(1)}} X_2^{\alpha_2^{(1)}} Y_2^{\beta_2^{(1)}}) \prec_{\mathbb{N}_0^2} w^{(2)}(X_1^{\alpha_1^{(2)}} Y_1^{\beta_1^{(2)}} X_2^{\alpha_2^{(2)}} Y_2^{\beta_2^{(2)}})$

2. $w^{(2)}(X_1^{\alpha_1^{(1)}} Y_1^{\beta_1^{(1)}} X_2^{\alpha_1^{(2)}} Y_2^{\beta_1^{(2)}}) = w^{(2)}(X_1^{\alpha_2^{(1)}} Y_1^{\beta_2^{(1)}} X_2^{\alpha_2^{(2)}} Y_2^{\beta_2^{(2)}})$
   but
   $$X_1^{\alpha_1^{(1)}} Y_1^{\beta_1^{(1)}} X_2^{\alpha_1^{(2)}} Y_2^{\beta_1^{(2)}} \prec_{\text{lex}} X_1^{\alpha_2^{(1)}} Y_1^{\beta_2^{(1)}} X_2^{\alpha_2^{(2)}} Y_2^{\beta_2^{(2)}}.$$

Here, $X_1 \succ_{\text{lex}} Y_1 \succ_{\text{lex}} X_2 \succ_{\text{lex}} Y_2$ is assumed. The set $\{X_1^{q+1} - Y_1^q - Y_1, X_2^{q+1} - Y_2^q - Y_2, X_1^{q^2} - X_1, X_2^{q^2} - X_2\}$ is a Gröbner basis for $I_{q^2}^{(2)}$ with respect to $\prec_{w^{(2)}}$ giving us the basis

$$\{X_1^{i_1} Y_1^{j_1} X_2^{i_2} Y_2^{j_2} + I_{q^2} \mid 0 \leq i_1, i_2 < q^2, 0 \leq j_1, j_2 < q\}$$

for $\mathbb{F}_{q^2}[X_1, Y_1, X_2, Y_2]/I_{q^2}^{(2)}$ as a vectorspace over $\mathbb{F}_{q^2}$. For the code construction we need the following bijective evaluation map

$$\text{EV} : \mathbb{F}_{q^2}[X_1, Y_1, X_2, Y_2]/I^{(2)} \to \mathbb{F}_{q^2}^{q^6}$$

given by $\text{EV}(F(X_1, Y_1, X_2, Y_2) + I_{q^2}^{(2)}) = (F(Q_1,), \ldots, F(Q_{q^6}))$. Define $\Lambda^{(2)} = \Lambda \times \Lambda$ and $(\Lambda^{(2)})^* = \Lambda^* \times \Lambda^*$. We have

$$(\Lambda^{(2)})^* = w^{(2)}(\Delta_{\prec_{w^{(2)}}}(I_{q^2}^{(2)}))$$

where no two monomials in $\Delta_{\prec_{w^{(2)}}}(I_{q^2}^{(2)})$ have the same weight. Similar to the situation of a Hermitian code we consider a codeword $c = \text{EV}(F(X_1, Y_1, X_2, Y_2) + I_{q^2}^{(2)})$ where without loss of generality we will assume that $F(X_1, Y_1, X_2, Y_2) \in \Delta_{\prec_{w^{(2)}}}(I_{q^2}^{(2)})$. We write $\lambda^{(2)} = (\lambda_1, \lambda_2) = w^{(2)}(\text{lm}(F))$. We can estimate

$$|\Delta_{\prec_{w^{(2)}}}(\langle F(X_1, Y_1, X_2, Y_2)\rangle + I_{q^2}^{(2)})| \leq |\Lambda^{(2)} - (\lambda^{(2)} + \Lambda^{(2)})|$$
$$\leq q^6 - (q^3 - \lambda_1)(q^3 - \lambda_2).$$

Hence, $w_H(c) \geq (q^3 - \lambda_1)(q^3 - \lambda_2)$.
Consider the code $\widetilde{E}(\delta)$ which is to Hermitian codes what Massey-Costello-Justesen codes [13] are to Reed-Solomon codes

$$\widetilde{E}(\delta) := \text{Span}_{\mathbb{F}_{q^2}} \Big\{ \text{EV}(X_1^{i_1} Y_1^{j_1} X_2^{i_2} Y_2^{j_2} + I_{q^2}^{(2)}) \mid 0 \leq i_1, i_2 < q^2, 0 \leq j_1, j_2 < q,$$
$$(q^3 - w(X_1^{i_1} Y_1^{j_1}))(q^3 - w(X_2^{i_2} Y_2^{j_2})) \geq \delta \Big\}.$$

From our discussion we conclude that the minimum distance satisfies $d(\widetilde{E}(\delta)) \geq \delta$. To estimate the dimension we make use of the characterization (7). The task is to estimate the number of (λ_1, λ_2)s that satisfies $(q^3 - \lambda_1)(q^3 - \lambda_2) \geq \delta$. For this purpose we can replace Λ^* with

$$\{g, g+1, \ldots, q^3 - 1\} \cup \{\lambda_{n-g+1}, \ldots, \lambda_n\}.$$

When estimating the dimension $k(\widetilde{E}(\delta))$ we shall furthermore ignore the elements in $\{\lambda_{n-g+1}, \ldots, \lambda_n\}$. Writing $T = q^3 - g$ we thereby get

$$k(\widetilde{E}(\delta)) \geq |\{(i,j) \mid 0 \leq i, j \leq T-1, (T-i)(T-j) \geq \delta\}|$$
$$\geq \int_0^{T-\frac{\delta}{T}} \int_0^{T-\frac{\delta}{T-i}} dj\,di = T^2 - \delta + \ln\left(\frac{\delta}{T^2}\right),$$

where the last inequality holds under the assumption $\delta \geq T$.

Proposition 1. *Assume $\delta \geq T$ where $T = q^3 - g$. The parameters of $\widetilde{E}(\delta)$ are $[n = q^6, k \geq T^2 - \delta + \delta \ln(\delta/T^2), d \geq \delta]$.*

In [8] Feng-Rao improved codes $\widetilde{C}(\delta)$ over $\mathbb{F}_{q^2}[X_1, Y_1, X_2, Y_2]/I_{q^2}^{(2)}$ were considered and a formula similar to the above proposition was derived under a stronger assumption on δ. Feng-Rao improved codes are described by means of their parity check matrix which is not very useful when the aim is to construct a small-bias space. This is why we included the description of $\widetilde{E}(\delta)$ in the present paper. We have a proof that $\widetilde{E}(\delta) = \widetilde{C}(\delta)$, however, we do not include it here as it has no implication for the construction of small-bias spaces. Observe that to derive Proposition 1 we did not use detailed information about the Weierstrass semigroup Λ but relied only on the genus and the number of roots of the Hermitian polynomial. Proposition 1 can be generalized to hold for not only two copies of Hermitian function fields but to arbitrary many such copies. Such constructions, however, are not useful when dealing with small-bias spaces so we do not treat them here.

From Proposition 1 and Theorem 2 we get a new class of ϵ-bias spaces:

Theorem 5. *For any ϵ, $0 < \epsilon < 1$ using codes $\widetilde{E}(\delta)$ as outer code in the construction of Theorem 2 one can construct ϵ-bias spaces with*

$$\mathcal{X} \subseteq \mathbb{F}_2^{\Omega(k)}, \quad |\mathcal{X}| = \mathcal{O}\left(\left(\frac{k}{\epsilon + (1-\epsilon)\ln(1-\epsilon)}\right)^{\frac{4}{3}}\right). \tag{8}$$

Proof. In the following we will use the substitution $1 - \epsilon = \delta/N$ which follows from $\epsilon = (N - \delta)/N$. Assume $\delta > \sqrt{N}$. We then have $\delta > T$ which is the condition in Proposition 1. Note that $\delta > \sqrt{N}$ is equivalent to $\epsilon < 1 - (1/\sqrt{N})$. For $N \to \infty$ this becomes $\epsilon < 1$ which is actually no restriction at all. From the proposition we get

$$\frac{K}{N} \geq \left(\frac{q^3 - g}{q^6}\right)^2 - \frac{\delta}{q^6} + \frac{\delta}{q^6} \ln\left(\frac{\delta}{(q^3-g)^2}\right)$$
$$\geq o(1) + 1 - (1-\epsilon) + (1-\epsilon)\ln(1-\epsilon)$$
$$= o(1) + \epsilon + (1-\epsilon)\ln(1-\epsilon).$$

With $q^2 = 2^s$ we have

$$|\mathcal{X}| \leq \frac{2^s}{s}\left(\frac{k}{o(1) + \epsilon + (1-\epsilon)\ln(1-\epsilon)}\right).$$

But $|\mathcal{X}| = (2^s)^4$ implies $2^s = |\mathcal{X}|^{1/4}$ and (8) has been demonstrated.

Theorem 6. *Consider the family of ϵ-bias spaces in Theorem 5. Given $\alpha \in \mathbb{R}^+$ choose $\epsilon = k^{-\alpha}$ and let $k \to \infty$. We have*

$$\log_k(|\mathcal{X}|) = \frac{4}{3} + \frac{8}{3}\alpha + o(1). \tag{9}$$

Proof. We have

$$\log_k(|\mathcal{X}|) \leq \frac{4}{3} - \frac{4}{3}\log_k(\epsilon + (1 - \epsilon)\ln(1 - \epsilon)).$$

We now apply Taylors formula to derive $\ln(1 - \epsilon) = -\epsilon - \epsilon^2/2(1 - c)^2$ for some $c \in [0, \epsilon]$. This produces

$$\log_k(|\mathcal{X}|) \leq \frac{4}{3} - \frac{4}{3}\log_k\left(\epsilon + (1 - \epsilon)(-\epsilon - \frac{\epsilon^2}{2(1 - \epsilon)^2})\right)$$

$$\leq \frac{4}{3} - \frac{4}{3}\log_k\left(\epsilon^2\left(\frac{2(1 - \epsilon)^2 - \epsilon^2}{(1 - \epsilon)^2}\right)\right).$$

With $\epsilon = k^{-\alpha}$ we arrive at (9).

4   Time Complexity Considerations

To build the multiset $\mathcal{X}$ in our construction we need to construct a generator matrix for the concatenated code. This involves the following tasks:

1. Build the generator matrix G_1 for $\widetilde{E}(\delta)$.
2. Express every entry of G_1 as a binary vector giving us G_2 (a matrix with binary vectors as entries).
3. For every row in G_2 we produce $s = \log_2(q^2)$ rows. This is done by taking cyclic shifts of all the vectors appearing in the row. We arrive at a matrix G_3.
4. Every entry in G_3 is a vector of length s and it must be multiplied with the $s \times 2^s$ generator matrix of the Walsh-Hadamard code producing G_4.

The total cost in binary operations is estimated as follows:

1. Determining functions and points for the code construction is inexpensive. To produce one entry costs $\mathcal{O}\left(\log(N)\log(\log(N))\right)$ operations. G_1 is a $K \times N$ matrix. Using $K \leq N - D + 1$, $\epsilon = (N - D)/N$, $\epsilon = k^{-\alpha}$, and $k = K\log_2(N)/6$ we arrive at $K \leq N^{\frac{1}{1+\alpha}}(\log_2(N))^{\frac{1}{1+\alpha}}6^{\frac{\alpha}{1+\alpha}}$. So the price for building G_1 is $\mathcal{O}\left(N^{\frac{2+\alpha}{1+\alpha}}(\log(N))^{\frac{1}{1+\alpha}}\log(\log(N))\right)$.

2. To produce one entry in G_2 costs $\mathcal{O}\left(N^{\frac{1}{3}}\log(N^{\frac{1}{3}})\log(\log(N^{\frac{1}{3}}))\right)$ operations. That is, to produce G_2 from G_1 amounts to $\mathcal{O}\left(N^{\frac{7+4\alpha}{3+3\alpha}}\log(N)^{\frac{1}{1+\alpha}}\log(\log(N))\right)$ operations.

3. There will be $\mathcal{O}\left(N^{\frac{2+\alpha}{1+\alpha}}(\log(N))^{\frac{1}{1+\alpha}}\right)$ entries in G_3 each coming with a cost of s operations. Altogether we have $\mathcal{O}\left(N^{\frac{2+\alpha}{1+\alpha}}(\log(N))^{\frac{2+\alpha}{1+\alpha}}\right)$ operations.

4. The price for multiplying with a generator matrix for the Walsh-Hadamard code is $N^{\frac{1}{3}} \log(N)$ giving a total cost of

$$\mathcal{O}\left(N^{\frac{7+4\alpha}{3+3\alpha}}(\log(N))^{\frac{2+\alpha}{1+\alpha}}\right) \tag{10}$$

operations for producing G_4 from G_3.

Clearly, the overall cost is that of (10). Note that (10) counts binary operations in contrast to (6) which counts operations in $\mathbb{F}_{q^2}$.

5   Small-Bias Spaces from Norm-Trace Codes

The method developed by Ben-Aroya and Ta-Shma for Hermitian codes in [2] were generalized to norm-trace codes by Matthews and Peachey in [12]. Given $r \geq 2$ consider the C_{ab} curve [11]

$$X^{\frac{q^r-1}{q-1}} - Y^{q^{r-1}} - Y^{q^{r-2}} - \cdots - Y^q - Y$$

known as the norm-trace curve over $\mathbb{F}_{q^r}$ [6]. Clearly, $r = 2$ corresponds to the Hermitian function field. The following theorem from [12] coincides with (3) when $l = 4$.

Theorem 7. *Given an integer l, $l \geq 4$, define $r = \lfloor (l+2)/3 \rfloor$. Let k be a positive integer and ϵ a real number, $0 < \epsilon < 1$ such that*

$$\frac{\epsilon}{\left(\log_v(1/\epsilon)\right)^{\frac{1}{\sqrt{l}}}} \leq k^{\frac{-1}{\sqrt{l}}} \tag{11}$$

holds. Here, v is any fixed real number larger than 1. Using the norm-trace function field over $\mathbb{F}_{q^r}$ one can construct an ϵ-bias space $\mathcal{X} \subseteq \mathbb{F}_2^{\Omega(k)}$ with

$$|\mathcal{X}| = \mathcal{O}\left(\left(\frac{k}{\epsilon^{l-\sqrt{l}}\log_v(1/\epsilon)}\right)^{\frac{l+1}{l}}\right).$$

In the above theorem it is not completely clear how well the cases $l \geq 5$ compete with the case $l = 4$. Below we address this question and also compare the small-bias spaces from Theorem 7 with those achieved by using the codes $\tilde{E}(\delta)$ as is done in the present paper.

We first translate Theorem 7 into the setting from Section 1 where for increasing k and fixed α we consider a sequence of ϵ-bias multisets with $\epsilon = k^{-\alpha}$. Condition (11) from Theorem 7 then translates into

$$k^{1-\alpha\sqrt{l}} \leq \alpha \log_v(k).$$

For fixed v, $\log_v(k) = \mathcal{O}\left(k^\beta\right)$ holds for any $\beta > 0$. Therefore we have

$$1 - \alpha\sqrt{l} \leq \log_k(\alpha).$$

Letting $k \to \infty$ we get the condition

$$\frac{1}{\sqrt{l}} \leq \alpha.$$

Theorem 7 therefore guarantees that for any $\alpha \geq 1/\sqrt{l}$ we can construct an infinite sequence of ϵ-bias spaces with $\epsilon = k^{-\alpha}$, $\mathcal{X} \subseteq \mathbb{F}_2^{\Omega(k)}$ such that

$$\log_k(|\mathcal{X}|) = \frac{l+1}{l}(1 + \alpha(l - \sqrt{l})) + o(1). \tag{12}$$

Given an α and two integers $l_1, l_2 \geq 4$ with $\alpha \geq 1/\sqrt{l_i}$, $i = 1,2$ it is clear from (12) that the best result is obtained by choosing the smallest l_i. So the advantage of Theorem 7 over (3) boils down to the fact that Theorem 7 allows for any α provided that the l is chosen accordingly while (3) requires $\alpha \geq 1/2$. Recall from Section 3 that using the code $\widetilde{E}(\delta)$ in the construction of Theorem 2 one achieves

$$\log_k(|\mathcal{X}|) = \frac{4}{3} + \frac{8}{3}\alpha + o(1) \tag{13}$$

for any choice of α. We now compare this result with (12) ignoring of course the $o(1)$ parts. For fixed l (12) is a linear expression in α which is smaller than the linear expression from (13) when $\alpha = 0$. We now show that for $\alpha = 1/\sqrt{l}$ (which

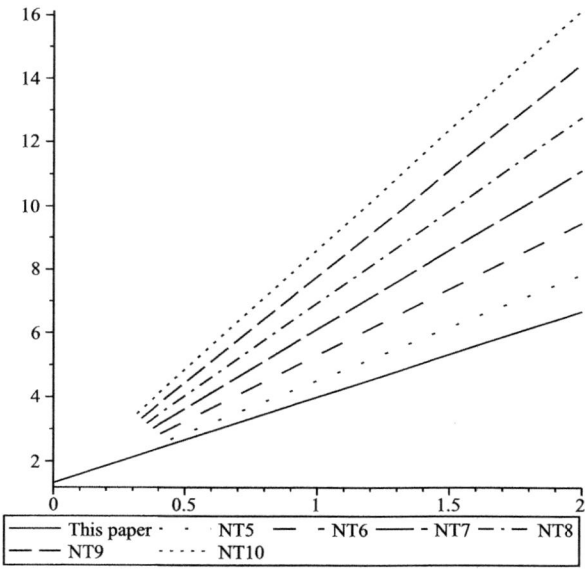

| | This paper | · · | NT5 | — · — NT6 | — · — NT7 | — · · — NT8 |
| | — — NT9 | · · · · · | NT10 | | | |

Fig. 2. Comparison of new construction with NT-construction where $l \in \{5, \ldots, 10\}$. First axis is α, second axis corresponds to $\log_k(|\mathcal{X}|)$ when $k \to \infty$.

is the smallest α allowed) (12) is larger than (13) when $l \geq 5$. It follows that none of the cases $l \geq 5$ can compete with the construction of the present paper. To show that (12) is larger than (13) for $\alpha = 1/\sqrt{l}$ we substitute $k = \sqrt{l}$ into (12)-(13) to get

$$\frac{1}{k^2}(k^3 - \frac{4}{3}k^2 - \frac{5}{3}k).$$

The function $k^3 - \frac{4}{3}k^2 - \frac{5}{3}k$ is positive for k belonging to the interval from 0 to approximately 2.119 and negative for higher values of k. Therefore for all $l \geq 5$ indeed (13) is better than (12). The situation is illustrated in Figure 2.

Acknowledgments. The present work was done while Ryutaroh Matsumoto was visiting Aalborg University as a Velux Visiting Professor supported by the Villum Foundation. The authors gratefully acknowledge this support. The authors also gratefully acknowledge the support from the Danish National Research Foundation and the National Science Foundation of China (Grant No. 11061130539) for the Danish-Chinese Center for Applications of Algebraic Geometry in Coding Theory and Cryptography.

References

1. Alon, N., Goldreich, O., Hastad, J., Peralta, R.: Simple constructions of almost k-wise independent random variables. Random Structures Algorithms 3(3), 289–303 (1992)
2. Ben-Aroya, A., Ta-Shma, A.: Constructing small-bias sets from algebraic-geometric codes. In: FOCS 2009, pp. 191–197 (2009)
3. Cox, D., Little, J., O'Shea, D.: Ideals, Varieties, and Algorithms, 2nd edn. Springer (1997)
4. Fitzgerald, J., Lax, R.F.: Decoding Affine Variety Codes Using Gröbner Bases. Des. Codes Cryptography 13, 147–158 (1998)
5. Garcia, A., Stichtenoth, H.: On the asymptotic behaviour of some towers of function fields over finite fields. J. Number Theory 61, 248–273 (1996)
6. Geil, O.: On codes from norm-trace curves. Finite Fields and their Applications 9, 351–371 (2003)
7. Geil, O., Høholdt, T.: Footprints or Generalized Bezout's Theorem. IEEE Trans. Inform. Theory 46, 635–641 (2000)
8. Geil, O., Høholdt, T.: On Hyperbolic Type Codes. In: Proceedings of 2003 IEEE International Symposium on Inf. Theory, Yokohama, p. 331 (2003)
9. Høholdt, T.: On (or in) Dick Blahut's 'footprint'. In: Vardy, A. (ed.) Codes, Curves and Signals, pp. 3–9. Kluwer Academic, Norwell (1998)
10. Høholdt, T., van Lint, J., Pellikaan, R.: Algebraic Geometry Codes. In: Pless, V.S., Huffman, W.C. (eds.) Handbook of Coding Theory, vol. 1, ch. 10, pp. 871–961. Elsevier, Amsterdam (1998)
11. Miura, S., Kamiya, N.: Geometric-Goppa codes on some maximal curves and their minimum distance. In: Proc. of 1993 IEEE Inf. Th. Workshop, Susonon-shi, Shizuoka, Japan, June 4-8, pp. 85–86 (1993)
12. Matthews, G.L., Peachey, J.: Small-bias sets from extended norm-trace codes. To appear in Proceedings of Fq10, Contemporary Mathematics. AMS

13. Massey, J., Costello, D.J., Justesen, J.: Polynomial Weights and Code Constructions. IEEE Trans. Inf. Theory 19, 101–110 (1973)
14. Meka, R., Zuckerman, D.: Small-Bias Spaces for Group Products. In: Dinur, I., Jansen, K., Naor, J., Rolim, J. (eds.) APPROX and RANDOM 2009. LNCS, vol. 5687, pp. 658–672. Springer, Heidelberg (2009)
15. Naor, J., Naor, M.: Small-bias probability spaces: eficient construction and applications. SIAM J. Comput. 22, 838–856 (1993)
16. Shum, K.W., Aleshnikov, I., Vijay Kumar, P., Stichtenoth, H., Deolalikar, V.: A Low-Complexity Algorithm for the Construction of Algebraic-Geometric Codes Better Than the Gilbert-Varshamov Bound. IEEE Trans. Inform. Theory 47, 2225–2241 (2001)
17. Vazirani, U.V.: Randomness, adversaries, and computation, Ph.D. thesis, EECS, UC Berkeley (1986)

An Improved Threshold Ring Signature Scheme Based on Error Correcting Codes

Pierre-Louis Cayrel[1], Sidi Mohamed El Yousfi Alaoui[2], Gerhrad Hoffmann[3], and Pascal Véron[4]

[1] Laboratoire Hubert Curien Université de Saint-Etienne, France
`pierre-louis.cayrel@univ-st-etienne.fr`
[2] CASED – Center for Advanced Security Research Darmstadt, Germany
`elyousfi@cased.de`
[3] Technische Universität Darmstadt, Germany
`hoffmann@mathematik.tu-darmstadt.de`
[4] IML/IMATH Université du Sud Toulon-Var. B.P.20132, F-83957 La Garde Cedex, France
`veron@univ-tln.fr`

Abstract. The concept of threshold ring signature in code-based cryptography was introduced by Aguilar et al. in [1]. Their proposal uses Stern's identification scheme as basis. In this paper we construct a novel threshold ring signature scheme built on the q-SD identification scheme recently proposed by Cayrel et al. in [14]. Our proposed scheme benefits of a performance gain as a result of the reduction in the soundness error from 2/3 for Stern's scheme to 1/2 per round for the q-SD scheme. Our threshold ring signature scheme uses random linear codes over the field $\mathbb{F}_q$, secure in the random oracle model and its security relies on the hardness of an error-correcting codes problem (namely the q-ary syndrome decoding problem). In this paper we also provide implementation results of the Aguilar et al. scheme and our proposal, this is the first efficient implementation of this type of code-based schemes.

Keywords: post-quantum cryptography, code-based cryptography, identification scheme, threshold ring signature scheme.

1 Introduction

The development in the field of quantum computing is a real threat to the security of many used public key cryptographic algorithms. Shor has demonstrated in 1994 that cryptographic schemes whose security relies on the difficulty of the factorization problem (e.g., RSA) and the difficulty of the discrete logarithm problem (e.g., DSA), could be broken using quantum computers. Consequently, it is necessary to have available alternative signature and identification schemes.

Coding based cryptography is one of the few alternatives supposed to be secure in a post quantum world. The most popular cryptosystems based on error-correcting codes are the McEliece [22] and Niederreiter [24] ones. The main

F. Özbudak and F. Rodríguez-Henríquez (Eds.): WAIFI 2012, LNCS 7369, pp. 45–63, 2012.

advantage of these two public key cryptosystems is the provision of a fast encryption and decryption procedure (about 50 times faster for encryption and 100 times faster for decryption than RSA).

Secure identification schemes were introduced by Feige, Fiat and Shamir [18]. These cryptographic schemes allow a prover to identify itself in polynomial time to a verifier without revealing any information of its secret key. Stern proposed at Crypto'93 [27] the first zero-knowledge identification scheme based on the syndrome decoding problem, this scheme could be turned into a digital signature via the Fiat-Shamir paradigm [17].

The concept of ring signatures was introduced first in 2001 by Rivest et al. [26]. Ring signatures permit any user from a set of intended signers to sign a message with no existing group manager and to convince the verifier that the author of the signature belongs to this set without revealing any information about its identity.

In 2002, Bresson et al. [12] extended ring signature schemes in a t-out-of-N threshold ring signature schemes, which enable any t participating users belonging to a set of N users to produce a signature in such a way that the verifier cannot determine the identity of the actual signers.

The concept of threshold ring signatures in code-based cryptography was introduced by Aguilar et al. in [1,2]. Their proposal is a generalization of Stern's identification scheme. The major advantage of this construction is that its complexity depends linearly on a maximum number of signers N, comparing with the complexity of threshold ring signature schemes based on number theory whose complexity is $\mathcal{O}(tN)$. However, the disadvantage of large public key size and signature length is still unsolved for this scheme.

Our Contribution: In this paper we propose an improved code-based threshold ring signature scheme. We achieve this by extending the five-pass zero-knowledge (q-SD) identification scheme proposed in [14] to a threshold ring identification scheme and applying the same idea as the Fiat-Shamir paradigm to transform it to a threshold ring signature. In this paper we provide also a first efficient implementation of the Aguilar et al. scheme and our scheme in order to show the advantage of our proposal in terms of performance.

Organization of the Paper: This paper is organized as follows: in Section 2 we recall some background on code-based cryptography. In Section 3 we present the q-SD identification scheme recently introduced by Cayrel et al. in [14], followed by some suggested improvements. We show in Section 4 how to use the q-SD identification scheme to construct our proposal, then we discuss the security and gives in detail the performance aspect of our construction by providing implementation results. Finally, we conclude the paper in Section 5.

2   Preliminaries

In this section, we recall useful notions of code-based cryptography. We refer to [9], for a general introduction to these issues.

2.1   Definitions

Linear codes are k-dimensional subspaces of an n-dimensional vector space over a finite field $\mathbb{F}_q$, where k and n are positive integers with $k < n$, and q a prime power. The error-correcting capability of such a code is the maximum number ω of errors that the code is able to decode. In short, linear codes with these parameters are denoted (n, k)-codes or $(n, n - r)$-codes, where r is the co-dimension of a code with $r = n - k$.

Definition 1 (Hamming weight). *The (Hamming) weight of a vector x is the number of non-zero entries. We use $\mathsf{wt}(x)$ to represent the Hamming weight of x.*

Definition 2 (Generator and Parity Check Matrix). *Let $\mathscr{C}$ be a linear code over $\mathbb{F}_q$. A generator matrix G of $\mathscr{C}$ is a matrix whose rows form a basis of $\mathscr{C}$:*

$$\mathscr{C} = \{xG : x \in \mathbb{F}_q^{n-r}\}$$

A parity check matrix H of $\mathscr{C}$ is defined by

$$\mathscr{C} = \{x \in \mathbb{F}_q^n : Hx^T = 0\}$$

and generates the dual space of $\mathscr{C}$.

Let n and r be two integers such that $n \geq r$, $\mathsf{Binary}(n, r)$ (resp. $q\text{-}\mathrm{ary}(n, r)$) be the set of binary (resp. q-ary) matrices with n columns and r rows of rank r. Moreover, denote by $x \xleftarrow{\$} A$ the random choice of x among the elements of a set A.

   We describe in the following the main hard problems on which the security of code-based schemes presented in this paper relies.

Definition 3 (Binary Syndrome Decoding (SD) problem).
Input : $H \xleftarrow{\$} \mathsf{Binary}(n, r)$, $y \xleftarrow{\$} \mathbb{F}_2^r$, *and an integer $\omega > 0$.*
Find : *a word $s \in \mathbb{F}_2^n$ such that $\mathsf{wt}(s) \leq \omega$ and $Hs^T = y$.*

This problem was proven to be NP-complete in 1978 [8]. It can be extended to arbitrary finite fields as follows.

Definition 4 (q-ary Syndrome Decoding (qSD) problem).
Input : $H \xleftarrow{\$} q\text{-}\mathrm{ary}(n, r)$, $y \xleftarrow{\$} \mathbb{F}_q^r$, *and an integer $\omega > 0$.*
Find : *a word $s \in \mathbb{F}_q^n$ such that $\mathsf{wt}(s) \leq \omega$ and $Hs^T = y$.*

In 1994, A. Barg proved that this last problem remains NP-complete [5, in russian].

Definition 5 (q-ary Minimum Distance (qMD) problem).
Input : $H \xleftarrow{\$} q\text{-}\mathrm{ary}(n, r)$, *and an integer $\omega > 0$.*
Find : *a word $s \in \mathbb{F}_q^n$ such that $\mathsf{wt}(s) \leq \omega$ and $Hs^T = 0$.*

Notice that the difficulties of solving the two problems (qSD and qMD) are equivalent [28]. The intractable assumptions associated to these problems are denoted by qSD assumption and qMD assumption, respectively.

Best known attack. The most efficient algorithm to attack code-based schemes is the Information Set Decoding (ISD) algorithm. Some improvements of this algorithm have been developed by Peters [25], Niebuhr et al. [23], and Bernstein et al. [10], and recently in [21] by Becker et al. and [7] by May et al.. The recent results of this attack are taken into account when choosing our parameters.

3   The q-SD Identification Scheme

An identification scheme is an interactive method for one party to prove to another that a statement is true, without revealing any additional information. In this section, we present the recent q-SD identification scheme based on error-correcting codes proposed by Cayrel et al. in [14]. The soundness error of approximately $1/2$ allows a performance gains when compared to Stern's scheme. For instance, one needs 16 rounds for the q-SD identification scheme and 28 rounds for the Stern's one to achieve the weak authentication probabilities of 2^{-16} according the norm ISO/IEC-9798-5.

3.1   Description of the q-SD Identification Scheme

In what follows, the elements of $\mathbb{F}_q^n$ are written as n blocks of size $\lceil \log_2(q) \rceil = m$ and each element of $\mathbb{F}_q$ is presented as m bits.

We first introduce a special transformation that will be used later.

Definition 6. *Let Σ be a permutation of $\{1, \ldots, n\}$ and $\gamma = (\gamma_1, \ldots, \gamma_n) \in \mathbb{F}_q^n$ such that $\forall i, \gamma_i \neq 0$. The transformation $\Pi_{\gamma, \Sigma}$ is defined as follows:*

$$\Pi_{\gamma, \Sigma} : \mathbb{F}_q^n \longrightarrow \mathbb{F}_q^n$$
$$v \mapsto (\gamma_{\Sigma(1)} v_{\Sigma(1)}, \ldots, \gamma_{\Sigma(n)} v_{\Sigma(n)})$$

Notice that $\forall \alpha \in \mathbb{F}_q$, $\forall v \in \mathbb{F}_q^n$, $\Pi_{\gamma, \Sigma}(\alpha v) = \alpha \Pi_{\gamma, \Sigma}(v)$, and $\mathsf{wt}(\Pi_{\gamma, \Sigma}(v)) = \mathsf{wt}(v)$.

The q-SD identification scheme is comprised of two algorithms: key generation and identification protocol. Given the security parameter κ, the key generation algorithm described in Fig. 1 picks a random $(r \times n)$ q-ary matrix H common to all users, a random vector s (secret key) with Hamming weight ω. It outputs the public key y by multiplication of the vector s by a matrix H. The identification protocol described in Fig. 2 corresponds to a zero-knowledge proof of knowledge that the prover $\mathcal{P}$ possesses a private key s with a given public key y.

In Fig. 1, $\mathrm{WF}_{\mathrm{ISD}}$ denotes the Work-Factor of ISD algorithm over $\mathbb{F}_q$ and in Fig. 2, h denotes a hash function, S_n the symmetric group of degree n and $\|$ the concatenation of two strings.

The authors of [14] proved that the protocol presented in Fig. 2 corresponds to a zero-knowledge interactive proof in the random oracle model, that means it satisfies the completeness, soundness, and zero-knowledge properties.

Key-Gen:

Choose n, r, ω, and q such that $\mathrm{WF}_{\mathrm{ISD}}(n, r, \omega, q) \geq 2^{\kappa}$

$H \xleftarrow{\$} \mathbb{F}_q^{r \times n}$

$s \xleftarrow{\$} \mathbb{F}_q^n$, s.t. $\mathsf{wt}(s) = \omega$.

$y \leftarrow H s^T$

Output $(\mathrm{sk}, \mathrm{pk}) = (s, (y, H, \omega))$

Fig. 1. Key generation algorithm: parameters n, r, ω, q are public

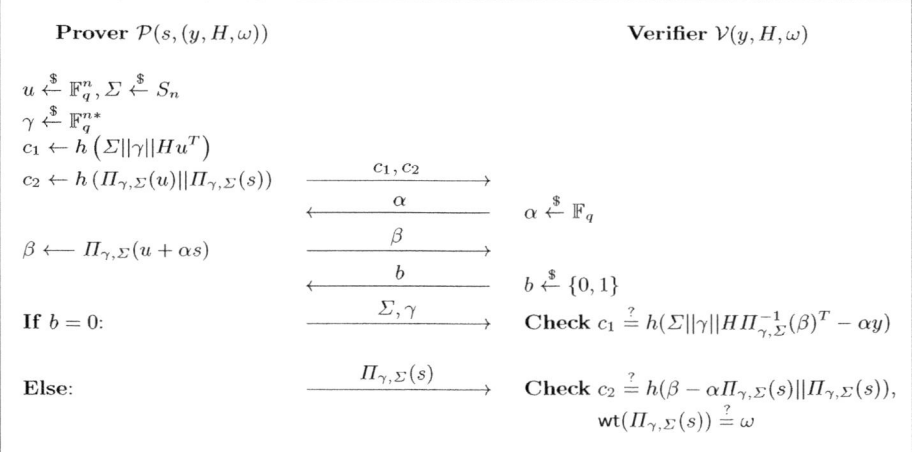

Fig. 2. The q-SD identification protocol

Some Improvements of the q-SD Scheme:

- To get better communication complexity in comparison to the original version of the protocol presented in Fig. 2, the prover could use a public function φ_q^{-1} by sending $\varphi_q^{-1}(\Pi_{\gamma, \Sigma}(s))$ instead of $\Pi_{\gamma, \Sigma}(s)$, where φ_q is an efficient bijective encoding which takes its input from the interval $[0, (q-1)^{\omega} \binom{n}{\omega}[$ and outputs a binary word of length n and Hamming weight ω. This function is described in Algorithm 2 (see Appendix).
- We could use the same random seed to generate the permutation Σ and the vector γ in order to reduce more the communication complexity.

Proposed parameters. According to the ISD algorithm, the suggested parameters for the q-SD scheme are: $q = 256, n = 128, r = 64, \omega = \mathsf{wt}(s) = 49$.

Table 1 shows the advantage regarding the communication cost and the size of the public key of the q-SD scheme in comparison with Stern's initial proposal and his five-pass variant, for the same security level of 2^{80} and an impersonation

Table 1. Stern's schemes vs. the q-SD scheme, security level 2^{80}, probability of cheating 2^{-16}

	Stern (3-pass)	Stern (5-pass)	q-SD
Rounds	28	16	16
Public data (bits)	122850	124950	33280
Secret key (bits)	700	4900	1024
Communication complexity (bits)	42019	62272	39056
Prover's Computation	$2^{22.7}$op. over $\mathbb{F}_2$	$2^{21.92}$op. over $\mathbb{F}_2$	2^{16}mult $+ 2^{16}$add op. over $\mathbb{F}_{256}$

resistance of 2^{-16}. It is considered that all seeds used are 128 bits long and that all hash values are 160 bits long.

Remark 1. To be fair, the two improvements above-mentioned according the scheme q-SD are not taken into account in the calculation of the communication complexity in table 1, since these improvements can be applied to Stern's schemes. However by using the same seed for Σ and γ, the communication complexity will be 30864 bits, and only 26760 bits if we use in addition an encoding function.

Remark 2. We want to mention that the authors in [3] propose a new five-pass identification scheme with small size of keys and an asymptotic cheating probability of $1/2$, this scheme is related to the Véron identification scheme and its security is based on the syndrome decoding Problem. The idea of this construction is based on deriving new challenges from the secret key through cyclic shifts of the initial public key syndrome.

3.2 Signature Schemes from Identification Schemes

One efficient method to derive a signature from an identification scheme is given via the Fiat-Shamir paradigm [17]. Pointcheval and Stern use the well-known forking lemma, in order to provide the security argument of signatures obtained from three-pass identification protocols. The authors in [4] extend the Fiat-Shamir transform and the Forking lemma to obtain secure signatures from identification protocols with multi-pass. For instance, in the case of a five-pass identification scheme, the signer replaces the two moves given from the verifier by the outputs of some random oracles, he sets the transcript $(\sigma_1||h_1||\sigma_2||h_2||\sigma_3)$ as a signature of a message M, where h_1 and h_2 the outputs of two hash functions $\mathcal{H}_1$ respectively $\mathcal{H}_2$ modeled as random oracles, and the σ_1, σ_2 and σ_3 the values given from the prover as in the identification scheme. Similarly to [20], the authors of [4] generalize the forking lemma even more for ring signatures schemes in order to give security arguments for such a class of signature.

4   Code-Based Threshold Ring Signature Schemes

The notion of threshold ring signature is introduced in 2002 by Bresson et al. [12]. We first define the formal definition of the threshold ring identification scheme which can be turned to get a threshold ring signature in accord with Section 3.2.

Definition 7. *Let $t < N$ be integers. We assume that each user P_i owns the pair of keys (sk_i, pk_i) corresponding respectively to the secret and the public keys. Let $P_1, \ldots, P_N$ be the N potential provers of the ring with their public keys $pk_1, \ldots, pk_N$. Let t of the N members form a group of provers, one of them is the leader L. The threshold identification scheme consists of the algorithms:*

- **Setup** *takes a secret parameter as input and outputs the public parameters and chooses the leader.*
- **Ring key generation** *takes public parameters as input and outputs a pair of keys corresponding to the secret and the public key.*
- **Commitment-challenge-answer and verification step** *is an interactive protocol between the t-users and the verifier consisting of the computation of the commitments, challenges and responses, following by a verification step which takes as input the answers of the challenges and verifies the honestly of the computation, and returns 1 (accept), and 0 (reject).*

For the security of a threshold ring signature scheme the basic criteria are:

- Unforgeablity: Without the knowledge of the t secret keys, it is infeasible to generate a valid (t, N) threshold ring signature.
- Anonymity: Given a message-signature pair, it should be infeasible for the verifier to reveal which t-subset of signers generated a signature.

See defintion 3 and 4 in [2] for a formal definition of the two above properties. In [1], Aguilar et al. introduced the first code-based threshold ring signature. The main idea of this scheme is to generalize the Stern's identification scheme and then convert this latter into a threshold ring signature using the Fiat-Shamir paradigm. Aguilar et al.'s scheme is proven to be a zero-knowledge protocol with soundness error of $2/3$ as in the Stern's protocol for each round. Its security relies on the hardness of the binary Minimum Distance problem.

A second code-based threshold ring signature scheme has been proposed by Dallot and Vergnaud in [16]. Their proposal is not derived from an identification scheme. It uses Goppa codes and combines the generic construction of Bresson et al. [12] and the CFS signature scheme [15]. The authors of [16] obtained a short signature but the required time to generate it, is too high, and the huge public key size is also a disadvantage of their proposal.

Using the Aguilar et al.'s approach, Cayrel et al. proposed in [13] a lattice-based threshold ring signature scheme based on the hardness of the SIS problem.

The soundness error of approximately $1/2$ for the q-SD identification scheme allows a performance gains when compared to Stern's scheme. In order to make use of this gain, we present in this section a novel threshold ring identification scheme and according to Section 3.2, we turn it into a threshold ring signature scheme.

To describe our scheme, we need the two notions of block permutations:

Definition 8. *Let n and N be two integers and let $\beta = (\beta_1, \ldots, \beta_n, \beta_{n+1}, \ldots, \beta_{2n}, \ldots, \beta_{nN})$ be a vector of length nN defined over*

some alphabet. Let us define for $i \in [1, N]$ the elements $\tilde{\beta}_i = (\beta_{(i-1)n+1}, \ldots, \beta_{in})$
such that β can be expressed as $(\tilde{\beta}_1, \ldots, \tilde{\beta}_N)$.

The constant (n, N)-block permutation Θ is a permutation over $\{1, \ldots, N\}$
which acts over vectors of length nN such that

$$\Theta(\beta) = \Theta(\tilde{\beta}_1, \ldots, \tilde{\beta}_N) = (\tilde{\beta}_{\Theta(1)}, \ldots, \tilde{\beta}_{\Theta(N)})$$

Let $\sigma = (\sigma_1, \ldots, \sigma_N)$ be a family of N permutations over $\{1, \ldots, n\}$, we define a
(n, N)-block permutation Π, as a permutation which acts over a vector of length
nN and which is the product of a constant n-block permutation Θ and the family
σ, i.e.

$$\Pi(\beta) = \Theta(\sigma_1(\tilde{\beta}_1), \ldots, \sigma_N(\tilde{\beta}_N))$$

Informally speaking a constant (n, N)-block permutation divides a vector of
length nN into N blocks of size n and permutes them. A (n, N)-block permuta-
tion permutes also for each block the components of the block.

Example 1. The permutation $(6, 5, 4, 3, 2, 1)$ is a $(2, 3)$-block permutation, and the
permutation
$(3, 4, 5, 6, 1, 2)$ is a constant $(2, 3)$-block permutation since the order on each
block $((1, 2), (3, 4)$ and $(5, 6))$ is preserved in the block permutation.

4.1 Description of Our Threshold Identification Protocol

We consider one set of N members. Let $(P_1, \ldots, P_t)$ be a subset of this set
consisting of the members who want to prove that they know some secret s,
whereas one of them is a leader L. The parameter t corresponding to the number
of provers has to be fixed at the beginning of the protocol.

Our protocol consists of the following steps: Setup, Ring public key generation,
Commitment-Challenge-Answer and Verification step. We can formally describe
each step as follows:

- **Setup** Given κ as security parameter, we generate the corresponding public
  parameters n, r, ω, and q such that $\mathrm{WF}_{\mathrm{ISD}}(n, r, \omega, q) \geq 2^\kappa$, where n and r
  are the parameters for each matrix H_i $(1 \leq i \leq N)$ which will be used to
  form the ring public matrix. Each matrix can be constructed as follows: we
  choose a random vector $s_i \in \mathbb{F}_q^n$ of weight ω, generate $n - r - 1$ random vectors and consider the code $\mathscr{C}_i$ obtained by these $n - r$ words (the operation
  can be repeated until the co-dimension of $\mathscr{C}_i$ is r). The matrix H_i is then a
  parity-check matrix of a code $\mathscr{C}_i$ and thus we have $H_i s_i^T = 0$, where $s_i \in \mathbb{F}_q^n$
  has a weight ω. The fact that we take a same syndrome and the same weight
  for the vectors s_i helps for conserving the anonymity in the group. For the
  $(N - t)$ other users, s_i are fixed at 0, because 0 is always a solution of the
  equation $H_i s_i^T = 0$.

- **Ring key generation** The leader collects all these matrices and forms
  among them a public key $(H, t\omega)$ called ring public key, the matrix H can
  be described as follows:

$$H = \begin{pmatrix} H_1 & 0 & \cdots & 0 \\ 0 & H_2 & 0 & 0 \\ \vdots & & \ddots & H_i & 0 \\ 0 & 0 & \cdots & H_N \end{pmatrix}$$

- **Commitment-challenge-answer and verification step** To simplify the description, we consider that the t provers correspond to the first matrices H_i $(1 \le i \le t)$. The leader L, member of the set of t provers among N members, want to prove to the verifier that he knows a secret key s, where s is a nN vector of weight tw. This will be achieved by performing the following steps:
  - Each member of the t provers (including L) creates local commitments using the secret keys s_i and sends them to L.
  - L collects all these commitments, simulates the missing ones for the $(N-t)$ other users by fixing all remaining s_i by 0, and create the master commitment using a random constant block permutation.
  - The master commitment are sent to the verifier V.
  - V chooses a random value α over $\mathbb{F}_q$ and sends it to L, the latter one forwards this value to the $(t-1)$ provers.
  - Each member of the t provers (including L) calculates the vectors β_i, L collects those values and creates a global vector β' using a constant block permutation and it will be sent to V.
  - V chooses a challenge from $\{0,1\}$ and sends it to L who forwards it to the $(t-1)$ provers.
  - L collects the answers from the $(t-1)$ provers, computes the responses for the other users and finally computes a global answer for V.
  - After receiving the global answer, V checks the correctness of the master commitments.

Algorithm 1 gives a full description of this interaction between the set of t provers and the verifier. This algorithm has to be performed in multi-rounds in order to reach the required impersonation resistance.

We stress that during the answer step (line 19 of Algorithm 1), the knowledge of the permutation ρ permits to recover Θ, Σ_i, and γ_i for $1 \le i \le N$. In addition, the verifier can easily obtain $\beta_i (1 \le i \le N)$ by applying the inverse of θ on the known vector β'.

4.2   Security

We first prove that the generalized q-SD identification protocol is an honest-verifier zero-knowledge proof of knowledge. The resulting threshold ring signature obtained from the application of the work presented in [4] on the generalized q-SD identification protocol is existentially unforgeable under chosen message attacks in the random oracle model.

Lemma 1. *Finding a vector s of length nN such that the global weight of s is tw, the weight of s for each of the N blocks of length n is 0 or ω, and such that s has a null syndrome for H, is hard under the assumption of hardness of the qMD problem.*

Algorithm 1. Generalized q-SD protocol

INPUT: $n, k, N, t \in \mathbb{N}$, where $k < n$ and $t < N$.
   $H \in \mathbb{F}_q^{rN \times nN}$, where $r = n - k$ and h a collision resistant hash function.

PRIVATE KEY: $s = (s_1, \ldots, s_N) \in \mathbb{F}_q^{nN}$, $\mathrm{wt}(s_j) = 0$ or $\mathrm{wt}(s_j) = \omega$ with $\mathrm{wt}(s) = t\omega$ and $Hs^T = 0$.

COMMITMENT STEP:
1: Each prover P_i chooses $u_i \xleftarrow{\$} \mathbb{F}_q^n$, $\Sigma_i \xleftarrow{\$} S_n$, $\gamma_i \xleftarrow{\$} \mathbb{F}_q^{n*}$ $(1 \leq i \leq t)$.
2: P_i constructs $c_{1,i} \leftarrow h\left(\Sigma_i || \gamma_i || H_i u_i^T\right)$ and $c_{2,i} \leftarrow h\left(\Pi_{\gamma_i, \Sigma_i}(u_i) || \Pi_{\gamma_i, \Sigma_i}(s_i)\right)$.
3: P_i sends $c_{1,i}$ and $c_{2,i}$ to leader L.
4: L fixes the secret keys s_i of the $N - t$ other users at 0 $(t+1 \leq i \leq N)$.
5: L chooses $N-t$ values $u_i \xleftarrow{\$} \mathbb{F}_q^n$ and $N-t$ permutations $\Sigma_i \xleftarrow{\$} S_n$ and $N-t$ values $\gamma_i \xleftarrow{\$} \mathbb{F}_q^{n*}$ $(t+1 \leq i \leq N)$.
6: L chooses $\Theta \xleftarrow{\$} S_N$ in order to obtain the master commitments.
7: L computes the master commitments $C_1 \leftarrow h(\Theta || c_{1,1} || \ldots || c_{1,N})$ and $C_2 \leftarrow h(\Theta(c_{2,1}, \ldots, c_{2,N}))$.
8: C_1 and C_2 are sent to the verifier V.
9: V sends back the value $\alpha \xleftarrow{\$} \mathbb{F}_q$ and L passes it to each P_i $(1 \leq i \leq t)$.
10: P_i computes $\beta_i \longleftarrow \Pi_{\gamma_i, \Sigma_i}(u_i + \alpha s_i)$ $(1 \leq i \leq t)$.
11: L computes $\beta_i \longleftarrow \Pi_{\gamma_i, \Sigma_i}(u_i)$ $(t + 1 \leq i \leq N)$.
12: $\beta' = \Theta(\beta) = \Theta(\beta_1, \cdots, \beta_N) = (\beta_{\Theta(1)}, \ldots, \beta_{\Theta(N)})$ is sent to V.

CHALLENGE STEP:
13: V sends a challenge $b \xleftarrow{\$} \{0, 1\}$

ANSWER STEP: ▷ The first part of this step is between each prover P_i $(1 \leq i \leq t)$ and the leader L.
14: **if** $b = 0$ **then**
15:    P_i sends γ_i and Σ_i to L.
16: **else if** $b = 1$ **then**
17:    P_i sends $\varphi_q^{-1}(\Pi_{\gamma_i, \Sigma_i}(s_i))$ to L.
18: **end if**   ▷ φ_q^{-1} is an implementation detail not necessary for the algorithm.

19: L simulates the $N - t$ other answers with $s_i = 0$ $(t + 1 \leq i \leq N)$.
20: L computes the answer for V:

21: **if** $b = 0$ **then**
22:    $\gamma = (\gamma_1, \ldots, \gamma_N)$, $\Sigma = (\Sigma_1, \ldots, \Sigma_N)$, and Θ are sent to V.
23: **else if** $b = 1$ **then**
24:    $\rho(s) = (\Pi_{\gamma_{\Theta(1)}, \Sigma_{\Theta(1)}}(s_{\Theta(1)}), \ldots, \Pi_{\gamma_{\Theta(N)}, \Sigma_{\Theta(N)}}(s_{\Theta(N)}))$ is sent to V.   ▷ NB: Not Θ.
25: **end if**

VERIFICATION STEP:
26: **if** $b = 0$ **then**
27:    V checks $C_1 \overset{?}{=} h(\Theta || h(\Sigma_1 || \gamma_1 || H_1 \Pi_{\gamma_1, \Sigma_1}^{-1}(\beta_1)^T) || \cdots || h(\Sigma_N || \gamma_N || H_N \Pi_{\gamma_N, \Sigma_N}^{-1}(\beta_N)^T))$ and
   $\Theta \overset{?}{\in} S_N$.
28: **else if** $b = 1$ **then**
29:    V checks

$$C_2 \overset{?}{=} h\left(\left(\begin{array}{c} h(\beta_{\Theta(1)} - \alpha \Pi_{\gamma_{\Theta(1)}, \Sigma_{\Theta(1)}}(s_{\Theta(1)}) || \Pi_{\gamma_{\Theta(1)}, \Sigma_{\Theta(1)}}(s_{\Theta(1)})) \\ h(\beta_{\Theta(2)} - \alpha \Pi_{\gamma_{\Theta(2)}, \Sigma_{\Theta(2)}}(s_{\Theta(2)}) || \Pi_{\gamma_{\Theta(2)}, \Sigma_{\Theta(2)}}(s_{\Theta(2)})) \\ \vdots \\ h(\beta_{\Theta(N)} - \alpha \Pi_{\gamma_{\Theta(N)}, \Sigma_{\Theta(N)}}(s_{\Theta(N)}) || \Pi_{\gamma_{\Theta(N)}, \Sigma_{\Theta(N)}}(s_{\Theta(N)})) \end{array}\right)^T\right),$$

   $\mathrm{wt}(\rho(s)) \overset{?}{=} t\omega$, and that $\rho(s)$ is formed of N blocks of length n and of weight ω or weight 0.
30: **end if**

Proof. The construction of the matrix H (described above) and the vector s implies that finding such a n-block of length nN is also equivalent to finding a solution of a local hard problem s_i of weight ω such that $H_i s_i = 0$, which is hard under our assumption.

Theorem 1. *Our scheme is an honest verifier zero-knowledge proof of knowledge, with soundness error bounded by $1/2$, that the group of t provers knows a vector s of length nN such that the global weight of s is $t\omega$, and such that the vector s has a null syndrome for H. The scheme is secure in the random oracle model under the assumption of the hardness of the qMD problem.*

Proof. We prove that our scheme satisfies the three properties: completeness, soundness and zero-knowledge.

Completeness: It is clear that each group of honest provers who has the knowledge of a valid secret key is able to answer correctly any of the honest leader's queries, which permit him to compute the master commitments. The leader, on his turn is able to reveal the information necessary to the honest verifier, in order to check the correctness of these commitments.

Soundness: In [14] it was proven that the regular q-SD scheme satisfies this property and that the soundness error is bounded by $1/2$, assuming that the qMD problem is hard. Because our protocol can be seen as a composition of t simultaneous executions of the q-SD scheme and given that the latter one can be reduced to our protocol by making all signing instances equal, this implies that this soundness error cannot be higher than $1/2$ for our protocol in one single round.

Zero-knowledge: The zero-knowledge property for our protocol can be proven in the random oracle model. In order to do that, we use the classical idea of resettable simulation. Let M be a polynomial-time probabilistic turing machine (simulator) using a dishonest verifier. Because of the two interactions with the leader (prover for our case), we have to assume that the dishonest verifier could contrive two strategies: $St_1(C_1, C_2)$ taking as input the leader's (master) commitments and generating a value $\alpha \in \mathbb{F}_q$, $St_2(C_1, C_2, \beta')$ taking as input the leader's commitments, the answer β and generating as output a challenge in the set $\{0, 1\}$. M will generate a communication tape representing the interaction between leader and verifier. The goal is to produce a communication tape whose distribution is indistinguishable from a real tape by an honest interaction. The simulator M is constructed as follows:

Step 1. M randomly picks a query b from $\{0, 1\}$.

- If $b = 0$, M randomly chooses: u_i, γ_i, Σ_i ($1 \leq i \leq N$) and Θ as a random constant block permutation on N blocks $\{1, 2, \ldots, N\}$, and solves the equation: $Hs'^T = y$ for some vector $s' = (s'_1, \ldots, s'_N)$ of length nN and not necessarily satisfying the condition $\mathsf{wt}(s') = t\omega$. The values $c_{1,i}$ ($1 \leq i \leq N$) can be computed as follows: $c_{1,i} = h\left(\Sigma_i || \gamma_i || H_i u_i^T\right)$, the master commitments are taken then as $C_1 = h(\Theta || c_{1,1} || \ldots || c_{1,N})$ and C_2 as a random string. By simulating the verifier, M applies $St_1(C_1, C_2)$ to get $\alpha \in \mathbb{F}_q$, and then computes β' as follows: $\beta' = \Theta(\Pi_{\gamma_1, \Sigma_1}(u_1 + \alpha s'_1), \ldots, \Pi_{\gamma_N, \Sigma_N}(u_N + \alpha s'_N))$, and has the information needed to derive the simulated communication data between leader and verifier. Therefore the candidates to be written in the communication tape consist of elements $A = C_1 || C_2$, β' and $ans= \rho = \Theta(\Pi_{\gamma_1, \Sigma_1}, \ldots, \Pi_{\gamma_N, \Sigma_N})$. Taking into account the uniform distribution of the random variables used in the computation of A, ans and β', it follows that the distribution of these elements is indistinguishable from those resulting from a fair interaction.

- If $b = 1$, M randomly chooses u_i, γ_i, Σ_i $(1 \leq i \leq N)$ and Θ as a random constant block permutation on N blocks $\{1, 2, \ldots, N\}$. This time it picks $s = (s_1, \ldots, s_N)$ as a random vector from the set $\mathbb{F}_q^{nN}$ with weight $t\omega$ and formed of N blocks of length n and of weight ω or 0. The commitments C_1 will be given uniformly at random values and $C_2 = h(\Theta(c_{2,1}, \ldots, c_{2,N}))$ such that each $c_{2,i} = h(\Pi_{\gamma_i, \Sigma_i}(u_i) || \Pi_{\gamma_i, \Sigma_i}(s_i))$. Again, from $St_1(C_1, C_2)$, M gets $\alpha \in \mathbb{F}_q$ and computes β' as follows: $\beta' = \Theta(\Pi_{\gamma_1, \Sigma_1}(u_1 + \alpha s_1), \ldots, \Pi_{\gamma_N, \Sigma_N}(u_N + \alpha s_N))$, and has the information needed to derive the simulated communication data. The communication set features elements $A = C_1 || C_2$, β' and $ans = \rho(s) = (\Pi_{\gamma_{\Theta(1)}, \Sigma_{\Theta(1)}}(s_{\Theta(1)}), \ldots, \Pi_{\gamma_{\Theta(N)}, \Sigma_{\Theta(N)}}(s_{\Theta(N)}))$. The uniformly random character of the choices made will render these elements indistinguishable from those resulting from a fair interaction.

Step 2. M applies the verifier's strategy obtaining b' as result.

Step 3. When $b = b'$, the machine M writes on its communication tape the values of A, α, β', b and ans. If the values differ, however, nothing is written and the machine returns to Step 1.

Therefore, in 2δ rounds on average, M produces a communication tape indistinguishable from one that corresponds to a fair interaction process execution that takes δ rounds. □

We derive the following corollary due to the results from [4].

Corollary 1. *The resulting threshold ring signature scheme obtained from the honest verifier zero-knowledge q-SD identification protocol is unforgeable under chosen message attacks in the random oracle model.*

Theorem 2. *Our threshold ring signature scheme obtained from the q-SD identification protocol is anonymous in the random oracle model.*

Proof. The second property we would like to examine is the anonymity property in the random oracle model of the resulting threshold ring signature scheme obtained from our q-SD identification protocol. In other words, the verifier must not be able to determine the identity of the real signers, a part from the fact that they were at least t among the n specified ring members. For the challenge 0 the response of both real signers and the non-signers are completely indistinguishable, since Θ, Σ_i, and γ_i are chosen uniformly at random and therefore the response is random. So the only possibility to identify non-signers is challenge 1. In this case the verifier receives a permuted value of the secret key without having access to the used permutation. As consequence, the anonymity of the signers is preserved. □

Table 2. Comparison code-based threshold ring signature schemes

Threshold ring Signatures	Pk size in KBytes	Sign. size in KBytes	Sign. cost in bops
Aguilar et. al's scheme	1470	2448	2^{30}
Dallot and Vergnaud's scheme	10137122	7	2^{35}
Our scheme	400	2384	2^{26}

4.3   Performance and Implementation

Theoretical Results. In general the signature length of signature schemes derived from identification schemes is constrained by the number of rounds. Our proposal is built by applying the q-SD identification scheme, which needs a smaller number of rounds to reach the same probability of cheating as Stern's identification scheme. For a probability of cheating of 2^{-80}, one needs about 140 rounds for Stern's identification scheme and only 80 rounds for our proposal. This fact has a positive effect in terms of signature length for our proposal as we will see below.

Dallot and Vergnaud's scheme uses a CFS signature scheme as basis, therefore it inherits the advantage to provide a shorter signature length, however it suffers from slow signature generation cost and large key sizes compared to our construction.

In Table 2 we give our results in terms of key sizes, signature length and the signing cost for the parameter set $(N, t) = (100, 50)$ in comparison with the code-based threshold ring signature schemes due to Aguilar et al. [1] and Dallot and Vergnaud [16].

Before we do that we have to mention that the signature length of our scheme is N times the signature length of one q-SD signature length, similarly is for the computation of the public key size and the complexity of the protocol.

For our scheme, taking into account the performances of the ISD algorithm, we suggest the following parameters: $q = 256$, $n = 128$, $r = 64$, and $\omega = 49$.

For the same security level, we need to take respectively: $q = 2$, $n = 694$, $r = 347$, $\omega = 69$ for Aguilar et al.'s scheme [1] and, $q = 2$, $n = 2^{22}$, $r = 198$, $\omega = 9$ for Dallot and Vergnaud's scheme [16].

We considered that all seeds used are 128 bits long, the hash outputs are 160 bits long, the security level is 2^{80}, and that the probability of cheating is bounded by 2^{-80}.

Table 2 shows that for the same level of security, the public key size for our scheme is almost four times smaller than Aguilar et. al's one and 25344 times smaller than Dallot and Vergnaud's one. The signature length of our construction is 2384 Kbytes, where that of Aguilar et al.'s scheme is 2448 Kbytes and almost 7 Kbytes for Dallot and Vergnaud's one. For signing cost, we obtain better results.

Using the two improvement presented in Section 3, the signature length of our threshold ring signature scheme is almost 1633 KBytes.

We also compare our scheme to the recent work [13], which is a lattice based threshold ring signature based on the hardness of SIS problem and is related to this work.

For a 111 bit-security, the authors of [13] obtain 45 Mbytes for the signature length. Using our scheme the signature length is only 4 Mbytes for the same security level and the same hash length (224 bits), using parameters $q = 256, n = 204, r = 102$, and $\omega = 71$.

Remark 3. To reduce more the public key size, we can use the proposal in [19] and [6] by replacing a random matrix H by a double circulant matrix respectively a quasi-dyadic matrix. In this case, we obtain a public key size in 12.5 Kbytes for our construction, 8.47 Kbytes for Aguilar et al.'s one. and 1146 Kbytes for Dallot and Vergnaud's scheme.

4.4   Practical Results

General Remarks. The following tables show the timings we have obtained for a C implementation. The test system was an Intel(R) Core(TM)2 Duo CPU E8400@3.00GHz, running Debian 6.0.3. The sources have been compiled using gcc 4.6.2.

In all cases, we used parity check matrices in systematic form. Due to the row-major order of C, the transposed matrices have been stored. The number of the ring members was always set to 100, the number of ring provers to 50. The tables show the setup time and the time running the protocol, where the setup time is consumed for the generation of the necessary public and private keys.

Finally, the use of quasi-dyadic matrices does not allow for all theoretically possible parameters. For instance, dyadic matrices have the dimension $2^p \times 2^p$ ($p \in \mathbb{N}$), which means that for quasi-dyadic matrices $r = d2^p$ for some $d \in \mathbb{N}$. In order to have comparable results and a uniform implementation, we have used this restriction for the random and the quasi-cyclic case as well.

Aguilar et. al Scheme. The number of rounds for the scheme has been set to 28 (probability of cheating 2^{-16}), the dimension of the parity check matrix H^T over $\mathbb{F}_2$ has been set to 704×352, but only the redundancy part has been stored in memory, which is of dimension 352×352. The weight of the secrets been set to 76 (Table 3).

Our Scheme. For our scheme the parity check matrices H^T have been chosen over $\mathbb{F}_{2^8}$, mainly because in this case a field element fits exactly in one byte. The

Table 3. Aguilar et al. timings for 28 protocol rounds, $H \in \mathbb{F}_2^{352 \times 704}$

Matrix Type	Dim. $[n \times r]$	Weight	Setup $[ms]$	Protocol $[ms]$	Total $[ms]$	Sec
Random	704×352	76	108.539	98.662	207.200	2^{80}
Quasi-dyadic	704×352	76	811.202	474.737	1285.939	2^{80}
Quasi-cyclic	704×352	76	476.796	302.935	779.731	2^{80}

Table 4. Timings for our scheme and 16 protocol rounds, $H \in \mathbb{F}_{2^8}^{72 \times 144}$

Matrix Type	Dim. $[n \times r]$	Weight	Setup $[ms]$	Protocol $[ms]$	Total $[ms]$	Sec
Random	144×72	54	32.979	18.499	51.477	2^{80}
Quasi-dyadic	144×72	54	44.331	29.109	73.439	2^{80}
Quasi-cyclic	144×72	54	38.747	26.550	65.298	2^{80}

number of rounds has been set to 16 (probability of cheating 2^{-16}), the weight of the secrets has been set to 54. The number of the ring members was again set to 100 and the number of ring provers to 50 (Table 4).

Remark 4. As one can see, the computational cost for quasi-dyadic/-cyclic cases is always higher than using random parity check matrices. The reason is that the vector-matrix product is more expensive in those cases, because the matrix has to be reconstructed on the fly during the multiplication without actually building the whole matrix in memory. The savings in memory have to be paid for with additional runtime.

Remark 5. The given implementation is given as a proof of concept. For instance, the communication between the leader and the provers takes place on the same machine, even inside the same executable. In reality, the provers would be located on different computers, having a different architecture, connected to the leader via network connections and the like. In such a heterogeneous scenario, the communication latency for those network connections had to be taken into account. It also might be possible that some provers use a very fast machine, whereas others use a very slow one. The interaction process would be dominated then by the slowest possible prover.

Transforming into a Signature Scheme. Similar to the general technique shown by Fiat-Shamir, we can transform our scheme into a signature scheme. The idea is that the signing and verifying part are handled separately, i.e. at first, the leader simulates the challenge step of the verifier using a public stream cipher or public hash function with some predefined starting value involving the document to sign. The protocol is then run without further verification, but all the data which are needed for verification, in particular the master commitments, has to be recorded. These data form the signature.

The verifier uses the recorded data to run the protocol without the signing side. For the challenge step, the verifier uses the same starting value for the stream cipher or hash function as the signing part did. The document is part of this starting value. A consequence of this approach is that the signatures become quite large as everything needed for verification has to be recorded in the signature.

In Table 5 we give some timings for the resulting signature scheme. We used the same settings as above, but run the protocol with random matrices only. The savings using other matrix types is negligible compared to the gained signature sizes.

Table 5. Timings for 80 protocol rounds, $H \in \mathbb{F}_{2^8}^{72 \times 144}$

Doc. [MiB]	Sig. [MiB]	Dim. [$n \times r$]	Weight	Signing [ms]	Verification [ms]	Total [ms]	Sec
1	4	144×72	54	544	454	998	2^{80}
10	13	144×72	54	3643	3551	7194	2^{80}
25	28	144×72	54	8803	8700	17503	2^{80}

The signature sizes are not fixed, but show a small variation depending on the values chosen during the challenge step. More specifically, the answers transmitted for the cases b=0,1,2 vary in size, which effectively leads to varying signature sizes as well. The values are therefore average values obtained while running the protocol 80 rounds (probability of cheating 2^{-80}).

5   Conclusion

Starting from the recently proposed q-SD zero-knowledge identification scheme [14], we presented in this work a novel threshold ring signature scheme based on error-correcting codes based on the q-SD identification scheme. Since the q-SD scheme has a low soundness error allowing a specified security to be reached in few rounds, our construction uses this fact to achieve a scheme which shorter signature length, smaller public key size and signature cost compared to Aguilar et al.'s which is based on Stern's identification scheme. We have confirmed our results by implementing the both schemes in C that shows clearly the advantage of our proposal. The source code of our implementation can be found here: http://www.cayrel.net/IMG/tgz/waifi-files.tgz.

References

1. Aguilar Melchor, C., Cayrel, P.-L., Gaborit, P.: A New Efficient Threshold Ring Signature Scheme Based on Coding Theory. In: Buchmann, J., Ding, J. (eds.) PQCrypto 2008. LNCS, vol. 5299, pp. 1–16. Springer, Heidelberg (2008)
2. Aguilar Melchor, C., Cayrel, P.-L., Gaborit, P., Laguillaumie, F.: A new efficient threshold ring signature scheme based on coding theory. IEEE Transactions on Information Theory 57(7), 4833–4842 (2011)
3. Aguilar Melchor, C., Gaborit, P., Schrek, J.: A new zero-knowledge code based identification scheme with reduced communication (2011), http://arxiv.org/PS_cache/arxiv/pdf/1111/1111.1644v1.pdf
4. El Yousfi Alaoui, S.-M., Dagdelen, Ö., Véron, P., Galindo, D., Cayrel, P.-L.: Extended security arguments for (ring) signature schemes. Cryptology ePrint Archive, Report 2012/068 (2012)
5. Barg, S.: Some new NP-complete coding problems. Problemy Peredachi Informatsii 30, 23–28 (1994)
6. Barreto, P.S.L.M., Cayrel, P.-L., Misoczki, R., Niebuhr, R.: Quasi-Dyadic CFS Signatures. In: Lai, X., Yung, M., Lin, D. (eds.) Inscrypt 2010. LNCS, vol. 6584, pp. 336–349. Springer, Heidelberg (2011)

7. Becker, A., Joux, A., May, A., Meurer, A.: Decoding Random Binary Linear Codes in $2^{(n/20)}$: How $1 + 1 = 0$ Improves Information Set Decoding. In: Pointcheval, D., Johansson, T. (eds.) EUROCRYPT 2012. LNCS, vol. 7237, pp. 520–536. Springer, Heidelberg (2012)

8. Berlekamp, E., McEliece, R., van Tilborg, H.: On the inherent intractability of certain coding problems. IEEE Transactions on Information Theory 24(3), 384–386 (1978)

9. Bernstein, D.J., Buchmann, J., Dahmen, E.: Post-Quantum Cryptography, 1st edn. Springer Publishing Company, Incorporated (2008)

10. Bernstein, D.J., Lange, T., Peters, C.: Smaller Decoding Exponents: Ball-Collision Decoding. In: Rogaway, P. (ed.) CRYPTO 2011. LNCS, vol. 6841, pp. 743–760. Springer, Heidelberg (2011)

11. Biswas, B., Sendrier, N.: McEliece Cryptosystem Implementation: Theory and Practice. In: Buchmann, J., Ding, J. (eds.) PQCrypto 2008. LNCS, vol. 5299, pp. 47–62. Springer, Heidelberg (2008)

12. Bresson, E., Stern, J., Szydlo, M.: Threshold Ring Signatures and Applications to Ad-hoc Groups. In: Yung, M. (ed.) CRYPTO 2002. LNCS, vol. 2442, pp. 465–480. Springer, Heidelberg (2002)

13. Cayrel, P.-L., Lindner, R., Rückert, M., Silva, R.: A Lattice-Based Threshold Ring Signature Scheme. In: Abdalla, M., Barreto, P.S.L.M. (eds.) LATINCRYPT 2010. LNCS, vol. 6212, pp. 255–272. Springer, Heidelberg (2010)

14. Cayrel, P.-L., Véron, P., El Yousfi Alaoui, S.M.: A Zero-Knowledge Identification Scheme Based on the q-ary Syndrome Decoding Problem. In: Biryukov, A., Gong, G., Stinson, D.R. (eds.) SAC 2010. LNCS, vol. 6544, pp. 171–186. Springer, Heidelberg (2011)

15. Courtois, N.T., Finiasz, M., Sendrier, N.: How to Achieve a McEliece-Based Digital Signature Scheme. In: Boyd, C. (ed.) ASIACRYPT 2001. LNCS, vol. 2248, pp. 157–174. Springer, Heidelberg (2001)

16. Dallot, L., Vergnaud, D.: Provably Secure Code-Based Threshold Ring Signatures. In: Parker, M.G. (ed.) Cryptography and Coding 2009. LNCS, vol. 5921, pp. 222–235. Springer, Heidelberg (2009)

17. Fiat, A., Shamir, A.: How to Prove Yourself: Practical Solutions to Identification and Signature Problems. In: Odlyzko, A.M. (ed.) CRYPTO 1986. LNCS, vol. 263, pp. 186–194. Springer, Heidelberg (1987)

18. Fiege, U., Fiat, A., Shamir, A.: Zero knowledge proofs of identity. In: Proceedings of the Nineteenth Annual ACM Symposium on Theory of Computing - STOC, pp. 210–217 (1987)

19. Gaborit, P., Girault, M.: Lightweight code-based authentication and signature. In: IEEE International Symposium on Information Theory–ISIT 2007, Nice, France, pp. 191–195. IEEE (2007)

20. Herranz, J., Sáez, G.: Forking Lemmas for Ring Signature Schemes. In: Johansson, T., Maitra, S. (eds.) INDOCRYPT 2003. LNCS, vol. 2904, pp. 266–279. Springer, Heidelberg (2003)

21. May, A., Meurer, A., Thomae, E.: Decoding random linear codes in $\tilde{\mathcal{O}}(2^{0.054n})$. In: Lee, D.H., Wang, X. (eds.) ASIACRYPT 2011. LNCS, vol. 7073, pp. 107–124. Springer, Heidelberg (2011)

22. McEliece, R.: A public-key cryptosystem based on algebraic coding theory. The Deep Space Network Progress Report, DSN PR 42–44 (1978), http://ipnpr.jpl.nasa.gov/progressreport2/42-44/44N.PDF

23. Niebuhr, R., Cayrel, P.-L., Bulygin, S., Buchmann, J.: On Lower Bounds for Information Set Decoding over $\mathbb{F}_q$. In: SCC 2010, RHUL, London, UK, pp. 143–157 (2010)
24. Niederreiter, H.: Knapsack-type cryptosystems and algebraic coding theory. Problems of Control and Information Theory 15(2), 159–166 (1986)
25. Peters, C.: Information-Set Decoding for Linear Codes over $\mathbb{F}_q$. In: Sendrier, N. (ed.) PQCrypto 2010. LNCS, vol. 6061, pp. 81–94. Springer, Heidelberg (2010)
26. Rivest, R.L., Shamir, A., Tauman, Y.: How to Leak a Secret: Theory and Applications of Ring Signatures. In: Goldreich, O., Rosenberg, A.L., Selman, A.L. (eds.) Theoretical Computer Science. LNCS, vol. 3895, pp. 164–186. Springer, Heidelberg (2006)
27. Stern, J.: A New Identification Scheme Based on Syndrome Decoding. In: Stinson, D.R. (ed.) CRYPTO 1993. LNCS, vol. 773, pp. 13–21. Springer, Heidelberg (1994)
28. Stern, J.: A new paradigm for public key identification. IEEE Transactions on Information Theory 42, 1757–1768 (1996)

A Encoding Function over $\mathbb{F}_q$

The constant weight encoding bijective function φ_q is described in Algorithm 2, this function takes its input from the interval $[0, (q-1)^\omega \binom{n}{\omega})[$ and outputs a q-ary word of length n and Hamming weight ω. Algorithm 2 uses a binary encoder method introduced by Biswas and Sendrier [11]. This function is a constant weight encoding function taking $s = \omega \log_2(n/\omega)$ input bits and outputing a binary word of length n and weight ω, and which is very efficient because of its linear time encoding.

Algorithm 2. q-ary EnumDecoding

Input: integers n, q, ω and $x \in [0, (q-1)^\omega \binom{n}{\omega})[, (\omega \le n)$
Output: q-ary word of length n and Hamming weight ω

1: $db \leftarrow \lfloor x/(q-1)^\omega \rfloor$
2: ret $\leftarrow$ binary encoder(db, n, ω)
3: $rest \leftarrow x \bmod (q-1)^\omega$
4: **for** i from 1 to n **do**
5:    if $0 < \text{ret}[i]$
6:    $\text{ret}[i] \leftarrow (rest \bmod (q-1)) + 1$
7:    $rest \leftarrow \lfloor rest/(q-1) \rfloor$
8:    **end if**
9: **end for**
10: **Return** ret

Sequences and Functions Derived from Projective Planes and Their Difference Sets

Alexander Pott, Qi Wang*, and Yue Zhou

Faculty of Mathematics, Otto-von-Guericke-University Magdeburg,
39106 Magdeburg, Germany
{alexander.pott,qi.wang}@ovgu.de, yue.zhou@st.ovgu.de

Abstract. Many interesting features of sequences and functions defined on finite fields are related to the interplay between the additive and the multiplicative structure of the finite field. In this paper, we survey some of these objects which are related to difference set representations of projective planes.

Keywords: projective plane, semifield, difference set, sequence, isomorphism of incidence structures.

1 Introduction

Many functions and sequences on finite fields which are important in cryptography and coding theory are based on the sometimes unexpected connection between the additive and the multiplicative structure of the field. Popular examples are for instance, m-sequences and bent functions.

In this paper, we will survey some connections to finite projective planes. It turns out that several functions and sequences occur as projections from bigger objects. Sometimes these bigger objects have been used to construct projective planes. It may be interesting, and that is what we suggest in this paper, to reverse this process: for instance, given a bent function or a certain sequence with good correlation properties, is it possible or not to find this sequence via projection. This can be seen as a "lifting problem". As a well known example, p-ary bent functions are the component functions of planar functions. But "in between" planar functions and bent functions, there are a lot more vectorial functions: not all bent functions are components of planar functions, but they are (in many cases) components of vectorial bent functions which are not necessarily planar. This particular question has been investigated in [1].

Let us begin with the basic definitions. Let G be an abelian additively written group of order v. If not stated otherwise, all the groups in this paper are abelian, and actually they are almost always either cyclic or elementary abelian. For background on difference sets, we refer the reader to [2,3,4]. A k-subset D of G is a (v, k, λ) *difference set* if the list of differences $d - d'$ with $d, d' \in D$,

---

* Q. Wang's research is supported by the Alexander von Humboldt (AvH) Stiftung/Foundation.

F. Özbudak and F. Rodríguez-Henríquez (Eds.): WAIFI 2012, LNCS 7369, pp. 64–80, 2012.

$d \neq d'$, covers all nonzero elements in G exactly λ times. In this paper, we need more general concepts. A *relative difference set* with parameters (m, n, k, λ) is a k-subset D of an abelian group G such that the list of differences $d - d'$ with $d, d' \in D$, $d \neq d'$, covers all the elements in $G \setminus N$ exactly λ times, and no nonzero element in N (the so called forbidden subgroup): here N is a subgroup of G of order n and index m, hence $|G| = mn$. It is obvious that one may extend this definition in various ways. For us, the notion of a difference set relative to several subgroups $N_1, \ldots N_r$ plays some role where $r = 2$ and $r = 3$. Let G be an abelian group of order v, and let N_i be subgroups of order n_i. We assume that the $N_1, \ldots, N_r$ intersect pairwise trivially. In order to avoid confusion with difference sets and relative difference sets, we will now speak about generalized difference sets. A *generalized* $(v; n_1, \ldots, n_r; k, \lambda; \lambda_1, \ldots, \lambda_k)$ *difference set* relative to the subgroups N_i is a k-subset D of G such that the list of differences $d - d'$ with $d, d' \in D$, $d \neq d'$, covers all the elements in $G \setminus (N_1 \cup N_2 \cup \ldots \cup N_r)$ exactly λ times, and the nonzero elements in N_i exactly λ_i times. The N_i's are the *exceptional subgroups*.

Example 1. 1. The set $\{1, 2, 4\}$ in $\mathbb{Z}_7$ is a $(7, 3, 1)$ difference set.
  2. The set $\{1, 2, 4\}$, now considered as a subset of $\mathbb{Z}_8$, is a relative $(4, 2, 3, 1)$ difference set relative to $\{0, 4\} \leq \mathbb{Z}_8$.
  3. The set $\{(0, 0), (1, 1), (2, 1)\}$ is a relative $(3, 3, 3, 1)$ difference set in $\mathbb{Z}_3 \times \mathbb{Z}_3$ relative to $\{0\} \times \mathbb{Z}_3$.
  4. The set $\{(1, 2), (2, 4), (3, 3), (0, 1)\}$ is a generalized $(20; 4, 5; 4, 1; 0, 0)$ difference set in $\mathbb{Z}_4 \times \mathbb{Z}_5$ relative to $\mathbb{Z}_4 \times \{0\}$ and $\{0\} \times \mathbb{Z}_5$.
  5. The set $\{(1, 2), (2, 0), (0, 3)\}$ is a relative $(16; 4, 4, 4; 3, 1; 0, 0, 0)$ difference set in $\mathbb{Z}_4 \times \mathbb{Z}_4$ relative to the three subgroups $\mathbb{Z}_4 \times \{0\}$, $\{0\} \times \mathbb{Z}_4$ and $\{(x, x) : x \in \mathbb{Z}_4\}$.

Note that the term "divisible difference set" is used in the literature to denote the case of $(v; n; k, \lambda; \mu)$ difference sets, i.e., the case where the elements outside a subgroup of order n are covered by the list of differences exactly λ times, and the nonzero elements in the subgroup exactly μ times, see [5].

We have the following easy equation that relates the parameters of a generalized difference set:

$$k(k-1) = \lambda(v - n_1 - \cdots - n_r + r - 1) + \lambda_1(n_1 - 1) + \lambda_2(n_2 - 1) + \cdots + \lambda_r(n_r - 1).$$

2   Difference Sets and Incidence Structures

Given a (generalized, relative) difference set D, we may construct an incidence structure out of it. We recall that an *incidence structure* is just a set of points $\mathcal{P}$ and blocks $\mathcal{B}$ together with an incidence relation defined on $\mathcal{P} \times \mathcal{B}$. In most cases, the blocks are just subsets of the point set, and the incidence relation is "is element of". The incidence structure (called the *development* dev(D) of D) that we construct from D is the following: the points of the incidence structure are the elements in G, and the blocks are the *translates* $D + g := \{d + g : d \in D\}$ of

D, where $g \in G$. It may be possible that $D + g = D + h$ for $g \neq h$, in which case we consider the block $D+g$ twice (or even more). The incidence structure related to a generalized difference set has the same number of points as blocks. If the generalized difference set has parameters $(v; n_1, \ldots, n_r; k, \lambda; \lambda_1, \ldots, \lambda_r)$ relative to subgroups N_i, we get an incidence structure with the following properties:

- The structure has v points and v blocks.
- Any block contains exactly k points.
- There are r equivalence relations $\sim_i$ on the point set, where the size of an equivalence class relative to $\sim_i$ is n_i.
- Any two equivalence classes of $\sim_i$ and $\sim_j$ with $i \neq j$ intersect in at most one point.
- Two distinct points which are not contained in one equivalence class are contained in exactly λ blocks, and points p, q with $p \sim_i q$ are contained in exactly λ_i blocks.

The equivalence classes of the incidence structure $\mathrm{dev}(D)$ are simply as follows: $g \sim_i h$ if and only if $g - h \in N_i$, and the equivalence classes are the cosets of the N_i. For more on incidence structures, we refer to [2].

Let us look more closely at the property that two points are contained in a constant μ blocks:

$$g, h \in D + x \text{ and } g \neq h \Leftrightarrow g = d + x \text{ and } h = d' + x \text{ for some } d, d' \in D, d \neq d'$$
$$\Leftrightarrow g - h = d - d' \text{ and } g = d + x$$

This shows that the number μ of x with $g, h \in D + x$ is exactly the number of possibilities to write $g - h$ as a difference with elements in D. This basically shows that $\mathrm{dev}(D)$ has the aforementioned properties.

The most interesting case for us is the case $\lambda = 1$. The incidence structures are "residuals" of projective planes.

A finite *projective plane* consists of v points and v blocks (lines) such that any two different points are joined by exactly one line. Moreover, any line has exactly k points. The number $n := k - 1$ is the *order* of the projective plane. It is well known and easy to see that $v = n^2 + n + 1$ if n is the order of the plane.

An $(n^2 + n + 1, n + 1, 1)$ difference set gives rise to a projective plane of order n, hence the first case in Example 1 describes a projective plane of order 2.

Theorem 1. *Let D be a (relative, generalized) difference set with one of the following parameters:*

1. *A difference set with parameters $(n^2 + n + 1, n + 1, 1)$.*
2. *A relative $(n + 1, n - 1, n, 1)$ difference set relative to some subgroup of order $n - 1$.*
3. *A relative $(n, n, n, 1)$ difference set relative to some subgroup of order n.*
4. *A generalized $(n(n - 1); n - 1, n; n - 1, 1; 0, 0)$ difference set relative to two subgroups of order $n - 1$ and n.*
5. *A generalized $((n - 1)^2; n - 1, n - 1, n - 1; n - 2, 1; 0, 0, 0)$ relative to three subgroups of order $n - 1$ which intersect pairwise trivially.*

Then dev(D) is a projective plane of order n (case (1.)) or it is an incidence structure which can be uniquely extended to a projective plane of order n.

Remark 1. These cases also occur in the classification of quasiregular collineation groups of projective planes in [6].

Remark 2. The (relative, generalized) difference sets in Theorem 1 are the *projective plane difference sets*. Those in (1.) are *planar*, those in (2.) *affine*.

We just sketch the proof of part (5.), which is the most involved case. We construct a projective plane of order n from a generalized $((n-1)^2; n-1, n-1, n-1; n-2, 1; 0, 0, 0)$ difference set relative to three subgroups N_1, N_2 and N_3 as follows (the groups N_i intersect pairwise trivially and have order $n-1$).

Points are the $(n-1)^2$ elements in G plus the cosets of the N_i (each coset $x + N_i$ is considered as a "new" point $[x]_i$). This gives another $3(n-1)$ points, hence together we have $(n-1)^2 + 3(n-1) = n^2 + n - 2$ points. In addition to that, we add three points "at infinity" ∞_1, ∞_2 and ∞_3. Now we come to the lines: the lines are the point sets $D + x$ plus three new points, namely exactly those cosets of N_i which are disjoint from $D + x$. That gives $(n-1)^2$ lines of size $n + 1$. Three more lines "at infinity" are the lines L_1, L_2 and L_3 whose points are the cosets of N_i $(i = 1, 2, 3)$ where L_i contains, additionally, the points ∞_j with $j \in \{1, 2, 3\} \setminus \{i\}$. Hence, these three lines also contain $n + 1$ points. The final $3(n-1)$ lines are the points of the cosets $N_i + x$ together with ∞_i and the "new" point $[x]_i$. We leave it to the reader to check that this incidence structure is, indeed, a projective plane.

The other cases in the list of Theorem 1 can be checked in a similar way. We refer to [3] and [7] for more details. Note that in Example 1 we gave instances for all the 5 cases listed above.

3   Equivalence of Difference Sets and Isomorphism of Incidence Structures

An important notion in the theory of projective planes and, more generally, incidence structures is the concept of isomorphisms: two incidence structures are *isomorphic* if there is a permutation π on points and a permutation σ on blocks such that for all points p and blocks B the following holds:

$$\pi(p) \text{ is incident with } \sigma(B) \quad \Leftrightarrow \quad p \text{ is incident with } B.$$

Note that σ is uniquely determined by π if no two different blocks consist of exactly the same points. Such incidence structures are simple, and in this paper we just consider simple incidence structures. The incidence structure $\mathcal{B}$ that we obtain by applying π is $\mathcal{B}^\pi$.

The set of point permutations π such that $\mathcal{B}^\pi = \mathcal{B}$ is the automorphism group of the incidence structure.

If D is a generalized difference set, then the incidence structure $\text{dev}(D)$ has the translations $\tau_g : G \to G$, $x \mapsto x + g$ as automorphisms. The translations form a group that acts regularly on points and blocks of dev(D). One can show that any incidence structure with an automorphism group acting regularly on points and blocks has a "difference set representation". We do not want to go into details here, but just refer to the literature ([2,3], for instance).

Two difference sets D_1 and D_2 in G are *equivalent* if there is a group automorphism ϕ such that $\phi(D_1) := \{\phi(d) : d \in D_1\} = D_2 + g$ for a suitable $g \in G$. It is obvious that equivalent generalized difference sets give rise to isomorphic incidence structures, but the converse is in general not true.

Problem 1 (Isomorphism problem). Does isomorphism of two incidence structures dev(D_1) and dev(D_2) derived from generalized difference sets D_1 and D_2 imply that the generalized difference sets are equivalent?

In our opinion, this is a fundamental (and in general difficult) question. We refer the reader to [8], for instance. In [9] it is shown that a deep theorem due to Ott [10], see also [11], is closely related to the isomorphism problem for cyclic planar difference sets. It is, in general, much easier to check whether difference sets are equivalent or not than checking for isomorphism. For semifields (which are basically projective planes derived from relative $(n, n, n, 1)$ difference sets), geometers introduced the notion of isotopism in order to describe when two planes are isomorphic.

In the next section, we will discuss several incidence structures obtained from projective plane difference sets via projection.

We think it is worthwhile not just to look at the equivalence problem for these difference sets, but also the isomorphism problem. For instance, several papers deal with the equivalence of bent functions, but not much seems to be known about the corresponding incidence structures.

There are only some invariants to distinguish incidence structures (rank of incidence matrix, Smith normal form, automorphism groups, intersection numbers), see [12] and the references quoted there. The papers [13] and [14] indicate the difficulty in distinguishing incidence structures derived from difference sets. In [13], the author used quite involved arguments to check the inequivalence of only a fairly small number of bent functions, and in [15] a long standing conjecture about the nonisomorphism of Gordon-Mills-Welch-type incidence structures has been proved using highly nontrivial arguments.

4   Projections of Generalized Difference Sets and a Lifting Problem

Let $\mathbb{C}[G]$ denote the set of formal sums $\sum_{g \in G} a_g g$, where $a_g \in \mathbb{C}$ and G is any (not necessarily abelian) group which we write here multiplicatively. The set $\mathbb{C}[G]$ is basically just the set of complex vectors whose basis is the set of group elements. We add these vectors componentwise, i.e.,

$$\sum_{g \in G} a_g g + \sum_{g \in G} b_g g = \sum_{g \in G} (a_g + b_g) g.$$

Multiplication is a bit more involved:

$$\sum_{g \in G} a_g g \cdot \sum_{g \in G} b_g g = \sum_{g \in G} \left(\sum_{h \in G} a_h b_{h^{-1}g} \right) g.$$

If $D = \sum_{g \in G} a_g g$, we define $D^{(-1)} = \sum_{g \in G} a_g g^{-1}$. If D is a subset of G, we identify D with the group ring element $\sum_{d \in D} d$. Then the following holds (the proof is straightforward).

Proposition 1. *The set D is a generalized $(v; n_1, \ldots, n_r; k, \lambda; \lambda_1, \ldots, \lambda_r)$ difference set relative to the subgroups N_i if and only if*

$$\begin{aligned} D \cdot D^{(-1)} = k &- (\lambda(1-r) + \lambda_1 + \cdots + \lambda_r) + \\ &+ \lambda(G - N_1 - N_2 - \ldots - N_r) + \lambda_1 N_1 + \cdots + \lambda_r N_r. \end{aligned} \quad (1)$$

From now on, we just consider abelian groups (if not stated otherwise). Let $H \leq G$ be a subgroup of G, and let ϕ_H denote the canonical epimorphism $G \to G/H$ with $\phi_H(g) = gH$. We may extend ϕ_H by linearity to an epimorphism $\mathbb{C}[G] \to \mathbb{C}[G/H]$. If $H \leq N_i$ for some N_i with $\lambda_i = 0$, then $\phi_H(D)$ has just $0, 1$-coefficients, hence we may interpret $\phi_H(D)$ as a subset of G/H.

In general, it is not clear what happens with the right hand side of (1) when we apply ϕ_H. But in all the cases we are interested in (namely the cases of projective plane difference sets), it turns out that one may easily compute the right hand side. In particular, if $r = 2$, we have the following proposition.

Proposition 2. *Let D be a generalized $(v; n_1, n_2; k, \lambda; 0, 0)$ difference set in G relative to N_1 and N_2. If $H \subseteq N_1$ is a subgroup of order m, then $\phi_H(D)$ is a generalized $(v/m; n_1/m, n_2; k, m\lambda; 0, \lambda(m-1))$ difference set.*

Proof. This can be easily seen by applying ϕ_H to $\lambda(G - N_1 - N_2)$.

If D is a generalized difference set relative to more than 2 subgroups, the situation is more involved since in that case $\phi_H(N_2)$ and $\phi_H(N_3)$ may have a nontrivial intersection when projecting onto G/H with $H \leq N_1$.

Proposition 3. *Let D be a generalized $(v; n_1, n_2, n_3; k, \lambda; 0, 0, 0)$ difference set in G relative to N_1, N_2 and N_3. Let $H \subseteq N_1$ be a subgroup of order m, and let $K \leq G/H$ be the intersection $K = \phi_H(N_2) \cap \phi_H(N_3)$. Then*

$$\begin{aligned} \phi_H(D)\phi_H(D)^{(-1)} = \lambda m(G - \phi_H(N_1) - \phi_H(N_2) + K) &+ \lambda(m-1)\phi_H(N_2) + \\ &+ \lambda(m-1)\phi_H(N_2) - \lambda_m K, \end{aligned}$$

hence every element outside of $\phi_H(N_i)$, $i = 1, 2, 3$, has λm representations as a difference with elements from $\phi_H(D)$, no element in $\phi_H(N_1)$ has such a representation, and the elements in $\phi_H(N_2)$ and $\phi_H(N_3)$ not in K have $\lambda(m-1)$ representations. Finally, the nonzero elements in K have $\lambda(m-2)$ representations.

For the sake of completeness, we also note the projections of relative difference sets.

Proposition 4. *Let D be a relative (m, n, k, λ) difference set in G relative to N. If $H \subseteq N$ is a subgroup of order t, then $\phi_H(D)$ is a relative $(m, n/t, k, t\lambda)$ difference set. If $t = n$, then $\phi_H(D)$ is a difference set.*

Note that the existence of a generalized or relative difference set D implies the existence of $\phi_H(D)$, which satisfies, as indicated above, certain group ring equations. Therefore, if we want to show that for certain groups and parameters a generalized or relative difference set does not exist, it is sufficient to show the nonexistence of the hypothetical projections. However, the existence of the projections does not give information about the original difference set. For instance, the existence of a relative $(n, 2, n, n/2)$ difference set does not imply the existence of an $(n, 4, n, n/4)$ or even a relative $(n, n, n, 1)$ difference set.

It turns out that some of the relative or generalized difference sets that one can construct from those in Theorem 1 via projection are of some interest in their own right. The most popular examples are the p-ary bent functions which occur as images from $(p^a, p^a, p^a, 1)$ difference sets: p-ary bent functions are precisely the same as (p^a, p, p^a, p^{a-1}) difference sets in elementary abelian groups. But there are many bent functions which can not be constructed from $(p^a, p^a, p^a, 1)$ difference sets. However, sometimes they can be constructed from (p^a, p^b, p^a, p^{a-b}) difference sets. We may formulate the following quite general lifting problem.

Problem 2 (Lifting problem). Assume that R is a subset of a group G' such that $RR^{(-1)} = \phi_H(DD^{(-1)})$ for some $D \subseteq G$, where $G' = \phi_H(G)$. Is it possible to choose D such that $\varphi(D) = R$?

In other words, R looks as if it can be obtained from D via projection. Does it mean that it really can be constructed as a projection? Good references for the existence of the aforementioned relative (p^a, p^b, p^a, p^{a-b}) difference sets are [16,17], see also [18] and [19].

5   The Prime Power Conjecture for Projective Planes

The difference sets in Theorem 1 correspond to projective planes. The classical projective planes are the Desarguesian planes, where the points are the one-dimensional subspaces of $\mathbb{F}_q^3$, and the lines are the two-dimensional subspaces. This is a $(q^2 + q + 1, q + 1, 1)$ difference set. The order of the plane is q. It turns out that these Desarguesian planes give rise to difference sets of any of the types in Theorem 1. However, there are also non-Desarguesian planes with such representations. For background on projective planes, we refer to [20]

One of the main problems about projective planes is to decide whether the order of a plane must be a prime power. This is a purely combinatorial problem. The situation is slightly different if we impose some structure on the plane, for instance it has a representation as a certain type of difference set.

In the rest of the paper, we would like to summarize some results about difference set representations of projective planes, where we focus on the following questions.

1. What is the difference set representation of the Desarguesian plane?
2. Can we answer the question about isomorphism of planes derived form difference sets?
3. What is known about the prime power conjecture for planes admitting difference set representations?
4. Are there interesting projections of the difference sets? What is known about the lifting problem?

6   Difference Sets and Affine Difference Sets

Now we are going to describe the "classical" difference sets which correspond to the Desarguesian projective planes.

For $n \geq 2$, let $G = \mathbb{F}_{q^n}^*$ be the multiplicative group of $\mathbb{F}_{q^n}$. Let $D = \{x \in \mathbb{F}_{q^n} : \mathsf{tr}(x) = 1\}$. It is not difficult to see that D is a relative $((q^n - 1)/(q - 1), q - 1, q^{n-1}, q^{n-2})$ difference set relative to $N = \mathbb{F}_q^*$, see [3]. We apply the projection construction Proposition 3 to obtain a difference set with parameters $((q^n - 1)/(q - 1), q^{n-1}, q^{n-1} - q^{n-2})$. Taking the complement in the case $n = 3$, we get a planar difference set with parameters $(q^2 + q + 1, q + 1, 1)$. The development of this is the Desarguesian plane. Similarly, if $n = 2$ we obtain a relative $(q + 1, q - 1, q, 1)$ difference set which also describes the Desarguesian projective plane of order q.

It is widely conjectured that the only projective planes which have a development as a $(q^2 + q + 1, q + 1, 1)$ difference set or an affine $(q + 1, q - 1, q, 1)$ difference set are the Desarguesian ones, in other words that there is just one example (up to isomorphism). In the case of Desarguesian planes, it is known that isomorphism implies equivalence, i.e., it is impossible that the Desarguesian plane is the development of two inequivalent (planar or affine) difference sets (see [9] for the planar case; the affine case is similar). We are not aware of a criteria which can be used to distinguish different planar or affine difference sets directly. Perhaps not so many people tried to find such a criteria since no examples are known that have to be distinguished!

A lot of work has been done to prove the prime power conjecture for planar or affine difference sets. There are many necessary conditions that the order n of a planar or affine difference set has to satisfy. Unfortunately, the criteria that we know are not sufficient for a proof of the prime power conjecture, i.e., for a proof that n has to be a prime power, see [21].

It should be mentioned that also non-abelian planar and non-abelian affine difference sets exist. The planes described by these non-abelian difference sets are also Desarguesian.

Now let us consider the "lifting" problem. A lifting of a planar difference set or an affine difference set can not exist since $\lambda = 1$. However, the complement of a planar difference set, hence a $(q^2 + q + 1, q^2, q^2 - q)$ difference set, may have a lifting. Actually, we did construct this difference set via projection of a relative difference set. We note that the problem whether a difference set occurs as a projection from a relative difference set is known under the name "Waterloo problem". We refer the reader to [22] and [23].

In the case of affine difference sets, we get cyclic relative difference sets as image sets under projections. So far, all cyclic relative difference sets which are known whose parameters are $(q + 1, (q - 1)/t, q, t)$ are constructed using projections from the classical affine difference sets. One reason may be that not many researchers tried to find non-classical examples which are different from projections of affine difference sets. However, we think this is an interesting topic of research since the projections give rise to interesting sequences, as we are now going to describe.

Let D be an affine difference set with parameters $(n + 1, n - 1, n, 1)$ relative to a subgroup N of order $n - 1$. We choose a subgroup H of N that has index p in N. Then $\phi_H(D)$ is a relative $(n + 1, p, n, (n - 1)/p)$ difference set. We have

$$\phi_H(D) \cdot \phi_H(D)^{(-1)} = n + (n - 1)/p \cdot (G/H - N/H),\tag{2}$$

where N/H is a subgroup of order p. We have $G/H \cong \mathbb{Z}_{n+1} \times \mathbb{Z}_p$. Let g be a generator of $\mathbb{Z}_p$. We replace g^i by ζ_p^i, where $\zeta = e^{\frac{2\pi i}{p}}$ is a complex primitive p-th root of unity. Then the right hand side of (2) becomes n, and $\phi_H(D) \in \mathbb{C}[\mathbb{Z}_{n+1}]$, where the coefficient of exactly one element in $\mathbb{Z}_{n+1}$ is 0, the remaining coefficients are powers of ζ_p. This can be interpreted as an (almost) p-ary sequence $(a_i)_{i=0,...,n}$ of period $n + 1$ where a_i is a power of ζ_p for all i but one, for which $a_i = 0$, see also [58]. That explains the term "almost". This sequence is perfect in the sense that all nontrivial autocorrelation coefficients $\sum_{i=0}^{n} a_i \overline{a_{i+t}}$ (i.e., $t = 1, \ldots, n$) are 0. For more details, we refer to [24,25]. If $p = 2$, we do not get a sequence of length $n + 1$ out of $\phi_H(D)$ since the group does not split $\mathbb{Z}_{n+1} \times \mathbb{Z}_2$.

Example 2. A relative $(5, 3, 4, 1)$ difference set in $\mathbb{Z}_{15}$ is $\{0, 1, 3, 7\}$. Using $\mathbb{Z}_{15} \cong \mathbb{Z}_5 \times \mathbb{Z}_3$, this set can be also written as $\{(0, 0), (2, 2), (1, 0), (4, 2)\}$, which gives rise to the complex sequence $(1, 1, \zeta^2, 0, \zeta^2)$ of period 5, where ζ is a 3-rd root of unity.

7   Difference Sets and Semifields

One of the most interesting examples of difference sets related to projective planes are relative $(n, n, n, 1)$ difference sets. First of all, for these difference sets the prime power conjecture has been proved, see [26]. Moreover, in contrast to planar and affine difference sets, many inequivalent examples are known. It is far beyond the scope of this article to give a survey of these difference sets, but we just want to highlight some interesting facts. These difference sets are closely related to semifields and Dembowski-Ostrom polynomials: a *semifield* is, roughly speaking, a (not necessarily commutative) field without associativity with respect to multiplication. For more, we refer to [27]. A *Dembowski-Ostrom* polynomial in $\mathbb{F}_q[x]$, $q = p^m$, p prime, is of the form $f(x) = \sum_{i,j} a_{i,j} x^{p^i + p^j}$. In other words, the mapping $x \mapsto f(x + a) - f(x) - f(a) + f(0)$ is $\mathbb{F}_p$-linear on $\mathbb{F}_q$.

7.1   The n Even Case

In this case, we do not just know that n is a power of 2, but we know that $G \cong \mathbb{Z}_4^a$ and the forbidden subgroup is $\mathbb{Z}_2^a$, see [28,29]. Such relative difference sets can be constructed from commutative semifields with 2^a elements. Many examples for n even are constructed in [30]. There is a systematic way how to go from semifields to relative difference sets, but as far as we know there is no "nice" difference set description of the semifields constructed in [30].

Projective planes constructed from semifields have certain additional properties: they are translation planes whose dual is again a translation plane. This holds for the case of even and of odd order. The question arises whether all relative $(n, n, n, 1)$ difference sets with n even are semifield planes. This seems to be unknown. A nice description of $(n, n, n, 1)$ difference sets in terms of finite fields has been obtained in [31]. The approach is similar to the "cocycle method" developed in [19].

Theorem 2. *Let n be a power of 2. Every relative $(n, n, n, 1)$ difference set can be expressed by a function $f : \mathbb{F}_{2^n} \to \mathbb{F}_{2^n}$ such that*

$$f(x + a) + f(x) + xa \tag{3}$$

is a permutation for every $a \neq 0$. Conversely, the existence of a function f with (3) gives rise to a relative difference set. The corresponding plane is a semifield plane if and only if f is a Dembowski-Ostrom polynomial.

Remark 3. If $f(x) = 0$ in this theorem, the corresponding plane is Desarguesian.

The isomorphism of planes that occur as developments of relative $(n, n, n, 1)$ difference sets can be decided on the level of the semifields (if the planes are semifield planes) using the concept of isotopism introduced in [32]. Isotopism is only slightly more general than equivalence of difference sets, see [33,34]

Question 1. Is it possible to construct relative $(2^a, 2^a, 2^a, 1)$ difference sets D such that the development $\mathrm{dev}(D)$ is not a semifield plane? In other words, is there a non-Dembowski-Ostrom polynomial f with (3)?

The projections of relative $(2^a, 2^a, 2^a, 1)$ difference sets are difference sets in groups of exponent 4. The question arises whether we may have relative $(2^a, 2^b, 2^a, 2^{a-b})$ difference sets also in other groups. The answer is yes, see [17]. Of particular interest are $(2^a, 2, 2^a, 2^{a-1})$ difference sets in $G \times \mathbb{Z}_2$ where the forbidden subgroup is a direct factor $\{0\} \times \mathbb{Z}_2$. Then D gives rise to a function $f : G \to \mathbb{Z}_2$ such that $\{(x, f(x)) : x \in G\} = D$. It turns out that f is a bent function, and any bent function gives rise to a relative difference set. We may use this as the definition of a bent function. The groups G which admit bent functions are completely classified, see [35]. The case of vectorial bent functions, which are relative $(2^a, 2^b, 2^a, 2^{a-b})$ difference sets in $G \times \mathbb{Z}_2^b$ relative to $\{0\} \times \mathbb{Z}_2^b$ is not yet completely solved. However, we have the following interesting theorem.

Theorem 3. *There is no abelian relative* $(2^{2a}, 2^b, 2^{2a}, 2^{2a-b})$ *difference set in* $G \times N$ *with* N *as the forbidden subgroup, if* $b > a$.

We refer to [36] for the elementary abelian case, and to [17] for the general case.

Bent functions, i.e., relative $(2^a, 2, 2^a, 2^{a-1})$ difference sets in $G \times \mathbb{Z}_2$ relative to $\{0\} \times \mathbb{Z}_2$ can exist only if $a = 2m$ is even. We cannot obtain them via projection of a relative $(2^a, 2^a, 2^a, 1)$ difference set, as discussed above, but perhaps as projection from a relative $(2^{2m}, 2^m, 2^{2m}, 2^m)$ difference set (also known as a vectorial bent function). There are many of such difference sets, we just describe two very general constructions, known as the Maiorana-McFarland and the partial-spread construction, see [37,38]. We formulate the constructions here for arbitrary characteristic.

Theorem 4 (Maiorana-McFarland [39]). *The sets*

$$\{(x, y, x \cdot \pi(y) + \sigma(y)) \ : \ x, y \in \mathbb{F}_{p^m}\}$$

are relative $(p^{2m}, p^m, p^{2m}, p^m)$ *difference sets in* $(\mathbb{F}_{p^m}^3, +)$, *i.e., the elementary abelian group of order* p^{3m}, *provided that* $\pi : \mathbb{F}_{p^m} \to \mathbb{F}_{p^m}$ *is a bijection, and* $\sigma : \mathbb{F}_{p^m} \to \mathbb{F}_{p^m}$ *is an arbitrary mapping.*

Theorem 5 (Partial Spreads [40]). *The sets*

$$\{(x, y, \pi(x \cdot y^{p^m - 2})) \ : \ x, y \in \mathbb{F}_{p^m}\}$$

are relative $(p^{2m}, p^m, p^{2m}, p^m)$ *difference sets in* $(\mathbb{F}_{p^m}^3, +)$, *i.e., the elementary abelian group of order* p^{3m}, *provided that* $\pi : \mathbb{F}_{p^m} \to \mathbb{F}_{p^m}$ *is a bijection.*

For the case $m = 2$, we also refer to [41].

7.2   The n Odd Case

Again, we know that the prime power conjecture is true, see [26]. We have restrictions on the type of the group that may contain the relative difference set, but these are not strong enough to ensure that the group must be elementary abelian. However, all known abelian examples are subsets of the elementary abelian group. In the situation that n is odd, all known relative $(p^a, p^a, p^a, 1)$ difference sets can be written as $D = \{(x, f(x)) : x \in H\}$, where $G = H \times N$ and N is the forbidden subgroup. Here f is a function $H \to N$. In the case n even, it was impossible to write the difference sets using a function since the forbidden subgroup had no complement. In order to emphasize that D is constructed using the function f, we denote the difference set D_f.

The function f such that D_f is a relative $(p^a, p^a, p^a, 1)$ difference set is *planar*. Many planar functions can be described easily using the fact that H and N are elementary abelian, hence they are the additive groups of finite fields. The example corresponding to the Desarguesian plane is $f(x) = x^2$. Other examples are x^{p^k} where $n/\gcd(k, n)$ is odd. The corresponding planes are non-Desarguesian (so-called *Albert semifields*). The planar functions whose corresponding planes are semifield planes are precisely those functions which are planar and Dembowski-Ostrom, see [33].

There is one example of a planar function which is not of Dembowski-Ostrom type.

Theorem 6 ([42]). *The function* $f(x) = x^{(3^k+1)/2}$ *is planar on* $\mathbb{F}_{3^n}$ *if* $\gcd(k, n) = 1$ *and* k *is odd.*

Question 2. Are there more planar functions which are not of Dembowski-Ostrom type?

Since this paper is not a survey about semifields, we cannot refer to all of the many interesting papers on semifields. Please consult [27] for a recent survey.

Relative (p^a, p^b, p^a, p^{a-b}) difference sets in elementary abelian groups are *vectorial bent functions*. Relative difference sets with these parameters are also known in non elementary abelian groups. The interest of research about these relative difference sets was to decide which groups contain difference sets: the equivalence or isomorphism problem for general groups did not attract a lot of attention. However, in the elementary abelian case, the question about the equivalence of different constructions of (p^a, p, p^a, p^{a-1}) has been studied recently, mainly due to the fact that quite a few new constructions of relative (p^a, p, p^a, p^{a-1}) difference sets have been given, see for instance [43,44,45,46,47]. A possibility to check inequivalence is given by the following Theorem whose proof is straightforward.

Theorem 7. *Let D_1 and D_2 be relative (p^a, p, p^a, p^{a-1}) difference sets in the elementary abelian group of order p^{a+1}. If D_1 is the projection of a (p^a, p^b, p^a, p^{a-b}) difference set, but D_2 not, then D_1 and D_2 are inequivalent.*

Therefore, it may be interesting to determine for p-ary (including binary) bent functions how far they can be lifted, i.e., what is the largest b such that the bent function occurs as the projection from a relative (p^a, p^b, p^a, p^{a-b}) difference set, see also [18].

8   Direct Product Difference Sets

The difference sets that we are going to describe in this and in the following section are less well known. Let G be a group. We study generalized difference sets with parameters $(n(n-1); n, n-1; n-1, 1; 0, 0)$ in G relative to subgroups N_1 of order n and N_2 of order $n-1$. Since in this case $G \cong N_1 \times N_2$ the difference sets are called *direct product difference sets*, see [48]. It is known that n has to be a prime power, see [49]. The classical example of such a difference set D where $\mathrm{dev}(D)$ is the Desarguesian plane is the set $\{(x, x) : x \in \mathbb{F}_q^*\} \subseteq (\mathbb{F}_q^*, \cdot) \times (\mathbb{F}_q, +)$. It is known that these are the only abelian examples. There are also non-abelian difference sets with these parameters. Therefore, to some extent, these planes are completely understood. The planes are nearfield planes, see also [50]. Via projections we obtain other interesting objects.

Let us begin with projections ϕ_H where $H \leq (\mathbb{F}_q, +)$. We consider the trace function, i.e., $\phi_H(\mathbb{F}_q) \cong \mathbb{F}_p$. If $f(x) = x$, hence if we start with the Desarguesian plane, we obtain the set $\{(x, \mathrm{tr}(x)) \ : \ x \in \mathbb{F}_q^*\}$. This is not a relative difference set, but the sequence $(\mathrm{tr}(\alpha^t))_t$ is a p-ary m-sequence and has a two-valued autocorrelation spectrum (here α is a primitive element in $\mathbb{F}_q^*$). For background on sequences, we refer to [51]. Other constructions of such sequences are in [52]. In our language of generalized difference sets, these constructions show that there are several generalized $(p(p^a - 1); p, p^a - 1; p^a - 1, p^{a-1}; 0, p^{a-1} - 1)$ difference sets in $\mathbb{Z}_{p^a - 1} \times \mathbb{Z}_p$, where p is prime, see Proposition 2. It may be interesting to try to find vectorial versions, or ask the question which of the known constructions are projections from generalized

$$(p^b(p^a - 1); p^b, p^a - 1; p^a - 1, p^{a-b}; 0, p^{a-b} - 1) \tag{4}$$

difference sets. An easy example is the Gordon-Mills-Welch construction [14].

Theorem 8. *Let $\mathbb{F}_{q'}$ be a subfield of $\mathbb{F}_q$, and let $\mathrm{tr}\,'$ denote the trace from $\mathbb{F}_q$ to $\mathbb{F}_{q'}$. Then the sets*

$$D_r := \{(x, [tr\,'(x)]^r) \ : \ x \in \mathbb{F}_q^*\} \subseteq (\mathbb{F}_q^*, \cdot) \times (\mathbb{F}_{q'}, +)$$

satisfy the same group algebra equation

$$D_r \cdot D_r^{(-1)} = p^a + p^{a-b}G - p^{a-b}N_1 - N_2$$

for all r with $\gcd(r, q' - 1) = 1$, i.e., they are generalized difference sets with parameters (4), where $q = p^a$ and $q' = p^b$ and $N_1 = (\mathbb{F}_{q'}, +)$ and $N_2 = \mathbb{F}_q^*$.

Note that D_r describes a vectorial function $f : \mathbb{F}_q^* \to \mathbb{F}_p^m$, where $q' = p^b$. Let α be a primitive element of $\mathbb{F}_q^*$. Then the component functions $\mathrm{tr}(a \cdot f(\alpha^t))_{t=0,\ldots,q-2}$ give rise to p-ary sequences $(\zeta_p^{\mathrm{tr}(a \cdot f(\alpha^t))})_t$, where $\zeta_p = \exp(\frac{2\pi i}{p})$ is a complex primitive p-th root of unity. These sequences are the well known Gordon-Mills-Welch sequences whose nontrivial autocorrelation coefficients are -1. We have stated the Gordon-Mills-Welch construction in "vectorial" form, i.e., the construction actually gives a vector space of functions with autocorrelation -1. The dimension of this vector space is b. In view of the "lifting problem", we ask the following question.

Question 3. Find more vectorial functions $\mathbb{F}_q^* \to \mathbb{F}_p^b$ such that all component functions give rise to p-ary sequences with autocorrelation -1, equivalently the set $D = \{(x, f(x)) \ : \ x \in \mathbb{F}_q^*\}$ is a generalized $(p^b(q - 1); p^b, q - 1; q - 1, q/p^b; 0, (q/p^b) - 1)$ difference set.

Finally, let us look at possible projections ϕ_H where $H \leq \mathbb{F}_q^*$ is a subgroup of the multiplicative group of the field $\mathbb{F}_q$. If q is odd, we may project onto a group of order 2. The following theorem can easily be proved. Here a *skew Hadamard difference set* is a $(q, (q-1)/2, (q-3)/4)$ difference set in G such that $D + D^{(-1)} = G - 0$ (using group ring notation). The proof is basically contained in [3,53].

Theorem 9. *Let D be a direct product generalized difference set with parameters $(q(q-1); q, q-1; q-1, 1; 0, 0)$ in $\mathbb{F}_q^* \times (\mathbb{F}_q, +)$. Let $q \equiv 3 \pmod{4}$ and let H be a subgroup of order $(q-1)/2$ in $\mathbb{F}_q^*$. Then the set $S := \{x \in \mathbb{F}_q : (1, x) \in \phi_H(x)\}$ is a skew Hadamard difference set in $(\mathbb{F}_q, +)$.*

A similar statement holds for the case $q \equiv 1 \pmod{4}$, in which case the set S must be a partial difference set. We refer to [3] and [53] for more details. If we use the classical Desarguesian plane, the set S is obviously the set of nonzero squares in $\mathbb{F}_q$. Recently, quite a few different skew Hadamard difference sets have been constructed ([54,55,56]). It would be interesting to decide whether these are also projections of some "bigger" difference sets, as the squares are the projections from a direct product difference set.

There are also some interesting p-ary sequences related to this type of difference sets, see [57,58].

9 Generalized Difference Sets Relative to Three Subgroups

This type of difference sets has parameters $((n-1)^2; n-1, n-1, n-1; n-2, 1; 0, 0, 0)$. The Desarguesian plane is the development of the difference set

$$\{(x, 1-x) : x \neq 0, 1, x \in \mathbb{F}_q\} \subseteq \mathbb{F}_q^* \times \mathbb{F}_q^*.$$

For these difference set types, the prime power conjecture is still open. It is also not known (though conjectured), whether the above construction is the only one, up to equivalence, see [59].

10 Conclusion

In this paper, we described five classes of difference sets which can be constructed from Desarguesian projective planes. In one of the cases, there are also other planes known which have such a difference set representation (planar functions, semifields). All these difference sets correspond to certain cases in the classification of projective planes with "large" quasiregular collineation groups. Using projections, these difference sets give rise to objects which are of independent interest: bent functions, autocorrelation -1-sequences, skew Hadamard difference sets. There are also examples of these objects which are different from those constructed from projective planes. It would be interesting to decide whether these non-classical examples can be also obtained via projection from some "bigger" difference sets.

References

1. Pott, A., Zhou, Y.: Switching Construction of Planar Functions on Finite Fields. In: Hasan, M.A., Helleseth, T. (eds.) WAIFI 2010. LNCS, vol. 6087, pp. 135–150. Springer, Heidelberg (2010)

2. Beth, T., Jungnickel, D., Lenz, H.: Design theory. Vol. I, 2nd edn. Encyclopedia of Mathematics and its Applications, vol. 69. Cambridge University Press, Cambridge (1999)
3. Pott, A.: Finite geometry and character theory. Lecture Notes in Mathematics, vol. 1601. Springer, Berlin (1995)
4. Schmidt, B.: Characters and cyclotomic fields in finite geometry. Lecture Notes in Mathematics, vol. 1797. Springer, Berlin (2002)
5. Jungnickel, D.: On automorphism groups of divisible designs. Canad. J. Math. 34(2), 257–297 (1982)
6. Dembowski, P., Piper, F.: Quasiregular collineation groups of finite projective planes. Math. Z. 99, 53–75 (1967)
7. Ghinelli, D., Jungnickel, D.: Finite projective planes with a large abelian group. In: Surveys in Combinatorics, 2003, Bangor. London Math. Soc. Lecture Note Ser., vol. 307, pp. 175–237. Cambridge Univ. Press, Cambridge (2003)
8. Muzychuk, M.: On the isomorphism problem for cyclic combinatorial objects. Discrete Math. 197/198, 589–606 (1999); 16th British Combinatorial Conference, London (1997)
9. Jungnickel, D.: The isomorphism problem for abelian projective planes. Appl. Algebra Engrg. Comm. Comput. 19(3), 195–200 (2008)
10. Ott, U.: Endliche zyklische Ebenen. Math. Z. 144(3), 195–215 (1975)
11. Ho, C.Y.: Finite projective planes with abelian transitive collineation groups. J. Algebra 208(2), 533–550 (1998)
12. Xiang, Q.: Recent progress in algebraic design theory. Finite Fields Appl. 11(3), 622–653 (2005)
13. Dempwolff, U.: Automorphisms and equivalence of bent functions and of difference sets in elementary abelian 2-groups. Comm. Algebra 34(3), 1077–1131 (2006)
14. Gordon, B., Mills, W.H., Welch, L.R.: Some new difference sets. Canad. J. Math. 14, 614–625 (1962)
15. Kantor, W.M.: Note on GMW designs. European J. Combin. 22(1), 63–69 (2001)
16. Ma, S.L., Schmidt, B.: Relative (p^a, p^b, p^a, p^{a-b})-difference sets: a unified exponent bound and a local ring construction. Finite Fields Appl. 6(1), 1–22 (2000)
17. Schmidt, B.: On (p^a, p^b, p^a, p^{a-b})-relative difference sets. J. Algebraic Combin. 6(3), 279–297 (1997)
18. Chen, Y.Q.: Divisible designs and semi-regular relative difference sets from additive Hadamard cocycles. J. Combin. Theory Ser. A 118(8), 2185–2206 (2011)
19. Horadam, K.J.: Hadamard matrices and their applications. Princeton University Press, Princeton (2007)
20. Hughes, D.R., Piper, F.C.: Projective planes. Graduate Texts in Mathematics, vol. 6. Springer, New York (1973)
21. Gordon, D.M.: The prime power conjecture is true for $n < 2,000,000$. Electron. J. Combin. 1, Research Paper 6, approx. 7 pp. (1994) (electronic)
22. Arasu, K.T., Dillon, J.F., Jungnickel, D., Pott, A.: The solution of the Waterloo problem. J. Combin. Theory Ser. A 71(2), 316–331 (1995)
23. Arasu, K.T., Dillon, J.F., Leung, K.H., Ma, S.L.: Cyclic relative difference sets with classical parameters. J. Combin. Theory Ser. A 94(1), 118–126 (2001)
24. Wolfmann, J.: Almost perfect autocorrelation sequences. IEEE Trans. Inform. Theory 38(4), 1412–1418 (1992)
25. Pott, A., Bradley, S.P.: Existence and nonexistence of almost-perfect autocorrelation sequences. IEEE Trans. Inform. Theory 41(1), 301–304 (1995)

26. Blokhuis, A., Jungnickel, D., Schmidt, B.: Proof of the prime power conjecture for projective planes of order n with abelian collineation groups of order n^2. Proc. Amer. Math. Soc. 130(5), 1473–1476 (2002) (electronic)
27. Lavrauw, M., Polverino, O.: Finite semifields. In: Storme, L., Beule, J.D. (eds.) Current Reserach Topics in Galois Geometry. Mathematics Research Developments, pp. 131–159. Nova Science Publishers, Inc., New York (2012)
28. Ganley, M.J.: On a paper of P. Dembowski and T. G. Ostrom: Planes of order n with collineation groups of order n^2. Math. Z. 103, 239–258 (1968); Arch. Math. (Basel) 27(1), 93–98 (1976)
29. Jungnickel, D.: On a theorem of Ganley. Graphs Combin. 3(2), 141–143 (1987)
30. Kantor, W.M.: Commutative semifields and symplectic spreads. J. Algebra 270(1), 96–114 (2003)
31. Zhou, Y.: $(2^n, 2^n, 2^n, 1)$-relative difference sets and their representations (2010) (manuscript)
32. Albert, A.A.: Finite division algebras and finite planes. In: Proc. Sympos. Appl. Math., vol. 10, pp. 53–70. American Mathematical Society, Providence (1960)
33. Coulter, R.S., Henderson, M.: Commutative presemifields and semifields. Adv. Math. 217(1), 282–304 (2008)
34. Zhou, Y., Pott, A.: A new family of semifields with 2 parameters. arXiv:1103.4555 (March 2011)
35. Kraemer, R.G.: Proof of a conjecture on Hadamard 2-groups. J. Combin. Theory Ser. A 63(1), 1–10 (1993)
36. Nyberg, K.: Differentially Uniform Mappings for Cryptography. In: Helleseth, T. (ed.) EUROCRYPT 1993. LNCS, vol. 765, pp. 55–64. Springer, Heidelberg (1994)
37. Carlet, C.: Vectorial boolean functions for cryptography. In: Crama, Y., Hammer, P.L. (eds.) Boolean Models and Methods in Mathematics, Computer Science, and Engineering, pp. 398–469. Cambridge University Press (2010)
38. Carlet, C.: Relating three nonlinearity parameters of vectorial functions and building APN functions from bent functions. Des. Codes Cryptogr. 59(1-3), 89–109 (2011)
39. Chabaud, F., Vaudenay, S.: Links between Differential and Linear Cryptanalysis. In: De Santis, A. (ed.) EUROCRYPT 1994. LNCS, vol. 950, pp. 356–365. Springer, Heidelberg (1995)
40. Dillon, J.F.: Elementary Hadamard Difference-Sets. ProQuest LLC, Ann Arbor (1974); Thesis (Ph.D.)–University of Maryland, College Park
41. Zhang, X., Guo, H., Gao, Z.: Characterizations of bent and almost bent function on $\mathbb{Z}_p^2$. Graphs Combin. 27(4), 603–620 (2011)
42. Coulter, R.S., Matthews, R.W.: Planar functions and planes of Lenz-Barlotti class II. Des. Codes Cryptogr. 10(2), 167–184 (1997)
43. Helleseth, T., Kholosha, A.: Monomial and quadratic bent functions over the finite fields of odd characteristic. IEEE Trans. Inform. Theory 52(5), 2018–2032 (2006)
44. Helleseth, T., Hollmann, H.D.L., Kholosha, A., Wang, Z., Xiang, Q.: Proofs of two conjectures on ternary weakly regular bent functions. IEEE Trans. Inform. Theory 55(11), 5272–5283 (2009)
45. Helleseth, T., Kholosha, A.: On the Dual of Monomial Quadratic p-ary Bent Functions. In: Golomb, S.W., Gong, G., Helleseth, T., Song, H.-Y. (eds.) SSC 2007. LNCS, vol. 4893, pp. 50–61. Springer, Heidelberg (2007)
46. Helleseth, T., Kholosha, A.: New binomial bent functions over the finite fields of odd characteristic. IEEE Trans. Inform. Theory 56(9), 4646–4652 (2010)
47. Carlet, C., Mesnager, S.: On Dillon's class H of bent functions, Niho bent functions and o-polynomials. J. Combin. Theory Ser. A 118(8), 2392–2410 (2011)

80   A. Pott, Q. Wang, and Y. Zhou

48. Ganley, M.J.: Direct product difference sets. J. Combinatorial Theory Ser. A 23(3), 321–332 (1977)
49. Jungnickel, D., de Resmini, M.J.: Another case of the prime power conjecture for finite projective planes. Adv. Geom. 2(3), 215–218 (2002)
50. Hiramine, Y.: Difference sets relative to disjoint subgroups. J. Combin. Theory Ser. A 88(2), 205–216 (1999)
51. Golomb, S.W., Gong, G.: Signal design for good correlation. Cambridge University Press, Cambridge (2005); For wireless communication, cryptography, and radar
52. Helleseth, T., Gong, G.: New nonbinary sequences with ideal two-level autocorrelation. IEEE Trans. Inform. Theory 48(11), 2868–2872 (2002)
53. Pott, A.: On projective planes admitting elations and homologies. Geom. Dedicata 52(2), 181–193 (1994)
54. Ding, C., Yuan, J.: A family of skew Hadamard difference sets. J. Combin. Theory Ser. A 113(7), 1526–1535 (2006)
55. Ding, C., Wang, Z., Xiang, Q.: Skew Hadamard difference sets from the Ree-Tits slice symplectic spreads in $PG(3, 3^{2h+1})$. J. Combin. Theory Ser. A 114(5), 867–887 (2007)
56. Feng, T., Xiang, Q.: Cyclotomic constructions of skew Hadamard difference sets. J. Combin. Theory Ser. A 119(1), 245–256 (2012)
57. Ma, S.L., Ng, W.S.: On non-existence of perfect and nearly perfect sequences. Int. J. Inf. Coding Theory 1(1), 15–38 (2009)
58. Chee, Y.M., Tan, Y., Zhou, Y.: Almost p-Ary Perfect Sequences. In: Carlet, C., Pott, A. (eds.) SETA 2010. LNCS, vol. 6338, pp. 399–415. Springer, Heidelberg (2010)
59. Kantor, W.M.: Projective planes of type $I - 4$. Geometriae Dedicata 3, 335–346 (1974)

On Formally Self-dual Boolean Functions in 2, 4 and 6 Variables

Lin Sok and Patrick Solé*

Telecom Paristech, Dept. Comelec, 46 rue Barrault, 75013 Paris, France
{lin.sok,patrick.sole}@telecom-paristech.fr

Abstract. In this paper, we classify all formally self-dual Boolean functions and self-dual bent functions under the action of the extended symmetric group in 2, 4 variables, and give a lower bound for the number of non-equivalent functions in 6 variables. There are exactly 2, 91 (1, 3 respectively) and at least 5535376 representatives from equivalence class of formally self-dual Boolean functions (self-dual bent functions respectively).

Keywords: Boolean functions, formally self-dual Boolean functions, self-dual bent functions.

1   Introduction

Bent functions were first introduced by Rothaus [7] in 1976. Bent functions are Boolean functions on $\mathbb{F}_2^n$ with n even and $\mathbb{F}_2$ being the two-element field $\{0, 1\}$, which are of highest non-linearity in the sense that their Walsh-Hadamard transform attains the values $\pm 2^{\frac{n}{2}}$. Because of their relations to coding theory and cryptology, bent functions are of great interest for research. In symmetric key cryptography, these functions can be used as building block of stream ciphers. If the number of variables is ≥ 10, the class of such functions is extremely large in terms of enumeration and classification so it is interesting to enumerate and classify its subclass called self-dual bent functions. Self-dual bent functions have been studied in [3], characterized in terms of the Rayleigh quotient of the Sylvester Hadamard matrix and classified for two, four and six variables as well as eight variables in the quadratic case [5].

Recently the authors of [4] have introduced the superclass, called formally self-dual Boolean functions, of self-dual bent functions and have given the exact number of such functions that is an upper bound for the number of self-dual bent functions.

Based on the result of [4], we can classify all formally self-dual Boolean functions and self-dual bent functions in two, four variables and give a lower bound for the number of non-equivalent formally self-dual Boolean functions in six variables. The idea of our classification is to apply the action of the symmetric group on the support of a function with a given near weight enumerator defined in [4].

---

* MECAA, Math Dept, King Abdulaziz University, Jeddah, Saudi Arabia

F. Özbudak and F. Rodríguez-Henríquez (Eds.): WAIFI 2012, LNCS 7369, pp. 81–91, 2012.
© Springer-Verlag Berlin Heidelberg 2012

In our algorithm, we successively classify the support subcode under the right action of the group of $n \times n$ permutation matrices.

The paper is organised as follows. Section 2 collects the notation and definition used through the paper. Section 3 describes the method that we use to classify formally self-dual Boolean functions and give some numerical results. Some conjectures are given in Section 4.

2   Notation and Definitions

In this section, we give some useful notation and definitions that will be used through the paper. For more details on Boolean functions, see [2].

Let $\mathbb{F}_2$ be a two-element field $\{0, 1\}$. A **Boolean function** f is a map from $\mathbb{F}_2^n$ to $\mathbb{F}_2$. The algebraic normal form (ANF) of a boolean function f is defined by

$$f(x_1, x_2, \cdots, x_n) = \sum_{u=(u_1, u_2, \cdots, u_n) \in \mathbb{F}_2^n} a_u x_1^{u_1} x_2^{u_2} \cdots x_n^{u_n}.$$

Its truth table is a binary vector $(f_0, f_1, \cdots, f_{2^n-1})$ of length 2^n, where $f_a := f(a)$ with $a := \sum_{i=1}^n a_i 2^i := a_1 a_2 \cdots a_n \in \mathbb{F}_2^n$ and its sign function is $F := (-1)^f := ((-1)^{f_0}, (-1)^{f_1}, \cdots, (-1)^{f_{2^n-1}}) \in \{-1, 1\}^{2^n}$.

The **support code** C_f of a Boolean function f is defined by

$$C_f := \{u \in \mathbb{F}_2^n : f(u) = 1\}.$$

For $u \in F_2^n$, the **Walsh-Hadamard transform** (WHT) of a Boolean function f is defined as

$$\hat{F}(u) := \sum_{v \in \mathbb{F}_2^n} (-1)^{f(v) + u.v},$$

where $u.v$ is the usual inner product of u and v.

A Boolean function f is said to be **bent** if $\forall u \in F_2^n, |\hat{F}(u)| = 2^{\frac{n}{2}}$. It follows that bent functions exist only when n is even.

Let $H_1 := \begin{pmatrix} 1 & 1 \\ 1 & -1 \end{pmatrix}$ and $H_n := \begin{pmatrix} H_{n-1} & H_{n-1} \\ H_{n-1} & -H_{n-1} \end{pmatrix}$, the n-fold tensor power of H_1. By considering $\hat{F}$ as a vector of length 2^n, we can express the WHT in matrix form

$$\hat{F} = F H_n.$$

If f is bent then there exists a function $\tilde{f}$, called the **dual** of f, whose sign function $\tilde{F}$ satisfies $F H_n = 2^{\frac{n}{2}} \tilde{F}$. Notice that $\tilde{f}$ is also bent and $\tilde{\tilde{f}} = f$. A bent function f is called **self-dual** (anti-self-dual) if $f = \tilde{f}$ ($f = 1 + \tilde{f}$) respectively.

Since our classification is based on the weight distribution of codes, we now introduce some definitions related to coding theory. For more details about coding theory, see [6].

A code C of length n over $\mathbb{F}_2$ is a subset of $\mathbb{F}_2^n$ and any vector u in C is called a codeword. The (Hamming) weight of a codeword $u = u_1 u_2 ... u_n$ is the number of non-zero coordinate u_i. To study the weight distribution of a code C, we associate a homogeneous polynomial $W_C(x, y) := \sum_{i=0}^{n} A_i(C) x^i y^{n-i}$, called the weight enumerator of C, where $A_i(C)$ is the number of codeword of weight i in C.

The **near weight enumerator** of a Boolean function f in n variables is defined in [4] as

$$W_{C_f}^N(x, y) = 2^{\frac{n}{2}-1} x^n + W_{C_f}(x, y). \tag{1}$$

A Boolean function f is called **formally self-dual** (fsd for short) with respect to $W_{C_f}^N(x, y)$ if

$$W_{C_f}^N(x, y) = W_{C_f}^N \left(\frac{x+y}{\sqrt{2}}, \frac{x-y}{\sqrt{2}} \right).$$

The following property is common to fsd Boolean functions and to bent functions.

Proposition 1. *If f in n variables is a fsd Boolean function with respect to the near weight enumerator then the size of its support code is $2^{n-1} \pm 2^{n/2-1}$. In particular such functions can only exist if n is even.*

Proof. Plug $x = 1$, $y = 0$ in the defining equation of a fsd Boolean function. The LHS is

$$W_{C_f}^N(1, 0) = 2^{\frac{n}{2}-1} + W_{C_f}(1, 0).$$

The RHS is, using the fact that W_{C_f} is homogeneous, given by

$$1/2 + W_{C_f}(1/\sqrt{2}, 1/\sqrt{2}) = 1/2 + W_{C_f}(1, 1)/\sqrt{2^n} = 1/2 + |C_f|/2^{\frac{n}{2}}.$$

Equating the two sides and using $W_{C_f}(1, 0) \in \{0, 1\}$, the first assertion follows. The second assertion is immediate since $2^{\frac{n}{2}}$ must be an integer. □

We will require the following two results from [4].

Proposition 2 ([4], lemma 3.1). *Let f be a formally self-dual Boolean function in n variables with respect to $W_{C_f}^N(x, y)$. Then*

$$W_{C_f}(x, y) = -2^{\frac{n}{2}-1} x^n + \sum_{j=0}^{\frac{n}{2}} a_j (x^2 + y^2)^{\frac{n}{2}-j} (xy - y^2)^j, \tag{2}$$

where the a_j's are integers.

Remark 1. By considering $W_{C_f}(x, y)$ of (2) as $\sum_{i=0}^{n} A_i(C_f) x^i y^{n-i}$ and using the fact, $0 \leq A_i(C_f) \leq \binom{n}{i}$ for $i = 0 \cdots n$, we can get all the possible values a_js.

Proposition 3 ([4], lemma 3.3). *Every self-dual bent function is formally self-dual with respect to $W_{C_f}^N(x, y)$.*

3   Classification of Formally Self-dual Boolean Functions

In this section, we classify formally self-dual Boolean functions using the action of the symmetric group on the support of a function with a given near weight enumerator. To do this classification, we need to partition the support of the function found in Proposition 2.

Let $\mathcal{P}_n$ be the group of $n \times n$ permutation matrices isomorphic to the symmetric group S_n. A right action of $\mathcal{P}_n$ on the support C_f is defined as

$$C_f \times \mathcal{P}_n \to C_f$$
$$(x, M) \mapsto xM.$$

In words we are just relabelling the Boolean variables.

For $x \in C_f$, the orbit of x is defined as $[x]_{C_f} := \{xM : M \in \mathcal{P}_n\}$. Let $C_f^{(i,k_i)}$ denote a subcode of C_f of weight i and with size k_i. Let $\mathcal{P}_n$ act on $C_f^{(i,k_i)}$ and denote $O^{(i,k_i)} := \cup\{[x]_{C_f^{(i,k_i)}}\}$.

By successively applying the right action of $\mathcal{P}_n$ on the support subcode $O^{(i,k_i)} \times C_f^{(i+1,k_{i+1})}$, we can finally classify all the support of a function with a given near weight enumerator and this leads to the classification of formally self-dual Boolean functions.

We say that two formally self-dual Boolean functions f and g are equivalent if there exists $M \in \mathcal{P}_n$ and $c \in \mathbb{F}_2$ such that $g(x) = f(xM) + c$ for all $x \in \mathbb{F}_2^n$.

Proposition 2, allows us to determine all the possible weight distributions of the support of formally self-dual Boolean functions. For instance, the number of formally self-dual Boolean functions with support size $2^{n-1} - 2^{n/2-1}$ is the same as that with support size $2^{n-1} + 2^{n/2-1}$ by complementation (case of M trivial and c =all-one vector). This will help us to easily list all representatives from the equivalence class by considering only those with support size $2^{n-1} - 2^{n/2-1}$.

Proposition 3 enables us to deduce all the self-dual bent functions.

We give a complete classification of all formally self-dual Boolean functions in two and four variables but not for six variables. The fact is that the size of $O^{(i,k_i)} \times C_f^{(i+1,k_{i+1})}$ becomes larger and larger especially when the size of the acting group is small. For six variables, we need a larger group than the symmetric group and what we can do here is just to give a lower bound on the number of representatives from equivalence class which can be derived from the product of the sizes of $O^{(i,k_i)}$. In fact, we can have a complete classification in some weight distributions, for example, the formally self-dual Boolean functions with weight distribution $[0, 0, 0, 16, 12, 0, 0]$ have 3513 representatives and as a result we give a complete classification of the functions with weight distributions $[0, 6, 14, 1, 1, 5, 1]$ and $[0, 6, 15, 0, 0, 6, 1]$ in Table 5.

All the computations were done with MAGMA [1]. From our computational results, we have the following theorem.

Theorem 1. *There are* 2, 91 *and at least* 5535376 *non-equivalent formally self-dual Boolean functions* (1, 3 *non-equivalent self-dual bent functions) in* 2, 4 *and* 6 *variables respectively.*

In the tables, we use the following notation:

- i: weight of codeword,
- A_i^k: number of weight-i codewords of the k-th weight distribution,
- 12.1: $x_1 x_2 + x_1$ for the column representative,
- $*$: bent but not self-dual bent case,
- $**$: self-dual bent case,
- $\prod |O^{(i,k_i)}|$: product of orbit size $|O^{(i,k_i)}|$.

There are four different weight distributions of formally self-dual Boolean funtions in two variables among which two of them are non equivalent and are listed in Table 1. For the weight distribution $A_i^1 = [0,0,1]$, there is only one weight-two codeword and only one orbit (represented by $\{11\}$) with size one. This formally self-dual Boolean function has its ANF, $f(x_1, x_2) = x_1 x_2$ and its truth table, $f = (0,0,0,1)$. Its sign function is $F = (1,1,1,-1)$ and its WHT is $FH_2 = 2(1,1,1,-1)$ so it is self-dual. Hence there are in total six formally self-dual Boolean functions in two variables among which two of them are self-dual bent.

In Tables 2 and 3, we list eight among the 16 weight distributions of the functions in four variables. For $A_i^2 = [0,0,2,4,0]$, there are two orbits of size 3 and 12 which are represented by

$$\{0110, 1001, 0111, 1011, 1101, 1110\}$$

and

$$\{1001, 1100, 0111, 1011, 1101, 1110\}.$$

The ANFs of the two representatives are $x_1 x_4 + x_2 x_3$ and $x_1 x_2 x_4 + x_1 x_2 + x_1 x_4 + x_2 x_3 x_4$. There are in total 3220 formally self-dual Boolean functions among which 20 are self-dual bent.

In Table 4, we list 98 among the 196 different weight distributions of the functions in six variables. In the column Size of Table 4, the (three) numbers in the multiplication are the sizes of $O^{(2,k_2)}$, $O^{(3,k_3)}$ and $O^{(4,k_4)}$ respectively. There are at least 5535376 representatives of equivalence class for formally self-dual Boolean functions in six variables.

In Table 5, we give a complete classification for two weight distributions. The first one has only one representative of size 1 and its function is self-dual bent while the second one has 78 representatives of total size 27000 but contains no bent function.

Table 1. Formally self-dual Boolean functions in two variables

i 0 1 2	Representative	Size
A_i^1 0 0 1	12	1^{**}
A_i^2 0 1 0	12.1	2^{*}
	Number of formally self-dual Boolean functions	3

Table 2. Formally self-dual Boolean functions in four variables

i 0 1 2 3 4	Representative	Size
A_i^1 0 0 3 2 1	123.124.12.13.34	24
	124.13.14.234.34	12
	123.124.14.23.24	24
	13.14.24	12*
	123.124.12.13.14	12
	123.124.12.134.234.24.34	12
	12.134.13.14.234	12
	123.124.12.13.23	12
A_i^2 0 0 2 4 0	14.23	3**
	124.12.14.234	12
A_i^3 0 1 3 1 1	123.12.134.14.1.23.34	12
	124.12.134.13.1.23.34	24
	123.124.12.13.14.23.24.34.3	12
	124.13.14.234.3	24
	12.13.14.23.24.34.3	4**
	123.124.134.13.14.234.4	24
	123.12.1.234.34	24
	124.13.14.1.234.24.34	24
	13.24.4	12*
	124.12.134.14.2	24
	123.124.134.234.4	4
	123.124.12.13.24.2.34	24
	123.124.134.13.14.234.23.24.4	12
	124.134.1	12
	123.14.234.23.4	12
	12.23.24.34.3	24*
	134.13.14.234.23.34.4	24
	123.12.13.234.3	24
A_i^4 0 1 2 3 0	123.124.12.14.24.2	24
	123.134.13.14.1.24	24
	123.124.13.24.34.4	24
	134.13.14.234.23.24.34.3	12
	123.234.24.4	24
	123.124.12.134.234.2	12
	123.12.134.14.1.24	12
	123.124.12.1	12
	12.13.1.24	24*
	123.124.12.13.14.24.34.4	24
	123.124.12.134.13.14.234.23.24.2	12
	123.124.12.134.14.234.24.2	12
	124.12.14.234.34.4	24
Number of formally self-dual Boolean functions		695

Table 3. Formally self-dual Boolean functions in four variables

i 0 1 2 3 4	Representative	Size
A_i^5 0 2 2 2 0	124.12.13.14.234.23.2.4	24
	123.12.234.24.2.3	12
	12.134.13.1.234.23.34.4	24
	123.124.12.13.1.23.24.2	24
	123.134.23.2.34.4	12
	12.13.14.1.23.24.34.4	6*
	124.134.13.1.34.4	12
	124.134.14.1.24.2	24
	134.1.234.23.24.2	12
	123.124.12.13.23.2.34.4	24
	123.124.134.13.14.1.234.23.34.3	24
	12.1.34.4	12*
	12.1.23.24.34.3	24*
	12.134.234.23.24.2.34.4	24
	124.13.14.234.24.34.3.4	24
	123.124.12.134.13.14.1.234.23.24.34.4	6
	123.124.134.14.1.234.23.2	12
	123.134.14.1.23.2	24
	123.124.12.134.1.234.23.24.34.3	24
	12.23.24.2.34.4	24*
	124.12.13.14.1.234.23.24.34.4	24
	123.14.1.234.34.3	24
	123.14.1.234.23.3	24
	123.124.1.23.34.3	24
	123.134.13.14.1.2	24
	124.12.134.13.14.24.2.3	24
	123.124.13.14.24.34.3.4	24
A_i^6 0 2 3 0 1	124.134.13.23.2.34.3	12
	134.13.1.234.4	24
	123.124.12.13.14.3.4	12
	12.14.1.34.4	12*
	123.124.13.14.1.2.34	12
	123.124.12.134.14.234.23.3.4	12
	123.13.14.234.23.24.2.34.4	24
	134.13.14.234.34.3.4	12
A_i^7 0 3 2 1 0	134.13.1.234.23.24.2.4	24
	123.124.12.134.13.14.234.24.2.34.3.4	12
	123.124.134.14.1.234.2.3	12
	123.12.134.13.14.1.23.2.34.4	12
	124.134.14.1.2.3	12
	124.12.134.14.1.24.2.3	24
	123.13.14.1.234.2.34.4	12
	124.12.134.1.23.34.3.4	24
	12.134.1.234.24.2.34.3	24
	12.13.2.34.3.4	24*
	123.124.12.134.13.1.234.23.3.4	12
	124.12.234.2.3.4	24
	123.124.12.13.14.1.24.2.34.3	24
A_i^8 0 4 2 0 0	123.124.14.1.24.2.3.4	12
	14.1.23.2.3.4	3**
	Number of formally self-dual Boolean functions	915

Table 4. Formally self-dual Boolean functions in six variables

A_0	A_1	A_2	A_3	A_4	A_5	A_6	$\prod |O^{(i,k_i)}|$
0	0	5	11	7	5	0	15.312.24
0	0	5	10	10	2	1	15.352.15
0	0	4	12	8	4	0	9.249.24
0	0	4	11	11	1	1	9.312.9
0	0	3	13	9	3	0	5.161.21
0	0	3	12	12	0	1	5.249.5
0	0	2	14	10	2	0	2.94.15
0	0	1	15	11	1	0	1.43.9
0	0	0	16	12	0	0	21.5
0	0	9	6	6	6	1	21.94.21
0	0	8	7	7	5	1	24.161.24
0	0	7	8	8	4	1	24.249.24
0	0	6	10	6	6	0	21.352.21
0	0	6	9	9	3	1	21.312.21
0	1	4	12	8	3	0	9.249.24
0	1	4	11	11	0	1	9.312.9
0	1	3	13	9	2	0	5.161.21
0	1	2	14	10	1	0	2.94.15
0	1	1	15	11	0	0	1.43.9
0	1	10	5	5	6	1	15.43.15
0	1	9	6	6	5	1	21.94.21
0	1	8	7	7	4	1	24.161.24
0	1	7	9	5	6	0	24.312.15
0	1	7	8	8	3	1	24.249.24
0	1	6	10	6	5	0	21.352.21
0	1	6	9	9	2	1	21.312.21
0	1	5	11	7	4	0	15.312.24
0	1	5	10	10	1	1	15.352.15
0	2	10	5	5	5	1	15.43.15
0	2	3	13	9	1	0	5.161.21
0	2	11	4	4	6	1	9.21.9
0	2	2	14	10	0	0	2.94.15
0	2	9	6	6	4	1	21.94.21
0	2	8	8	4	6	0	24.249.9
0	2	8	7	7	3	1	24.161.24
0	2	7	9	5	5	0	24.312.15
0	2	7	8	8	2	1	24.249.24
0	2	6	10	6	4	0	21.352.21
0	2	6	9	9	1	1	21.312.21
0	2	5	11	7	3	0	15.312.24
0	2	5	10	10	0	1	15.352.15
0	2	4	12	8	2	0	9.249.24
0	3	9	6	6	3	1	21.94.21
0	3	9	7	3	6	0	21.161.5
0	3	10	5	5	4	1	15.43.15
0	3	11	4	4	5	1	9.21.9
0	3	12	3	3	6	1	5.7.5
0	3	8	8	4	5	0	24.249.9
0	3	8	7	7	2	1	24.161.24
0	3	7	9	5	4	0	24.312.15
0	3	7	8	8	1	1	24.249.24
0	3	6	10	6	3	0	21.352.21
0	3	6	9	9	0	1	21.161.21
0	3	5	11	7	2	0	15.312.24
0	3	4	12	8	1	0	9.249.24
0	3	3	13	9	0	0	5.161.21
0	4	8	8	4	6	0	24.249.9
0	4	8	7	7	1	1	24.161.24
0	4	9	6	6	2	1	21.94.21
0	4	9	7	3	5	0	21.161.5
0	4	10	5	5	3	1	15.43.15
0	4	10	6	2	6	0	15.94.2
0	4	11	4	4	4	1	9.21.9
0	4	12	3	3	5	1	5.7.5
0	4	13	2	2	6	1	2.3.2
0	4	7	9	5	3	0	24.312.15
0	4	7	8	8	0	1	24.249.24
0	4	6	10	6	2	0	21.352.21
0	4	5	11	7	1	0	15.312.24
0	4	4	12	8	0	0	9.249.24
0	5	7	9	5	2	0	24.312.15
0	5	8	8	4	3	0	24.249.9
0	5	8	7	7	0	1	24.161.24
0	5	9	6	6	1	1	21.94.21
0	5	9	7	3	4	0	21.161.5
0	5	10	5	5	2	1	15.43.15
0	5	10	6	2	5	0	15.94.2
0	5	11	4	4	3	1	9.21.9
0	5	11	5	1	6	0	9.43.1
0	5	12	3	3	4	1	5.7.5
0	5	13	2	2	5	1	2.3.2
0	5	14	1	1	6	1	1.1.1
0	5	6	10	6	1	0	21.352.21
0	5	5	11	7	0	0	15.312.24
0	6	6	10	6	0	0	21.352.21
0	6	7	9	5	1	0	24.312.15
0	6	8	8	4	2	0	24.249.9
0	6	9	6	6	0	1	21.94.21
0	6	9	7	3	3	0	21.161.5
0	6	10	5	5	1	1	15.43.15
0	6	10	6	2	4	0	15.94.2
0	6	11	4	4	2	1	9.21.9
0	6	11	5	1	5	0	9.43.1
0	6	12	3	3	3	1	5.7.5
0	6	12	4	0	6	0	5.21
0	6	13	2	2	4	1	2.3.2
0	6	14	1	1	5	1	**1.1.1**
0	6	15	0	0	6	1	**1**

Total number of representatives from equivalence classes ≥ 5535376

Table 5. Complete classification for two weight distributions

Representatives in the weight distribution [0, 6, 15, 0, 0, 6, 1] 12,13,14,15,16,1,23,24,25,26,2,34,35,36,3,45,46,4,56,5,6	Size 1**
12346.1234.1235.1236.123.1245.1256.12.13.14.15.16.1.23456.234.2356.235.236.2456.245.24.25.26.2.34.35.36.3.45.46.4.56.5.6	720
Representative in the weight distribution [0, 6, 14, 1, 1, 5, 1]	
12345.12356.1235.12456.1246.12.13.14.15.16.1.23456.2346.234.2356.235.236.2456.245.24.25.26.2.34.35.36.3.45.46.4.56.5.6	720
12345.12346.12356.12456.1246.12.13.14.15.16.1.23456.236.13.146.14.156.15.1.2345.2356.235.23.2456.245.24.256.26.2.34.35.36.3.45.46.4.56.5.6	180
12346.123.12456.1245.12.13456.1346.13.14.15.16.1.2345.2356.235.23.2456.245.24.25.26.2.34.35.36.3.45.46.4.56.5.6	180
1234.1236.123.12456.125.12.13456.1346.13.14.15.16.1.2345.2356.235.23.2456.245.24.25.26.2.34.35.36.3.45.46.4.56.5.6	180
1236.1246.126.12.136.13.146.14.15.1.23.24.25.26.2.34.35.36.3.45.46.4.56.5.6	180
12345.12346.1235.12356.1236.123.12456.1245.1246.124.1256.126.13.14.15.16.1.23456.234.23.24.25.26.2.3456.34.356.35.36.3.45.46.4.56.5.6	120
12345.12346.12356.12456.1246.124.12.13456.1345.1346.134.13.14.15.16.1.23456.236.23.2456.245.24.25.26.2.3456.34.25.26.2.34.35.36.3.45.46.44.56.5.6	360
12345.12346.12356.12456.1245.12.13456.1346.1356.136.13.14.15.16.1.23456.2346.2356.236.23.2456.245.24.25.26.2.3456.346.34.356.35.3.45.46.4.56.5.6	120
12345.1234.12356.1235.124.12.13456.135.13.14.15.16.1.23456.2345.2356.235.23.24.25.26.2.3456.345.34.356.36.3.45.46.4.56.5.6	60
12345.12346.1234.1235.1256.125.12.13456.1345.1346.13.14.15.16.1.23456.2346.234.23.24.25.26.2.3456.345.346.34.356.35.3.45.46.4.56.5.6	360
12346.12356.1235.12456.1256.12.13456.13.1456.14.156.15.16.1.23456.2346.2356.236.23.24.25.26.2.3456.345.346.34.356.35.3.45.46.4.56.5.6	360
12356.1236.12.13.14.15.16.1.23456.2345.2346.2346.234.23.24.25.26.2.3456.345.346.35.36.3.45.46.4.56.5.6	360
12345.12346.1234.12356.1236.12.13456.1346.1356.136.13.1456.14.156.15.16.1.23.24.25.26.2.34.35.36.3.45.46.4.56.5.6	360
1235.1236.123.1245.1246.124.125.12.13.1456.14.156.15.16.1.23.24.25.26.2.34.35.36.3.45.46.4.56.5.6	180
12346.12456.1245.1246.125.12.13.14.156.15.16.1.2345.2346.2356.235.23.245.246.24.256.26.2.34.356.35.3.45.46.4.56.5.6	360
12345.1256.12.13456.13.14.156.15.16.1.2356.23.24.256.25.26.2.34.356.35.36.3.45.46.4.56.5.6	360
12345.12346.1234.1235.1246.1245.125.12.13.14.15.16.1.2356.23.2456.245.24.25.26.2.3456.345.34.356.35.36.3.45.46.4.56.5.6	360
1234.12356.1236.1246.1256.12.13.1245.135.13.1456.14.15.16.1.23456.2346.2356.236.23.24.25.26.2.34.35.36.3.45.46.4.56.5.6	180
12345.1234.12356.1235.1346.1356.135.13.1456.145.14.15.16.1.23456.2346.23.2456.245.24.256.26.2.3456.34.356.35.36.3.456.46.4.56.5.6	180
1235.1236.1256.126.12.13456.135.13.14.15.16.1.23456.2346.236.23.24.25.26.2.34.35.36.3.45.46.4.56.5.6	360
12345.12346.1234.1235.1246.124.1256.126.12.13456.1346.13.14.15.16.1.23456.2346.23.2456.245.24.256.26.2.3456.345.34.35.36.3.45.46.4.56.5.6	720
12345.1234.12356.12456.1246.124.12.13.14.156.145.14.15.16.1.23456.2346.23.24.25.26.2.3456.345.34.356.35.36.3.45.46.4.56.5.6	360
12346.1236.12456.1256.126.12.13456.1346.13.14.15.16.1.23456.2346.236.23.2456.246.24.256.25.26.2.34.35.36.3.45.46.4.56.5.6	360
12356.1245.1246.12.13456.1345.134.13.14.15.16.1.2345.2346.234.23.245.246.24.25.26.2.34.35.36.3.45.46.4.56.5.6	180
12346.12356.1246.1256.126.12.13456.1346.13.14.15.16.1.2345.2346.235.236.23.24.25.26.2.34.35.36.3.456.45.46.4.56.5.6	360
12346.1234.12356.1235.12456.1245.12.13456.134.13.14.15.16.1.2345.2346.235.136.14.15.16.1.23.24.25.26.2.34.35.36.3.45.46.4.56.5.6	360
12346.12356.12456.1245.12.13456.1346.13.14.15.16.1.2345.2346.235.236.23.24.25.26.2.34.35.36.3.45.46.4.56.5.6	120
12346.1234.12.12356.1246.126.12.13456.1345.1356.136.13.1456.14.15.16.1.2345.23.245.24.25.26.2.3456.345.346.35.36.3.45.46.4.56.5.6	360
12345.1245.12.13456.1356.136.13.145.14.156.15.16.1.2345.23.2456.245.24.25.26.2.3456.345.346.35.36.3.45.46.4.56.5.6	180
12346.12356.1246.126.12.13456.1356.136.13.14.156.15.1.23.24.25.26.2.3456.345.346.34.356.35.3.45.46.4.56.5.6	720
12345.1245.1246.1256.126.12.13456.136.13.14.156.15.1.23.24.25.26.2.3456.345.34.25.26.2.34.35.36.3.456.46.4.56.5.6	180
12346.1236.1246.12.13456.13.1456.146.14.156.15.1.23.24.25.26.2.3456.346.23.24.25.26.2.34.35.36.3.45.46.4.56.5.6	180
12345.12346.1245.1246.12.13456.136.13.14.15.16.1.23456.235.23.2456.246.24.25.26.2.34.356.36.3.45.46.4.56.5.6	360
12345.1235.1256.125.12.13456.1345.13.1456.145.14.15.16.1.23456.235.23.24.25.26.2.34.356.36.3.45.46.4.56.5.6	360
12345.1235.1256.12.13456.135.13.14.156.15.16.1.23456.235.23.24.25.26.2.34.356.36.3.45.46.4.56.5.6	720
12346.1235.1236.125.12.13456.1345.13.1456.145.14.15.16.1.2345.2356.235.23.2456.245.24.256.26.2.34.35.36.3.45.46.4.56.5.6	360

Table 5. *Continued*

12346.12356.1235.12456.1246.1256.125.12.13.14.15.16.1.23456.2345.2346.2356.235.23.245.246.24.256.26.2.34.35.36.3.45.46.4.56.5.6	720
12346.12356.1235.1236.12456.1245.1256.125.12.13456.1345.1356.135.13.1456.145.14.156.16.1.2345.2346.2356.236.23.24.25.26.2.34.35.36.3.45.46.4.56.5.6	360
12345.12356.1235.12456.13.14.15.16.1.23456.2345.24.256.25.26.2.34.356.36.3.456.45.46.4.56.5.6	360
12345.12356.1235.1236.123.12456.1346.13.14.15.16.1.2345.234.2356.236.2456.246.24.25.26.2.34.35.36.3.45.46.4.56.5.6	360
12345.12356.1235.1236.123.12456.1246.12.13456.1356.135.13.14.15.16.1.2345.2346.2356.236.2456.245.24.25.26.2.34.35.36.3.45.46.4.56.5.6	360
12345.12356.1235.1236.123.12456.1245.1246.1256.126.12.13456.1356.135.13.14.15.16.1.23456.2356.235.23.24.25.26.2.345.346.34.356.36.3.45.46.4.56.5.6	720
12345.12356.1235.1236.12.1345.1346.1356.135.13.14.15.16.1.2345.2356.235.23.24.25.26.2.345.346.34.356.35.3.45.46.4.56.5.6	120
12345.1234.12356.1236.12456.1346.1356.136.13.14.15.16.1.23456.2345.2346.2356.236.23.24.25.26.2.3456.346.34.356.36.3.45.46.4.56.5.6	360
12346.12356.1236.12456.1346.1356.136.13.14.15.16.1.23456.2345.2346.2356.236.23.24.25.26.2.3456.346.34.356.35.3.45.46.4.56.5.6	360
12346.12456.1246.12.13456.1356.136.13.14.15.16.1.23456.2345.2346.234.23.24.25.26.2.34.35.36.3.45.46.4.56.5.6	720
12345.12346.1235.1236.12456.1245.12.13456.1345.13.14.15.16.1.2346.234.2356.235.2456.245.24.25.26.2.3456.346.34.35.36.3.456.46.4.56.5.6	360
1234.1235.1236.12.12456.1246.12.13.24.25.26.2.34.356.35.3.45.46.4.56.5.6	360
1234.1236.123.12456.1246.12.13456.134.13.146.14.156.16.1.23456.2346.23.24.25.26.2.34.35.36.3.45.46.4.56.5.6	120
12345.1234.12356.1246.124.12.13456.1345.1346.134.13.1456.14.15.16.1.2345.234.23.24.25.26.2.34.35.36.3.45.46.4.56.5.6	720
12345.12356.1235.12456.1246.124.12.13456.1356.13.14.15.16.1.23.24.25.26.2.3456.345.346.34.35.36.3.45.46.4.56.5.6	720
12345.1235.1236.124.12456.1246.124.12.13456.1356.13.14.15.16.1.23.24.256.26.2.3456.345.346.34.35.36.3.45.46.4.56.5.6	180
12346.1234.12356.12456.1246.124.12.13456.134.1356.13.14.15.16.1.2345.2346.2356.236.23.24.25.26.2.34.356.36.3.45.46.4.56.5.6	180
12346.1234.12356.12456.1246.124.12.1356.135.13.14.15.16.1.23456.2346.2356.2456.245.246.25.26.2.34.356.34.356.35.3.45.46.4.56.5.6	360
1234.1235.123.12456.12.13456.13.1456.14.15.16.1.23.24.25.26.2.3456.345.346.34.35.36.3.45.46.4.56.5.6	60
1234.12356.123.12.13456.1356.136.13.145.14.15.16.1.23456.2345.2356.235.24.256.2.3456.346.34.356.35.3.45.46.4.56.5.6	720
12345.12346.1235.12456.1246.1256.125.12.13456.13.1456.14.15.16.1.2345.2346.2356.235.23.245.246.24.256.26.2.34.35.36.3.45.46.4.56.5.6	720
12345.12356.1236.12456.1256.125.12.13.14.15.16.1.2345.2346.2356.236.23.245.23.24.25.26.2.34.35.36.3.45.46.4.56.5.6	360
12346.1234.12356.12456.1246.12.13456.134.1356.13.14.15.16.1.23456.2346.2356.236.23.24.25.26.2.34.356.34.356.36.3.45.46.4.56.5.6	360
12346.1234.12356.12456.1246.12.13456.1356.13.14.15.16.1.23456.2346.2356.236.2456.245.246.25.26.2.34.356.34.356.35.3.45.46.4.56.5.6	360
1234.12356.123.12.13456.1356.136.13.1456.14.15.16.1.23.24.25.26.2.3456.346.34.356.3.45.46.4.56.5.6	720
1234.12356.123.12456.123.12.13456.135.13.1456.135.13.145.14.15.16.1.23456.2356.235.23.24.25.26.2.3456.346.34.356.36.3.45.46.4.56.5.6	360
12346.1234.1245.12.13456.1346.1356.136.13.145.14.156.13.14.156.15.16.1.23456.2346.23.245.24.25.26.2.345.34.356.35.3.45.46.4.56.5.6	360
12346.1234.1245.12.13456.1346.1356.136.13.145.14.156.13.14.156.15.16.1.2345.234.236.23.24.25.26.2.345.34.356.35.3.45.46.4.56.5.6	120
12346.1234.1245.12.13456.1346.1356.136.13.1456.135.13.145.14.15.16.1.23.24.25.26.2.3456.346.34.35.36.3.45.46.4.56.5.6	120
12345.1234.12356.1256.12.13456.1345.13.14.15.16.1.23456.2346.234.2356.235.23.245.246.25.26.2.34.35.36.3.45.46.4.56.5.6	360
12356.1235.1236.124.12.13.14.15.16.1.23456.2346.234.2356.236.23.24.25.26.2.34.35.36.3.456.45.46.4.56.5.6	180
12345.1234.12356.12456.124.12.13456.1346.134.13.1456.14.15.16.1.2345.2346.2356.236.23.24.25.26.2.3456.346.34.356.35.3.45.46.4.56.5.6	360
12356.1235.1236.12456.1245.1246.124.12.13456.1346.13.14.15.16.1.23456.236.23.2456.245.246.25.26.2.3456.346.34.35.36.3.45.46.4.56.5.6	360
12356.1235.1236.12456.123.12456.1256.12.13456.1346.134.1356.135.136.13.14.15.16.1.23456.2346.23.2456.245.246.25.26.2.3456.346.34.35.36.3.45.46.4.56.5.6	180
1234.1236.123.12.1346.13.1456.145.14.156.16.1.23456.235.13.1456.145.14.156.16.1.23456.235.23.24.25.26.2.34.35.36.3.45.46.4.56.5.6	120
12356.1235.1236.12456.124.1256.126.12.13456.146.14.15.16.1.23456.236.23.2456.246.24.256.25.26.2.3456.236.23.2456.25.26.2.34.35.36.3.45.46.4.56.5.6	180
12345.12346.12356.1235.1236.1245.1256.125.12.13456.1345.1356.135.13.1456.145.14.156.16.1.23456.2345.235.23.24.25.26.2.34.35.36.3.45.46.4.56.5.6	360
1234.12356.1245.124.12.13456.1346.13.1456.146.14.15.16.1.2345.2346.23.2456.246.24.25.26.2.3456.346.34.35.36.3.456.45.46.4.56.5.6	360
12346.1234.12356.123.12456.1245.1246.124.12.13456.1345.1346.13.14.15.16.1.2345.234.35.34.356.345.34.356.35.3.45.46.4.56.5.6	360
12345.1235.1236.12456.124.12.13456.1346.1356.136.13.14.15.16.1.2345.234.2356.236.2456.246.24.25.26.2.34.35.36.3.45.46.4.56.5.6	180
12356.1235.1236.12456.124.1256.12.13456.1346.135.13.14.15.16.1.23456.2346.23.2456.235.236.23.2456.25.26.2.3456.346.34.35.36.3.45.46.4.56.5.6	180
12346.12356.1235.12456.125.126.12.13456.1356.13.14.15.16.1.23456.2346.23.2456.246.24.256.25.26.2.34.35.36.3.45.46.4.56.5.6	360
Number of formally self-dual Boolean functions	27000

Remark 2. From our computer program test, the formally self-dual Boolean functions are not invariant under the action of the orthogonal group $\mathcal{O}_n$ of $n \times n$ matrices, a supergroup of $\mathcal{P}_n$. The fact is that, in the case of six variables, the weight distribution $[0, 6, 13, 2, 2, 4, 1]$ contains a formally self-dual Boolean function with support

{100000, 010000, 001000, 000100, 000010, 000001, 101000, 000011, 011000, 100100, 010100, 100010, 010010, 001010, 000110, 010001, 001001, 000101, 001100, 110100, 100101, 101011, 111010, 111011, 110111, 101111, 111110, 111111}   and its orbit under the action of $\mathcal{O}_n$ contains a function with support {010000, 000100, 000011, 101000, 010100, 100010, 001010, 100001, 010001, 001001, 000101, 110100, 011100, 101010, 011010, 010110, 001110, 101001, 010101, 101101, 110011, 100111, 111010, 101110, 111001, 111011, 101111, 111111}   which   is not a formally self-dual Boolean function since its weight distribution is $[0, 2, 9, 8, 6, 2, 1]$ and is not among the 196 weight distributions.

4   Conclusion

We have provided a complete classification of formally self dual Boolean functions in 2 and 4 variables and a partial classification in 6 variables. The number of classes of formally self dual Boolean functions is higher than that of self-dual bent functions, a special case. Since the latter cannot be, as of now, classified completely in 8 variables [3,5], it is natural to stop in 6 variables for the former. We will be able to have a complete classification for all formally self-dual Boolean functions in six variables if we can find another invariant suitable group larger than the group $\mathcal{P}_n$, that is, an intermediate group between $\mathcal{P}_n$ and $\mathcal{O}_n$.

Acknowledgements. We would like to thank A. Wassermann and his colleagues for their warm hospitality and helpful discussions. We thank the referees for their careful reading and helpful remarks that greatly improved the presentation of the material.

References

1. Bosma, W., Cannon, J.: Handbook of Magma Functions, Sydney (1995)
2. Carlet, C.: Boolean Functions for Cryptography and Error Correcting Codes. In: Crama, Y., Hammer, P.L. (eds.) Boolean Models and Methods in Mathematics, Computer Science, and Engineering. Encyclopedia of Mathematics and its Applications, vol. 134, ch. 8, pp. 257–397. Cambridge University Press, Cambridge (2010)
3. Carlet, C., Danielsen, L.E., Parker, M.G., Solé, P.: Self-dual Bent functions. International Journal of Information and Coding Theory 1(4), 384–399 (2010)
4. Hyun, J.Y., Lee, H., Lee, Y.: MacWilliams Duality and Gleason-type Theorem on Self-dual Bent Functions. Designs, Codes and Cryptography 63(3), 295–304 (2012)
5. Feulner, T., Sok, L., Solé, P., Wasserman, A.: Towards the Classification of Self-dual Bent Functions in Eight Variables. Designs, Codes and Cryptography (to appear)
6. MacWilliams, F.J., Sloane, N.J.: Theory of Error Correcting Codes. North-Holland, Amsterdam (1998)
7. Rothaus, O.S.: On Bent Functions. J. Combinatorial Theory Ser. A 20(3), 300–305 (1976)

On the Algebraic Normal Form and Walsh Spectrum of Symmetric Functions over Finite Rings

Boris Batteux

CASSIDIAN
Cyber Security, France
Boris.Batteux@cassidian.com

Abstract. A function over finite rings is a function from a ring E_q^n to a ring E_r, where E_k is $\mathbb{Z}/k\mathbb{Z}$. These functions are well used in cryptography: cipher design, hash function design and in theoretical computer science. In this paper, we are especially interested in symmetric functions. We give practical ways of computing their ANF and their Walsh Spectrum in $\mathcal{O}\left(\binom{n+q-1}{q-1}^2\right)$ using linear algebra. Thus, we achieve a better complexity both in time and memory than the fast Fourier transform which is in $\mathcal{O}\left(q^n n \log(q)\right)$.

1 Introduction

In information science, it is a challenge to find functions with a large number of variables, which have good properties and can be implemented efficiently. For instance, good balanced non-linear functions are needed to ensure the quality of a cryptographic algorithm [5] [13].

The problem is that there exists trade-off between some important cryptographic properties [5]. For instance, Camion *et al.* [3] give a relation linking the correlation order t and the algebraic degree d of q-ary functions of n variables: $d + t < (q-1)n$. Furthermore, in order to resist to algebraic attacks, a function must have a high algebraic immunity. Bounds have been issued on the algebraic immunity with a given number of variables [2]. Meier *et al.* [12] show that the algebraic immunity of random Boolean function with n variables is almost always at least equal to $0.22n$. So increasing the number of variables allows us to find better cryptographic functions. So, if you want to find functions with higher algebraic degree and correlation order, you have to either increase the number of variables n or the domain size q.

But from a practical point of view, we cannot increase the number of variables as much as we want. Indeed, a Boolean function of more than 38 input variables, which is rather small, cannot fit into the largest available physical memory block.

In order to be used, functions with many variables must admit a representation in a concise form. Symmetric functions represent really good candidat on this point. Indeed, a symmetric function over finite rings can be implemented in

F. Özbudak and F. Rodríguez-Henríquez (Eds.): WAIFI 2012, LNCS 7369, pp. 92–107, 2012.

hardware with a number of gates which is polynomial in the number of input variables. More precisely, the number of gates is in $\mathcal{O}\left(\binom{n+q-1}{q-1}\right)$.

Besides, another reason for choosing symmetric functions is that their input variables are not distinguishable which can be very useful for instance if we want to filter the outputs of some LFSRs. Of course, a cryptographic algorithm cannot be built only with symmetric functions, but it can be worth using one of them in a larger scheme.

Moreover, if we are going to use q-ary functions, we need ways to analyse their cryptographic properties [6] [7] [8] [9] [11]. Although Boolean functions are well studied [4], there are not many existing algorithms for testing the quality of a q-ary function.

In this paper, we focus on symmetric functions from $(\mathbb{Z}/q\mathbb{Z})^n$ to $\mathbb{Z}/r\mathbb{Z}$. For those functions, we will give two algorithms which can be used to compute Algebraic Normal Form (ANF) and Walsh Spectrum of any symmetric function over finite rings. The complexity of these algorithms is in $\mathcal{O}\left(\binom{n+q-1}{q-1}^2\right)$, i.e. polynomial in the number of variables. Knowing ANF and Walsh Spectrum of a function makes us able to compute many cryptographic properties, e.g. algebraic degree and non-linearity.

This paper is organized as follows. In section 2, we introduce definitions about symmetric functions over finite rings and partitions. In section 3, we describe a way to compute the ANF of a symmetric function. In section 4, we define the Walsh transform over E_k and we give a way to compute the Walsh Spectrum of a symmetric function. Section 5 explains how our algorithms can be used to search good symmetric functions. Section 6 concludes the paper. We also add three sections in appendices where we give examples of the construction introduced in the paper and compare the complexity obtained in this paper with the previous ones.

2   Symmetric Functions over Finite Rings

Let us consider $E_k := \mathbb{Z}/k\mathbb{Z}$ a ring of k elements and n, q, r positive integers. The cardinal of a set E is denoted by $\mathrm{Card}(E)$.

Definition 1. *Let $\boldsymbol{x} = (x_1, \ldots, x_n) \in E_q^n$, we denote by $wt(\boldsymbol{x})$ the Hamming weight of $\boldsymbol{x}$ which is the number of components of $\boldsymbol{x}$ distinct from zero.*

$$wt(\boldsymbol{x}) = \mathrm{Card}\left(\{i \in \{1, \ldots, n\} \mid x_i \neq 0\}\right) . \tag{1}$$

2.1   Functions over Finite Rings

Definition 2. *Given three integers $n \geq 1$, $q \geq 2$, $r \geq 2$, we call function over finite rings any function from E_q^n to E_r.*

The set of functions over finite rings from E_q^n to E_r is denoted $\mathcal{M}_n(q, r)$.

A function $f \in \mathcal{M}_n(q,r)$ is characterized by a vector $\boldsymbol{f_v} \in E_r^{q^n}$ called its value vector, consisting in the evaluations of the function at every possible input:

$$\boldsymbol{f_v} = (f(0,\ldots,0), f(1,\ldots,0), f(2,\ldots,0),\ldots,f(q-1,\ldots,q-1))^T \ . \qquad (2)$$

2.2   Partitions

Definition 3. *A partition* $\boldsymbol{\lambda} = (\lambda_1,\ldots,\lambda_k)$ *of an integer* N *in* k *parts, whose largest part is* b*, is a sequence of* k *integers* $b \geq \lambda_1 \geq \lambda_2 \ldots \geq \lambda_k \geq 0$ *such as* $N = \sum_{i=1}^k \lambda_i$.

We denote by $\mathrm{Part}(b,k,N)$ the set of all partitions of all integers lower or equal to N in k parts whose largest part is b [14].

We say that a partition $(\lambda_1,\ldots,\lambda_n)$ is constant if $\lambda_1 = \cdots = \lambda_n$.

For all $\boldsymbol{x} = (x_1,\ldots,x_n) \in E_q^n$, there exists $\sigma \in S_n$ and a single partition denoted $\boldsymbol{\lambda(x)} \in \mathrm{Part}(q-1,n,n(q-1))$ such as $\boldsymbol{\sigma(x)} = \boldsymbol{\lambda(x)}$ (where $\boldsymbol{\sigma(x)} := (x_{\sigma(1)},\ldots,x_{\sigma(n)})$).

Example 1. $n = 7$, $q = 4$.

Let $\boldsymbol{x} = (1,3,2,0,1,0,3) \in E_4^7$, then $\boldsymbol{\lambda(x)} = (3,3,2,1,1,0,0)$.

Definition 4. *Let* $\boldsymbol{x} = (x_1,\ldots,x_n) \in E_q^n$, *for all* $\ell \in E_q$, *we denote by* $x[\ell]$ *the number of components of* $\boldsymbol{x}$ *which are equal to* ℓ.

$$x[\ell] := \mathrm{Card}\left(\{i \mid x_i = \ell\}\right) \ . \qquad (3)$$

Lemma 1. *Let* $\boldsymbol{x} \in E_q^n$ *and* $\boldsymbol{y} \in E_q^n$,

$$\boldsymbol{\lambda(x)} = \boldsymbol{\lambda(y)} \iff \forall \ell \in E_q, \ x[\ell] = y[\ell] \ . \qquad (4)$$

Proof. $\boldsymbol{\lambda(x)} = \boldsymbol{\lambda(y)}$ if and only if there exists $\sigma \in S_n$ such as $\boldsymbol{\sigma(x)} = \boldsymbol{y}$ which is equivalent to $\{x_i \mid i \in \{1,\ldots,n\}\} = \{y_i \mid i \in \{1,\ldots,n\}\}$. Which is equivalent to $x[\ell] = y[\ell]$ for all $\ell \in E_q$. $\square$

Andrew's Lemma [1] gives us the number of elements of the set $\mathrm{Part}(b,a,ab)$.

Lemma 2. *[1]* $\mathrm{Card}(\mathrm{Part}(b,a,ab)) = \binom{a+b}{a} = \binom{a+b}{b}$.

2.3   Symmetric Functions over Finite Rings

Definition 5. *A function over finite rings* $f : E_q^n \longrightarrow E_r$, *is symmetric, if* f *is invariant under any permutation of its input's variables:*

$$\forall \sigma \in S_n, \ f(x_1,\ldots,x_n) = f(x_{\sigma(1)},\ldots,x_{\sigma(n)}) \ . \qquad (5)$$

We denote by $\mathcal{SM}_n(q,r)$ the set of symmetric functions over finite rings from E_q^n to E_r.

Definition 6. *[14] We call symmetry class of* $\boldsymbol{x} \in E_q^n$, *the set* $P_{n,q}(\boldsymbol{x})$ *of vectors obtained by permuting the coordinates of* $\boldsymbol{x}$:

$$P_{n,q}(\boldsymbol{x}) := \{\boldsymbol{y} \in E_q^n \mid \text{there exists a partition } \boldsymbol{\lambda} \text{ such that } \boldsymbol{y} = \lambda(\boldsymbol{x})\}$$
$$= \{\boldsymbol{y} \in E_q^n \mid \exists \sigma \in S_n \text{ such that } \boldsymbol{y} = \sigma(\boldsymbol{x})\} \ . \qquad (6)$$

According to the Andrew's lemma, we deduce that there are $\binom{n+q-1}{q-1}$ symmetry classes in E_q^n. For each symmetry class, we designate as representative the smallest element in the lexicographical order. We call j^{th} symmetry class of E_q^n the class of symmetry whose representative s_j (which is a partition) is classified j^{th} among all representatives according to the lexicographical order. Notice that s_j is a partition [14].

A symmetric function over finite rings can be represented by a vector in $E_r^{\binom{n+q-1}{q-1}}$. Components of the vector are evaluations of the function for each representative of a symmetry class. We call this vector a *simplified value vector* of the function:

$$\boldsymbol{f_{sv}} = (f(0,\ldots,0),\ldots,f(q-1,0,\ldots,0),f(1,1,\ldots,0),\ldots,f(q-1,\ldots,q-1))^T$$
$$= \left(f(s_1),\ldots,f(s_q),f(s_{q+1}),\ldots,f(s_{\binom{n+q-1}{q-1}}) \right)^T \ . $$
$$(7)$$

To simplify notation, we denote $\binom{n+q-1}{q-1}$ by N.

3 Algebraic Normal Form

3.1 Definition

We first recall the definition of the algebraic normal form over a finite ring.

Definition 7. *Let $f \in \mathcal{M}_n(q,r)$, f can be expressed as a polynomial, called its Algebraic Normal Form (ANF):*

$$f(x_1,\ldots,x_n) = \sum_{(a_1,\ldots,a_n)\in E_q^n} h_f(a_1,\ldots,a_n)x_1^{a_1}\cdots x_n^{a_n} \pmod{r} \ , \qquad (8)$$

where $h_f(a_1,\ldots,a_n)$ is constant in E_r.

Let us focus on symmetric functions over finite rings. We first introduce symmetric polynomials:

Definition 8. *[15] If $\boldsymbol{\lambda}$ is a partition, let us denote $m_{\boldsymbol{\lambda}}$ a symmetric polynomial defined by:*

$$\forall \boldsymbol{x} \in E_q^n, \ m_{\boldsymbol{\lambda}}(\boldsymbol{x}) = \sum x_{i_1}^{\lambda_1}\cdots x_{i_n}^{\lambda_n} \ , \qquad (9)$$

where the sum is over all distinct monomials whose exponent is $\boldsymbol{\lambda}$.

Example 2. For all $\boldsymbol{x} = (x_1,x_2,x_3) \in E_3^3$,

$$m_{(2,1,0)}(\boldsymbol{x}) = x_1^2 x_2 + x_1^2 x_3 + x_2^2 x_1 + x_2^2 x_3 + x_3^2 x_1 + x_3^2 x_2 \ . \qquad (10)$$

Theorem 1. *[15] Let $f \in \mathcal{M}_n(q, r)$, f is symmetric if and only if its ANF can be written as follows:*

$$f(\boldsymbol{x}) = \sum_{j=1}^{N} h_f(\boldsymbol{s_j}) m_{\boldsymbol{s_j}}(\boldsymbol{x}) \pmod{r} \ . \tag{11}$$

So, the coefficient of the ANF of a symmetric function over finite rings can be expressed by a simplified ANF vector denoted $\boldsymbol{f_{sANF}} \in E_r^N$. Our goal is to find the relation between the two vectors $\boldsymbol{f_{sv}}$ and $\boldsymbol{f_{sANF}}$.

Our method consists in two steps. First, we compute the simplified value vector of each $m_{\boldsymbol{\lambda}}$. Thus, we obtain a matrix which allows us to compute the simplified value vector of a symmetric function knowing its simplified ANF. Then, if it is possible, we calculate the "inverse" of the matrix which gives us a way to compute the simplified ANF from a simplified value vector.

One naive way to obtain this matrix is computing every value vector of each monomial and then add those who are on the same symmetry class. Yet, there are q^n distinct monomials, so we will have to compute a matrix of size $q^n \times q^n$ which will be impossible when n is high.

In order to solve this problem, we describe a method for computing the simplified value vector of each $m_{\boldsymbol{\lambda}}$ without computing the value vector of each of its monomials.

3.2   Computing the Simplified Value Vector of $m_{\boldsymbol{\lambda}}$

We will compute the simplified value vector of $m_{\boldsymbol{\lambda}}$ for all $\boldsymbol{\lambda}$ of weight 0, then for all $\boldsymbol{\lambda}$ of weight 1,..., then for all $\boldsymbol{\lambda}$ of weight n.

So when we compute the simplified value vector of $m_{\boldsymbol{\lambda}}$, we assume we already know the simplified value vector of all $m_{\boldsymbol{\mu}}$ such that $wt(\boldsymbol{\mu}) < wt(\boldsymbol{\lambda})$. Besides, we know that the simplified value vector of $m_{(0,\ldots,0)}$ is $(1, \ldots, 1)$.

Assume that $wt(\boldsymbol{x}) < wt(\boldsymbol{\lambda})$.

Proposition 1. *Let $\boldsymbol{\lambda}$ be a partition of weight k, then for all $\boldsymbol{x} \in E_q^n$ such that $wt(\boldsymbol{x}) < k$, $m_{\boldsymbol{\lambda}}(\boldsymbol{x}) = 0$.*

Proof. Each monomial of $m_{\boldsymbol{\lambda}}(\boldsymbol{x})$ is a product of k distinct x_i. So if $wt(\boldsymbol{x}) < k$, all products will be null. □

Assume that $wt(\boldsymbol{x}) = wt(\boldsymbol{\lambda})$.

Lemma 3. *Let $\boldsymbol{\lambda}$ be a partition. $m_{\boldsymbol{\lambda}}$ is a monomial if and only if $\boldsymbol{\lambda}$ is constant. Moreover, if $\boldsymbol{\lambda} = (\lambda_1, \ldots, \lambda_n)$ is constant, for all $\boldsymbol{x} = (x_1, \ldots, x_n) \in E_q^n$, $m_{\boldsymbol{\lambda}}(\boldsymbol{x}) = x_1^{\lambda_1} \cdots x_n^{\lambda_1}$.*

Lemma 4. *Let $\boldsymbol{\lambda} = (\lambda_1, \ldots, \lambda_n)$ be a partition of weight k and $\boldsymbol{x} \in E_q^n$ of weight k. Then,*

$$m_{\boldsymbol{\lambda}}(\boldsymbol{x}) = m_{(\lambda_1, \ldots, \lambda_k)}((x_{i_1}, \ldots, x_{i_k})) \ , \tag{12}$$

where for all $j \in \{1, \ldots, k\}$, $x_{i_j} \neq 0$.

Proposition 2. *Let* $\boldsymbol{\lambda} = (\lambda_1, \ldots, \lambda_n)$ *be a non constant partition with* $\lambda_n \neq 0$ *and* $\boldsymbol{x} \in E_q^n$ *such that* $wt(\boldsymbol{x}) = n$, *then:*

$$m_{\boldsymbol{\lambda}}(\boldsymbol{x}) = m_{\boldsymbol{\mu_c}}(\boldsymbol{x}) \times m_{\boldsymbol{\mu_d}}(\boldsymbol{x}) \ , \tag{13}$$

where $\boldsymbol{\mu_c} = (\lambda_n, \ldots, \lambda_n)$ *is a constant partition and* $\boldsymbol{\mu_d} = (\lambda_1 - \lambda_n, \ldots, \lambda_{n-1} - \lambda_n, 0)$ *is the difference between* $\boldsymbol{\lambda}$ *and* $\boldsymbol{\mu_c}$.
    Moreover, $wt(\boldsymbol{\mu_d}) < wt(\boldsymbol{\lambda})$.

Proof. Let $\boldsymbol{x} = (x_1, \ldots, x_n) \in E_q^n$ and $\boldsymbol{\lambda} = (\lambda_1, \ldots, \lambda_n)$ be two non constant partition. We can write $\boldsymbol{\lambda} = \boldsymbol{\mu_c} + \boldsymbol{\mu_d}$, where $\boldsymbol{\mu_c} = (\lambda_n, \ldots, \lambda_n)$ is a constant partition and $\boldsymbol{\mu_d} = (\lambda_1 - \lambda_n, \ldots, \lambda_{n-1} - \lambda_n, 0)$ is the difference between $\boldsymbol{\lambda}$ and $\boldsymbol{\mu_c}$.

We will show that every monomial of $m_{\boldsymbol{\mu_c}}(\boldsymbol{x}) m_{\boldsymbol{\mu_d}}(\boldsymbol{x})$ is a monomial of $m_{\boldsymbol{\lambda}}(\boldsymbol{x})$ and conversely.

We know from lemma 3 that $m_{\boldsymbol{\mu_c}}(\boldsymbol{x}) = x_1^{\lambda_n} \cdots x_n^{\lambda_n}$. Let $M'(\boldsymbol{x})$ be a monomial of $m_{\boldsymbol{\mu_d}}(\boldsymbol{x})$, there exists $1 \leq j_1 < \ldots < j_n \leq n$ such that $M'(\boldsymbol{x}) = \prod_{i=1}^n x_{j_i}^{\lambda_i - \lambda_n}$.

As a monomial of $m_{\boldsymbol{\mu_c}}(\boldsymbol{x}) \times m_{\boldsymbol{\mu_d}}(\boldsymbol{x})$ is the product of $\prod_{i=1}^n x_i^{\lambda_n}$ and a monomial of $m_{\boldsymbol{\mu_d}}(\boldsymbol{x})$, if $M(\boldsymbol{x})$ is a monomial of $m_{\boldsymbol{\mu_c}}(\boldsymbol{x}) \times m_{\boldsymbol{\mu_d}}(\boldsymbol{x})$, there exists $1 \leq j_1 < \ldots < j_n \leq n$ such that $M(\boldsymbol{x}) = \prod_{i=1}^n x_{j_i}^{\lambda_i - \lambda_n} x_{j_i}^{\lambda_n} = \prod_{i=1}^n x_{j_i}^{\lambda_i}$. So, by definition of $m_{\boldsymbol{\lambda}}$, $M(\boldsymbol{x})$ is a monomial of $m_{\boldsymbol{\lambda}}$.

Conversely, if $M(\boldsymbol{x})$ is a monomial of $m_{\boldsymbol{\lambda}}$, then there exists $1 \leq j_1 < \ldots < j_n \leq n$ such that $M(\boldsymbol{x}) = \prod_{i=1}^n x_{j_i}^{\lambda_i} = \prod_{i=1}^n x_{j_i}^{\lambda_i - \lambda_n} x_{j_i}^{\lambda_n}$ which is a monomial of $m_{\boldsymbol{\mu_c}}(\boldsymbol{x}) \times m_{\boldsymbol{\mu_d}}(\boldsymbol{x})$.   □

Example 3. Assume that $n = 3$ and $q = 4$. Let $\boldsymbol{\lambda} = (3, 2, 1) = \underbrace{(1, 1, 1)}_{\mu_c} + \underbrace{(2, 1, 0)}_{\mu_d}$,

we have:

$$\begin{aligned}
m_{(3,2,1)}(\boldsymbol{x}) &= x_1^2 x_2^3 x_3 + x_1^3 x_2^2 x_3 + x_1^2 x_2 x_3^3 + x_1^3 x_2 x_3^2 + x_1 x_2^2 x_3^3 + x_1 x_2^3 x_3^2 \\
&= x_1 x_2 x_3 \left(x_1 x_2^2 + x_1^2 x_2 + x_1 x_3^2 + x_1^2 x_3 + x_2 x_3^2 + x_2^2 x_3 \right) \\
&= m_{(1,1,1)}(\boldsymbol{x}) m_{(2,1,0)}(\boldsymbol{x}) \ .
\end{aligned} \tag{14}$$

Assuming that we know the simplified value vector of each $m_{\boldsymbol{\mu}}$ if $\boldsymbol{\mu} < \boldsymbol{\lambda}$, proposition 2 allows us to compute $m_{\boldsymbol{\lambda}}(\boldsymbol{x})$ for all $\boldsymbol{x} \in E_q^n$ such that $wt(\boldsymbol{x}) = wt(\boldsymbol{\lambda})$.

Assume that $\boldsymbol{wt(x) > wt(\lambda)}$.
    We assume that we know $m_{\boldsymbol{\lambda}}(\boldsymbol{x})$ for all $\boldsymbol{x} \in E_q^n$ such that $wt(\boldsymbol{x}) = wt(\boldsymbol{\lambda})$. We recall that s_j is the representative of the jth symmetry class.

Lemma 5. *The number of* s_j *such that* $wt(s_j) = k$ *is* $\binom{k+q-2}{q-2}$.

Proof.

$$\begin{aligned}
\text{Card}\left(\{s_j \mid wt(s_j) = k\}\right) &= \text{Card}\left(\{s_j \mid wt(s_j) \leq k\}\right) \\
&\quad - \text{Card}\left(\{s_j \mid wt(s_j) \leq k - 1\}\right) \\
&= \binom{k+q-1}{q-1} - \binom{k+q-2}{q-1} \\
&= \binom{k+q-2}{q-2} \ .
\end{aligned} \tag{15}$$

□

Let $k := wt(\boldsymbol{\lambda})$ and $K := \binom{k+q-2}{q-2}$ be the number of $\boldsymbol{s_j}$ such that $wt(\boldsymbol{s_j}) = k$.

Theorem 2. *Let $\boldsymbol{\lambda}$ be a partition of weight k, $K := \binom{k+q-2}{q-2}$ the number of $\boldsymbol{s_j}$ such that $wt(\boldsymbol{s_j}) = k$.*
Then, for all $\boldsymbol{x} \in E_q^n$ such that $wt(\boldsymbol{x}) > k$, we have:

$$m_{\boldsymbol{\lambda}}(\boldsymbol{x}) = \sum_{\{i \ | \ wt(\boldsymbol{s_i}) = k\}} \left(m_{\boldsymbol{\lambda}}(\boldsymbol{s_i}) \prod_{j \in E_q^*} \binom{x[j]}{s_i[j]} \right) . \tag{16}$$

We can already notice that theorem 2 allows us to compute all the simplified value vector of $m_{\boldsymbol{\lambda}}$ assuming we know the value of $m_{\boldsymbol{\lambda}}(\boldsymbol{x})$ for all $\boldsymbol{x} \in E_q^n$ such that $wt(\boldsymbol{x}) = wt(\boldsymbol{\lambda})$.

Lemma 6. *Let $\boldsymbol{\lambda} = (\lambda_1, \ldots, \lambda_{wt(\boldsymbol{\lambda})}, 0, \ldots, 0)$ be a partition and $\boldsymbol{x} \in E_q^n$, then:*

$$m_{\boldsymbol{\lambda}}(\boldsymbol{x}) = \sum_{1 \leq i_1 < \cdots < i_{wt(\boldsymbol{\lambda})} \leq n} m_{(\lambda_1, \ldots, \lambda_{wt(\boldsymbol{\lambda})})}(x_{i_1}, \ldots, x_{i_{wt(\boldsymbol{\lambda})}}) . \tag{17}$$

Lemma 7. *Let $\boldsymbol{\lambda} = (\lambda_1, \ldots, \lambda_{wt(\boldsymbol{\lambda})}, 0, \ldots, 0)$ be a partition, $\ell \in \mathbb{N}$ such that $wt(\boldsymbol{s_\ell}) = wt(\boldsymbol{\lambda})$ and $\boldsymbol{x} \in E_q^n$ such that $wt(\boldsymbol{x}) > wt(\boldsymbol{\lambda})$. s Then, the number of choices of $1 \leq i_1 < \cdots < i_{wt(\boldsymbol{\lambda})} \leq n$ such that $m_{(\lambda_1, \ldots, \lambda_{wt(\boldsymbol{\lambda})})}(x_{i_1}, \ldots, x_{i_{wt(\boldsymbol{\lambda})}}) = m_{(\lambda_1, \ldots, \lambda_{wt(\boldsymbol{\lambda})})}(\boldsymbol{s_\ell})$ is:*

$$\prod_{j \in E_q^*} \binom{x[j]}{s_\ell[j]} . \tag{18}$$

Proof. Let $\boldsymbol{\lambda} = (\lambda_1, \ldots, \lambda_{wt(\boldsymbol{\lambda})}, 0, \ldots, 0)$ be a partition, $\ell \in \mathbb{N}$ such that $wt(\boldsymbol{s_\ell}) = wt(\boldsymbol{\lambda})$ and $\boldsymbol{x} \in E_q^n$ such that $wt(\boldsymbol{x}) > wt(\boldsymbol{\lambda})$.
Then, $m_{(\lambda_1, \ldots, \lambda_{wt(\boldsymbol{\lambda})})}(x_{i_1}, \ldots, x_{i_{wt(\boldsymbol{\lambda})}}) = m_{(\lambda_1, \ldots, \lambda_{wt(\boldsymbol{\lambda})})}(\boldsymbol{s_\ell})$ if and only if:

$$\forall j \in E_q^*, \text{Card}\left(\{i \in \{i_1, \ldots, i_k\} \mid x_i = j\}\right) = s_\ell[j] . \tag{19}$$

Hence, we can view this problem as a draw without replacement in an urn which contains $wt(\boldsymbol{x})$ balls of $q-1$ colors. We number the colors from 1 to $q-1$ and our urn contains $x[i]$ balls of color i. To conclude, we just have to notice that the number of ways to draw $s_\ell[j]$ balls of color j for $j \in E_q^*$ is equal to $\prod_{j=1}^{q-1} \binom{x[j]}{s_\ell[j]}$. $\square$

Proof (of theorem 2). The theorem is a simple consequence of the two previous lemmas. $\square$

3.3   Link between Simplified ANF and Simplified Value Vector

Let us define the matrix ANF_to_TT of size $N \times N$ by:

$$\forall (i,j) \in \{1, \ldots, N\}^2, \ \text{ANF\_to\_TT}[i][j] = m_{\boldsymbol{s_j}}(\boldsymbol{s_i}) . \tag{20}$$

There is an example of how to construct ANF_to_TT in appendix A.

Let $f \in \mathcal{SM}_n(q, r)$, then $\boldsymbol{f_{sv}}$ and $\boldsymbol{f_{sANF}}$ verify $\boldsymbol{f_{sv}} = \text{ANF\_to\_TT} \boldsymbol{f_{sANF}}$. Moreover, if ANF_to_TT is invertible we denote by TT_to_ANF its inverse which verifies $\boldsymbol{f_{sANF}} = \text{TT\_to\_ANF} \boldsymbol{f_{sv}}$. Remember that ANF_to_TT has its coefficients in E_m which is a ring, so it is an inversion in an A-module.

Nevertheless, depending of the choice of (q, r, n), ANF_to_TT could be not invertible:

Example 4. Assume that $n = 4$, $q = 3$, $r = 2$ and $x \in E_3^4$, then:

$$
\begin{aligned}
m_{(2,0,0,0)}(x) &= x_1^2 + x_2^2 + x_3^2 + x_4^2 \bmod 2 \\
&= x_1 + x_2 + x_3 + x_4 \bmod 2 \\
&= m_{(1,0,0,0)}(x) \ .
\end{aligned}
\tag{21}
$$

So the 2nd row and the 3rd row of ANF_to_TT are equal. This implies that ANF_to_TT is singular.

In this case, to compute $\boldsymbol{f_{sANF}}$ we have to solve the system:

$$
\boldsymbol{f_{sv}} = \text{ANF\_to\_TT} \times \boldsymbol{X} \ ,
\tag{22}
$$

which will have more than one solution. The simplified ANF of f will be either of those solutions.

So, knowing either simplified ANF or simplified value vector, the complexity of computing the other is $\mathcal{O}(N^2)$ (if we assume that ANF_to_TT and TT_to_ANF are pre-computed).

4 Walsh Spectrum

4.1 Definition

For all $\boldsymbol{a} \in E_q^n$, we denote by $\varphi_{\boldsymbol{a}}$ the linear function $x \mapsto \boldsymbol{a} \cdot \boldsymbol{x}$.

Definition 9. *Let $f \in \mathcal{M}_n(q, r)$ and $\boldsymbol{a} \in E_q^n$. The Walsh coefficient of f in point $\boldsymbol{a}$ corresponds to*

$$
\mathcal{F}(f + \varphi_{\boldsymbol{a}}) = \sum_{\boldsymbol{x} \in E_q^n} w^{f(\boldsymbol{x}) - \boldsymbol{a} \cdot \boldsymbol{x}} \ ,
\tag{23}
$$

where $w = e^{\frac{2i\pi}{r}}$.
We call Walsh spectrum of f the set $\{\mathcal{F}(f + \varphi_{\boldsymbol{a}}), \ a \in E_q^n\}$ of all Walsh coefficients of f.

Proposition 3. *The Walsh transform of a symmetric function over finite rings is a complex symmetric function from E_q^n to $\mathbb{C}$, verifying:*

$$
\forall \boldsymbol{a} \in E_q^n, \ \mathcal{F}(f + \varphi_{\boldsymbol{a}}) = \sum_{i=1}^{N} \left(w^{f(\boldsymbol{s_i})} \sum_{\boldsymbol{x} \in P_{n,q}(\boldsymbol{s_i})} w^{-\boldsymbol{a} \cdot \boldsymbol{x}} \right) \ .
\tag{24}
$$

We call the simplified Walsh spectrum of f the vector $f_{ws} \in \mathbb{C}^N$ defined by:

$$
f_{ws} := (\mathcal{F}(f + \varphi_{\boldsymbol{s_1}}), \ldots, \mathcal{F}(f + \varphi_{\boldsymbol{s_N}})) \ .
\tag{25}
$$

Proof. Let $f \in \mathcal{SM}_n(q, r)$, a and b two elements of a same symmetry class. We are going to prove that $\mathcal{F}(f + \varphi_a) = \mathcal{F}(f + \varphi_b)$:

$$\begin{aligned} \mathcal{F}(f + \varphi_a) &= \sum_{\boldsymbol{x} \in E_q^n} w^{f(\boldsymbol{x}) - \boldsymbol{a} \cdot \boldsymbol{x}} \\ &= \sum_{i=1}^{N} \left(w^{f(\boldsymbol{s_i})} \sum_{\boldsymbol{x} \in P_{n,q}(\boldsymbol{s_i})} w^{-\boldsymbol{a} \cdot \boldsymbol{x}} \right) \ . \end{aligned} \quad (26)$$

We will prove that if $i \in \{1, \ldots, N\}$, $\sum_{\boldsymbol{x} \in P_{n,q}(\boldsymbol{s_i})} w^{-\boldsymbol{a} \cdot \boldsymbol{x}} = \sum_{\boldsymbol{x} \in P_{n,q}(\boldsymbol{s_i})} w^{-\boldsymbol{b} \cdot \boldsymbol{x}}$.

Saying that $\boldsymbol{a}$ and $\boldsymbol{b}$ are in the same symmetry class means that there exists $\sigma \in S_n$ such that $\boldsymbol{\sigma}(\boldsymbol{a}) = \boldsymbol{b}$. Besides,

$$\begin{aligned} \boldsymbol{b} \cdot \boldsymbol{x} = \boldsymbol{\sigma}(\boldsymbol{a}) \cdot \boldsymbol{x} &= \sum_{i=1}^{n} a_{\sigma(i)} x_i \\ &= \sum_{i=1}^{n} a_i x_{\sigma^{-1}(i)} \\ &= \boldsymbol{a} \cdot \boldsymbol{\sigma}^{-1}(\boldsymbol{x}) \ . \end{aligned} \quad (27)$$

Noticing that for all $\boldsymbol{x} \in E_q^n$, $\boldsymbol{x}$ and $\boldsymbol{\sigma}^{-1}(\boldsymbol{x})$ are in the same symmetry class, we have $\sum_{\boldsymbol{x} \in P_{n,q}(\boldsymbol{s_i})} w^{-\boldsymbol{a} \cdot \boldsymbol{x}} = \sum_{\boldsymbol{x} \in P_{n,q}(\boldsymbol{s_i})} w^{-\boldsymbol{b} \cdot \boldsymbol{x}}$.

Eventually, $\mathcal{F}(f + \varphi_a) = \mathcal{F}(f + \varphi_b)$ and so $\mathcal{F}(f + \varphi_\cdot)$ is a symmetric function from E_q^n to $\mathbb{C}$. □

Our goal is now to compute $\sum_{\boldsymbol{x} \in P_{n,q}(\boldsymbol{s_i})} w^{-\boldsymbol{a} \cdot \boldsymbol{x}}$ for all $i \in \{1, \ldots, N\}$ and $\boldsymbol{a} \in E_q^n$. To do this, we introduce Walsh matrix.

Definition 10. *Let $q \geq 2$, $r \geq 2$ and $n \geq 1$, we recall that $w = e^{\frac{2i\pi}{r}}$ and $N = \binom{n+q-1}{q-1}$. We call Walsh matrix of parameters (q, r, n), denoted $W_{q,r,n}$, the matrix of size $N \times N$ defined by:*

$$\forall (i, j) \in \{1, \ldots, N\}^2, \ W_{q,r,n}[i][j] = \sum_{\boldsymbol{x} \in P_{n,q}(\boldsymbol{s_i})} w^{-\boldsymbol{s_j} \cdot \boldsymbol{x}} \ . \quad (28)$$

We denote by $W_{q,r,n}[i][\cdot]$ (resp. $W_{q,r,n}[\cdot][j]$) the ith row (resp. the jth column).

4.2   Computing the Matrix $W_{q,r,n}$

Cardinal of a Symmetry Class
We state the following lemma which will be used later.

Lemma 8. *Let $\boldsymbol{\lambda}$ be a partition, then*

$$\text{Card}(P_{n,q}(\boldsymbol{\lambda})) = \frac{n!}{\lambda[0]! \cdots \lambda[q-1]!} \ . \quad (29)$$

Proof. Recall that $P_{n,q}(\boldsymbol{\lambda}) = \{\boldsymbol{x} \in E_q^n \mid \exists \sigma \in S_n \text{ such that } \boldsymbol{x} = \boldsymbol{\sigma}(\boldsymbol{\lambda})\}$, so we want to calculate the cardinal of $\boldsymbol{\lambda}$'s orbit under the action of S_n. To do that, we will calculate the cardinal of $\boldsymbol{\lambda}$'s stabilizer and then apply the orbit-stabilizer theorem.

$\boldsymbol{\lambda}$ being a partition, we know that:

$$\boldsymbol{\lambda} = (\underbrace{q-1, \ldots, q-1}_{\lambda[q-1]}, q-2, \ldots, \underbrace{0, \ldots, 0}_{\lambda[0]}) \ . \quad (30)$$

So, if $\mathrm{Stab}_{\boldsymbol{\lambda}} := \{\sigma \in S_n \mid \sigma(\boldsymbol{\lambda}) = \boldsymbol{\lambda}\}$ then $\sigma \in \mathrm{Stab}_{\boldsymbol{\lambda}}$ if and only if

$$\sigma = \sigma_{q-1} \circ \cdots \circ \sigma_0 \ , \tag{31}$$

where σ_i is a permutation of the set $\left\{1 + \sum_{j=i+1}^{q-1} \lambda[j], \ldots, \lambda[i] + \sum_{j=i+1}^{q-1} \lambda[j]\right\}$ which has $\lambda[i]$ elements.

So each σ_i is in $S_{\lambda[i]}$. This implies that $\mathrm{Card}(\mathrm{Stab}_{\boldsymbol{\lambda}}) = \prod_{i=0}^{q-1} \lambda[i]!$. To conclude, we just have to use the orbit-stabilizer theorem:

$$\mathrm{Card}(\mathrm{Part}_{\boldsymbol{\lambda}}) = \frac{\mathrm{Card}(S_n)}{\mathrm{Card}(\mathrm{Stab}_{\boldsymbol{\lambda}})} = \frac{n!}{\prod_{i=0}^{q-1} \lambda[i]!} \ . \tag{32}$$

□

Relation between $W_{q,r,n}$ Rows and Columns
Let $\boldsymbol{v} = (v_1, \ldots, v_n) \in E_q^n$, we denote by $\boldsymbol{v} \bmod \boldsymbol{r}$ the vector $(v_1 \bmod r, \ldots, v_n \bmod r)$.

Proposition 4. *Let* $(j_1, j_2) \in \{1, \ldots, N\}^2$ *such that* $\boldsymbol{s}_{j_1} \bmod \boldsymbol{r} \in P_{q,n}(\boldsymbol{s}_{j_2})$, *then:*

$$W_{q,r,n}[\cdot][j_1] = W_{q,r,n}[\cdot][j_2] \ . \tag{33}$$

Proof. Let $(j_1, j_2) \in \{1, \ldots, N\}^2$ such that $(\boldsymbol{s}_{j_1} \bmod \boldsymbol{r}) \in P_{q,n}(\boldsymbol{s}_{j_2})$. There exists $\boldsymbol{y} \in P_{q,n}(\boldsymbol{s}_{j_2})$ such that $\boldsymbol{y} = \boldsymbol{s}_{j_1} \bmod \boldsymbol{r}$. So, there exists $\boldsymbol{v} \in E_q^n$, such that $\boldsymbol{y} - \boldsymbol{s}_{j_1} = m\boldsymbol{v}$.

Therefore, for all $i \in \{1, \ldots, N\}$, we have:

$$
\begin{aligned}
W_{q,r,n}[i][j_1] &= \sum_{\boldsymbol{x} \in P_{q,n}(\boldsymbol{s}_i)} w^{-\boldsymbol{s}_{j_1} \cdot \boldsymbol{x}} \\
&= \sum_{\boldsymbol{x} \in P_{q,n}(\boldsymbol{s}_i)} w^{-\boldsymbol{y} \cdot \boldsymbol{x}} w^{(\boldsymbol{y} - \boldsymbol{s}_{j_1}) \cdot \boldsymbol{x}} \\
&= \sum_{\boldsymbol{x} \in P_{q,n}(\boldsymbol{s}_i)} w^{-\boldsymbol{y} \cdot \boldsymbol{x}} \underbrace{e^{\frac{2i\pi}{r} r(\boldsymbol{v} \cdot \boldsymbol{x})}}_{=1} \\
&= \sum_{\boldsymbol{x} \in P_{q,n}(\boldsymbol{s}_i)} w^{-\boldsymbol{y} \cdot \boldsymbol{x}} = \sum_{\boldsymbol{x} \in P_{q,n}(\boldsymbol{s}_i)} w^{-\boldsymbol{s}_{j_2} \cdot \boldsymbol{x}} = W_{q,r,n}[i][j_2] \ .
\end{aligned}
\tag{34}
$$

□

Proposition 5. *Let* $(i_1, i_2) \in \{1, \ldots, N\}^2$ *such that* $\boldsymbol{s}_{i_1} \bmod \boldsymbol{r} \in P_{q,n}(\boldsymbol{s}_{i_2})$. *Then:*

$$W_{q,r,n}[i_1][\cdot] = \frac{\mathrm{Card}\left(P_{q,n}(\boldsymbol{s}_{i_1})\right)}{\mathrm{Card}\left(P_{q,n}(\boldsymbol{s}_{i_2})\right)} W_{q,r,n}[i_2][\cdot] \ . \tag{35}$$

Proof. Let $(i_1, i_2) \in \{1, \ldots, N\}^2$ such that $(\boldsymbol{s}_{i_1} \bmod \boldsymbol{r}) \in P_{q,n}(\boldsymbol{s}_{i_2})$. There exists $\boldsymbol{y} \in P_{q,n}(\boldsymbol{s}_{j_2})$ such that $\boldsymbol{y} = \boldsymbol{s}_{j_1} \bmod \boldsymbol{r}$. So, there exists $\boldsymbol{v} \in E_q^n$, such that $\boldsymbol{y} - \boldsymbol{s}_{j_1} = m\boldsymbol{v}$.

Assume that $\mathrm{Card}\left(P_{q,n}(\boldsymbol{s}_{i_1})\right) \geq \mathrm{Card}\left(P_{q,n}(\boldsymbol{s}_{i_2})\right)$. We are going to show that for all $\boldsymbol{y} \in P_{q,n}(\boldsymbol{s}_{i_2})$ there exists $\frac{\mathrm{Card}\left(P_{q,n}(\boldsymbol{s}_{i_1})\right)}{\mathrm{Card}\left(P_{q,n}(\boldsymbol{s}_{i_2})\right)}$ distinct elements of $P_{q,n}(\boldsymbol{s}_{i_1})$ whose reductions modulo r are $\boldsymbol{y}$.

Let $\boldsymbol{x} \in P_{q,n}(\boldsymbol{s}_{i_1})$ and $\boldsymbol{y} \in P_{q,n}(\boldsymbol{s}_{i_2})$, for $k \in \{0, \ldots, q-1\}$, we have:

- $y[0] = \displaystyle\sum_{\{\ell \in \{0,\ldots,q-1\} \ | \ r \ \text{divide} \ \ell\}} x[\ell]$.

- If $k > 0$
    - If r divide k, $y[k] = 0$;
    - Else, $y[k] = x[k]$.

So, given $\boldsymbol{y}$, we can find $\dfrac{y[0]!}{\prod_{\{\ell \in \{0,\ldots,q-1\} \ | \ r \ \text{divide} \ \ell\}} s_{i_1}[\ell]!}$ distinct elements of $P_{q,n}(\boldsymbol{s_{i_1}})$ whose reductions modulo r are $\boldsymbol{y}$.

We obtain this result by modelling the problem as a draw without replacement in an urn of $y[0]$ balls which contains $s_{i_1}[\ell]$ balls of color ℓ for each ℓ verifying r divide ℓ.

Applying lemma 8, we obtain:

$$
\begin{aligned}
\frac{\text{Card}\big(P_{q,n}(\boldsymbol{s_{i_1}})\big)}{\text{Card}\big(P_{q,n}(\boldsymbol{s_{i_2}})\big)} &= \frac{\frac{n!}{\prod_{\ell=0}^{q-1} s_{i_1}[\ell]!}}{\frac{n!}{\prod_{\ell=0}^{q-1} s_{i_2}[\ell]!}} = \frac{\prod_{\ell=0}^{q-1} s_{i_2}[\ell]!}{\prod_{\ell=0}^{q-1} s_{i_1}[\ell]!} \\
&= \frac{s_{i_2}[0]! \prod_{\{\ell \in \{0,\ldots,q-1\} \ | \ r \ \text{do not divide} \ \ell\}} s_{i_1}[\ell]!}{\prod_{\ell=0}^{q-1} s_{i_1}[\ell]!} \\
&= \frac{s_{i_2}[0]!}{\prod_{\{\ell \in \{0,\ldots,q-1\} \ | \ r \ \text{divide} \ \ell\}} s_{i_1}[\ell]!} \ .
\end{aligned}
\tag{36}
$$

The last step is straightforward. For $j \in \{1,\ldots,N\}$,

$$
\begin{aligned}
W_{q,r,n}[i_1][j] &= \sum_{\boldsymbol{x} \in P_{q,n}(\boldsymbol{s_{i_1}})} w^{-s_j \cdot \boldsymbol{x}} \\
&= \sum_{\boldsymbol{y} \in P_{q,n}(\boldsymbol{s_{i_2}})} \sum_{\{\boldsymbol{x} \in P_{q,n}(\boldsymbol{s_{i_1}}) \ | \ \boldsymbol{x} \ \text{mod} \ r = \boldsymbol{y}\}} w^{s_j \cdot \boldsymbol{x}} \\
&= \sum_{\boldsymbol{y} \in P_{q,n}(\boldsymbol{s_{i_2}})} \sum_{\{\boldsymbol{x} \in P_{q,n}(\boldsymbol{s_{i_1}}) \ | \ \boldsymbol{x} \ \text{mod} \ r = \boldsymbol{y}\}} w^{-s_j \cdot \boldsymbol{y}} \underbrace{w^{-s_j \cdot (\boldsymbol{x} - \boldsymbol{y})}}_{=1} \\
&= \sum_{\boldsymbol{y} \in P_{q,n}(\boldsymbol{s_{i_2}})} \frac{\text{Card}\big(P_{q,n}(\boldsymbol{s_{i_1}})\big)}{\text{Card}\big(P_{q,n}(\boldsymbol{s_{i_2}})\big)} w^{-s_j \cdot \boldsymbol{y}} \\
&= \frac{\text{Card}\big(P_{q,n}(\boldsymbol{s_{i_1}})\big)}{\text{Card}\big(P_{q,n}(\boldsymbol{s_{i_2}})\big)} W_{q,r,n}[i_2][j] \ .
\end{aligned}
\tag{37}
$$

$\square$

Building $W_{q,r,n}$ Knowing $W_{r,r,n}$

Here, our aim is to give a quick way to build $W_{q,r,n}$ assuming we know $W_{r,r,n}$. We recall that $w = e^{\frac{2i\pi}{r}}$. There is an example of this construction in appendix B.

Let $N_r = \binom{n+r-1}{r-1}$ and $N_q = \binom{n+q-1}{q-1}$. We will denote $s_1,\ldots,s_{N_r}$ the representatives of the symmetry classes of E_r^n and $s'_1,\ldots,s'_{N_q}$ the representatives of the symmetry classes of E_q^n.

Let $\psi : \{1,\ldots,N_q\} \to \{1,\ldots,N_r\}$, the function defined by:

$$
\psi(i) = j \iff s'_i \ \text{mod} \ r \in P_{q,n}(s_j) \ .
\tag{38}
$$

We need an auxiliary matrix in order to build $W_{q,r,n}$. Let W' be a matrix of size $N_r \times N_q$ defined by:

$$
\forall (i,j) \in \{1,\ldots,N_r\} \times \{1,\ldots,N_q\}, \ W'[i][j] = \sum_{\boldsymbol{x} \in P_{q,n}(\boldsymbol{s_i})} w^{-s'_j \cdot \boldsymbol{x}} \ .
\tag{39}
$$

Using the same method as in the demonstration of proposition 4, we have:

$$W'[i][j] = \sum_{\boldsymbol{x} \in P_{q,n}(\boldsymbol{s_i})} w^{-\boldsymbol{s'_j} \cdot \boldsymbol{x}} = \sum_{\boldsymbol{x} \in P_{q,n}(\boldsymbol{s_i})} w^{-(\boldsymbol{s'_j} \bmod r) \cdot \boldsymbol{x}} w^{(\boldsymbol{s'_j} \bmod r - \boldsymbol{s'_j}) \cdot \boldsymbol{x}}$$
$$= \sum_{\boldsymbol{x} \in P_{q,n}(\boldsymbol{s_i})} w^{-\boldsymbol{s}_{\psi(j)} \cdot \boldsymbol{x}} = W_{r,r,n}[i][\psi(j)] \;,$$
(40)

for all $(i,j) \in \{1, \ldots, N_r\} \times \{1, \ldots, N_q\}$.

Applying the same method, we get for all $i \in \{1, \ldots, N_q\}$,

$$W_{q,r,n}[i][\cdot] = \frac{\mathrm{Card}\left(P_{q,n}(\boldsymbol{s'_i})\right)}{\mathrm{Card}\left(P_{q,n}(\boldsymbol{s}_{\psi(i)})\right)} W'[\psi(i)][\cdot] \;.$$
(41)

4.3   Computing $W_{r,r,n}$

In the previous subsection, we have shown how to compute $W_{q,r,n}$ assuming we know $W_{r,r,n}$.

The last step that we have to do is to give a way of building $W_{r,r,n}$ for any r and n.

Boolean Case

When $q = r = 2$, this is linked to Krawtchouk polynomial [10].

Definition 11. *The Krawtchouk polynomial of degree i, denoted K_i, is defined by:*

$$\forall (x,n) \in \mathbb{N}, \; K_i(x,n) = \sum_{j=0}^{i} (-1)^j \binom{x}{j} \binom{n-x}{i-j} \;.$$
(42)

We call Krawtchouk matrix of parameter n, the matrix K_n defined by:

$$\forall (i,j) \in \{0, \ldots, n\}^2, \; K_n[i][j] = K_i(j,n) \;.$$
(43)

It happens that we have $W_{2,2,n} = K_n$ for all n in $\mathbb{N}^*$. And there are some papers stating how to generate those matrices [16].

Extension for Any r

At this point, we do not have a method similar to Krawtchouk's one to generate the matrix $W_{r,r,n}$ for any r and n. Thus, we use fast Fourier Transform to compute for all $i \in \{1, \ldots, N_r\}$, $\mathcal{F}(0 + \varphi_{s_i})$ which can be directly used to build $W_{r,r,n}$.

The complexity of building $W_{r,r,n}$ this way is in $\mathcal{O}\left(N_r r^n n \log(r)\right)$.

4.4   Computing the Walsh Spectrum of a Symmetric Function

Now that we have matrix $W_{q,r,n}$, we can easily compute the simplified Walsh spectrum of a symmetric function. Indeed, if we denote f_{ws} the simplified Walsh spectrum of $f \in \mathcal{SM}_n(q,r)$, we have:

$$f_{ws} = W_{q,r,n}^T \times \boldsymbol{\phi}(\boldsymbol{f_{sv}}) \;,$$
(44)

where $\boldsymbol{\phi}(\boldsymbol{f_{sv}}) := (w^{f(s_1)}, \ldots, w^{f(s_N)})^T$.

5   Using These Algorithms to Test Large Set of Symmetric Functions

Assume we want to test T symmetric functions of $\mathcal{SM}_n(q,m)$, where T is a big number. If we are going to test each function individually the complexity of this analysis would be in $\mathcal{O}\left(T \times N^2\right)$, where $N = \binom{n+q-1}{q-1}$. For instance if we want to test all symmetric functions, the complexity will be in $\mathcal{O}\left(r^N N^2\right)$.

But, we can easily lower this complexity by noticing that if the simplified value vector of f differs from the simplified value vector of g from 1 bit, then we can get the simplified Walsh spectrum of g from the simplified Walsh spectrum of f only in $\mathcal{O}(N)$. This property can be extended easily:

Proposition 6. *Let* $f,\ g \in \mathcal{SM}_n(q,m)$ *and* $d = wt(f - g)$. *We denote by* $1 \leq i_1 < \cdots < i_d \leq N$ *the partition's numbers where* $f(s_{i_j}) \neq g(s_{i_j})$. *Then:*

$$g_{ws} = f_{ws} + \sum_{j=1}^{d} \left(w^{g(s_{i_j})} - w^{f(s_{i_j})} \right) W_{q,r,n}[i_j][\cdot] \ . \tag{45}$$

So, if we know f_{ws}, *we are able to compute* g_{ws} *in* $\mathcal{O}\left(dN\right)$ *operations.*

Proof. Let $f,\ g \in \mathcal{SM}_n(q,m)$ and $d = wt(f - g)$. We denote by $1 \leq i_1 < \cdots < i_d \leq N$ the partition's numbers where $f(s_{i_j}) \neq g(s_{i_j})$. Let $k \in \{1, \ldots, N\}$.

$$
\begin{aligned}
\mathcal{F}(g + \varphi_{s_k}) - \mathcal{F}(f + \varphi_{s_k}) &= \sum_{i=1}^{N} \left(w^{g(s_i)} \sum_{x \in P_{n,q}(s_i)} w^{-s_k \cdot x} \right) \\
&\quad - \sum_{i=1}^{N} \left(w^{f(s_i)} \sum_{x \in P_{n,q}(s_i)} w^{-s_k \cdot x} \right) \\
&= \sum_{i=1}^{N} \left(\left(w^{g(s_i)} - w^{f(s_i)} \right) \sum_{x \in P_{n,q}(s_i)} w^{-s_k \cdot x} \right) \\
&= \sum_{j=1}^{d} \left(\left(w^{g(s_{i_j})} - w^{f(s_{i_j})} \right) \sum_{x \in P_{n,q}(s_{i_j})} w^{-s_k \cdot x} \right) \\
&= \sum_{j=1}^{d} \left(\left(w^{g(s_{i_j})} - w^{f(s_{i_j})} \right) W_{q,r,n}[i_j][k] \right) \ .
\end{aligned}
\tag{46}
$$

So, to compute g_{ws} we have to do N times d multiplications and d additions. Hence the complexity is in $\mathcal{O}(dN)$.   □

Using Gray codes, we are able to enumerate all vectors of length N on the alphabet E_r. Each vector can be viewed as a function of $\mathcal{SM}_n(q,r)$. Thus, using proposition 6, we can compute the Walsh spectrum of each function in $\mathcal{O}(N)$, because the Hamming distance between two consecutive functions is only one by properties of Gray codes. So, to compute the Walsh spectrum of all of them, the complexity is in $\mathcal{O}(r^N N)$.

6   Conclusion

In this paper, we introduced two algorithms that can be used to compute efficiently the ANF and the Walsh spectrum of any symmetric function from E_q^n

to E_r. Their complexities are in $\mathcal{O}\left(\left(\binom{n+q-1}{q-1}\right)^2\right)$ so they can be used for symmetric functions with a lot of variables. Once Walsh spectrum is computed, we are able to get many information about the functions such as balancedness, non-linearity, order of correlation and order of resiliency. Moreover, our method is really efficient if we want to test many symmetric functions. Indeed the complexity for computing all simplified Walsh Spectrum of all symmetric functions of $\mathcal{SM}_n(q,r)$ is in $\mathcal{O}\left(r^{\binom{n+q-1}{q-1}}\binom{n+q-1}{q-1}\right)$, whereas if we use fast Fourier transform over the same set, complexity will be in $\mathcal{O}\left(r^{\binom{n+q-1}{q-1}}nq^n\log(q)\right)$.

References

1. Andrews, G.E.: The Theory of Partitions. Encyclopedia of Mathematics and its Applications, vol. 2. Addison-Wesley Publishing Co., Reading (1976); Reprinted by Cambridge University Press, Cambridge (1998)
2. Ars, G., Faugère, J.-C.: Algebraic immunities of functions over finite fields. Research Report RR-5532, INRIA (2005)
3. Camion, P., Canteaut, A.: Generalization of Siegenthaler Inequality and Schnorr-Vaudenay Multipermutations. In: Koblitz, N. (ed.) CRYPTO 1996. LNCS, vol. 1109, pp. 372–386. Springer, Heidelberg (1996), 10.1007/3-540-68697-5_28
4. Canteaut, A., Videau, M.: Symmetric boolean functions. IEEE Transactions on Information Theory 51(8), 2791–2811 (2005)
5. Carlet, C.: The complexity of boolean functions from cryptographic viewpoint. In: Krause, M., Pudlák, P., Reischuk, R., van Melkebeek, D. (eds.) Complexity of Boolean Functions, Dagstuhl, Germany. Dagstuhl Seminar Proceedings, vol. 06111. Internationales Begegnungs- und Forschungszentrum für Informatik (IBFI), Schloss Dagstuhl, Germany (2006)
6. Cusick, T.W., Li, Y., Stanica, P.: Balanced symmetric functions over GF(p). IEEE Transactions on Information Theory 54(3), 1304–1307 (2008)
7. Fu, S., Li, C., Sun, B.: Enumeration of Homogeneous Rotation Symmetric Functions over f_p. In: Franklin, M.K., Hui, L.C.K., Wong, D.S. (eds.) CANS 2008. LNCS, vol. 5339, pp. 278–284. Springer, Heidelberg (2008)
8. Gopalakrishnan, K., Stinson, D.R.: Three characterizations of non-binary correlation-immune and resilient functions. Designs, Codes and Cryptography 5, 241–251 (1997)
9. Hu, Y., Xiao, G.: Resilient functions over finite fields. IEEE Transactions on Information Theory 49(8), 2040–2046 (2003)
10. Krawtchouk, M.: Sur une généralisation des polynômes d'Hermite. C.R. Acad. Sci. Paris 189, 620–622 (1929)
11. Li, Y., Cusick, T.W.: Strict avalanche criterion over finite fields, submitted. Journal of Mathematical Cryptology 1, 65–78 (2005)
12. Meier, W., Pasalic, E., Carlet, C.: Algebraic Attacks and Decomposition of Boolean Functions. In: Cachin, C., Camenisch, J.L. (eds.) EUROCRYPT 2004. LNCS, vol. 3027, pp. 474–491. Springer, Heidelberg (2004)
13. Mouffron, M.: Transitive q-Ary Functions over Finite Fields or Finite Sets: Counts, Properties and Applications. In: von zur Gathen, J., Imaña, J.L., Koç, Ç.K. (eds.) WAIFI 2008. LNCS, vol. 5130, pp. 19–35. Springer, Heidelberg (2008)
14. Rovetta, C., Mouffron, M.: De Bruijn sequences and complexity of symmetric functions. Cryptography and Communications, 1–19 (2011), 10.1007/s12095-011-0054-2

15. Sagan, B.E.: The symmetric group - representations, combinatorial algorithms, and symmetric functions. Wadsworth & Brooks/Cole mathematics series. Wadsworth (1991)
16. Sarkar, S., Maitra, S.: Efficient search for symmetric boolean functions under constraints on walsh spectra values. In: Michon, J.-F., Valarcher, P., Yunès, J.-B. (eds.) Proceedings of BFCA 2006 Conference, Rouen, France, March 13-15, pp. 29–50 (2006)

A   Computing ANF_to_TT

Assume that $n = 3$, $q = 3$ and $r = 3$. The steps to obtain ANF_to_TT are:

$$
\begin{pmatrix} 1 \\ 1 \\ 1 \\ 1 \\ 1 \\ 1 \\ 1 \\ 1 \\ 1 \end{pmatrix}
\rightarrow
\begin{pmatrix} 1 \\ 1\ 1\ 1 \\ 1\ 2\ 1 \\ 1 \\ 1 \\ 1 \\ 1 \\ 1 \\ 1 \end{pmatrix}
\rightarrow
\begin{pmatrix} 1 \\ 1\ 1\ 1 \\ 1\ 2\ 1 \\ 1\ 2\ 2 \\ 1\ 0\ 2 \\ 1\ 1\ 2 \\ 1\ 0\ 0 \\ 1\ 1\ 0 \\ 1\ 2\ 0 \\ 1\ 0\ 0 \end{pmatrix}
\rightarrow
\begin{pmatrix} 1 \\ 1\ 1\ 1 \\ 1\ 2\ 1 \\ 1\ 2\ 2\ 1\ 2\ 1 \\ 1\ 0\ 2\ 2\ 0\ 1 \\ 1\ 1\ 2\ 1\ 1\ 1 \\ 1\ 0\ 0 \\ 1\ 1\ 0 \\ 1\ 2\ 0 \\ 1\ 0\ 0 \end{pmatrix}
\rightarrow
\begin{pmatrix} 1 \\ 1\ 1\ 1 \\ 1\ 2\ 1 \\ 1\ 2\ 2\ 1\ 2\ 1 \\ 1\ 0\ 2\ 2\ 0\ 1 \\ 1\ 1\ 2\ 1\ 1\ 1 \\ 1\ 0\ 0\ 0\ 0\ 0 \\ 1\ 1\ 0\ 2\ 2\ 0 \\ 1\ 2\ 0\ 2\ 1\ 0 \\ 1\ 0\ 0\ 0\ 0\ 0 \end{pmatrix}
\rightarrow
\begin{pmatrix} 1\ 0\ 0\ 0\ 0\ 0\ 0\ 0\ 0 \\ 1\ 1\ 1\ 0\ 0\ 0\ 0\ 0\ 0 \\ 1\ 2\ 1\ 0\ 0\ 0\ 0\ 0\ 0 \\ 1\ 2\ 2\ 1\ 2\ 1\ 0\ 0\ 0 \\ 1\ 0\ 2\ 2\ 0\ 1\ 0\ 0\ 0 \\ 1\ 1\ 2\ 1\ 1\ 1\ 0\ 0\ 0 \\ 1\ 0\ 0\ 0\ 0\ 0\ 1\ 0\ 0\ 1 \\ 1\ 1\ 0\ 2\ 2\ 0\ 2\ 2\ 1\ 1 \\ 1\ 2\ 0\ 2\ 1\ 0\ 1\ 2\ 2\ 1 \\ 1\ 0\ 0\ 0\ 0\ 0\ 2\ 0\ 0\ 1 \end{pmatrix}.
$$
(47)

B   Computing $W_{3,2,4}$

We show how we can apply our method to build any $W_{3,2,4}$, knowing $W_{2,2,4}$.

Example 5. Here is the Krawtchouk matrix of parameter 4:

$$
K_4 = W_{2,2,4} = \begin{pmatrix}
1 & 1 & 1 & 1 & 1 \\
4 & 2 & 0 & -2 & -4 \\
6 & 0 & -2 & 0 & 6 \\
4 & -2 & 0 & 2 & -4 \\
1 & -1 & 1 & -1 & 1
\end{pmatrix}
\rightarrow W' = \begin{pmatrix}
1 & 1\ 1 & 1 & 1\ 1 & 1\ 1 & 1\ 1 & 1 & 1\ 1 & 1\ 1 \\
4 & 2\ 4 & 0 & 2\ 4 & -2 & 0 & 2\ 4 & -4\ -2 & 0 & 2\ 4 \\
6 & 0\ 6 & -2 & 0\ 6 & 0\ -2 & 0\ 6 & 6 & 0\ -2 & 0\ 6 \\
4 & -2\ 4 & 0 & -2\ 4 & 2 & 0 & -2\ 4 & -4 & 2 & 0 & -2\ 4 \\
1 & -1\ 1 & 1 & -1\ 1 & -1 & 1 & -1\ 1 & 1 & -1 & 1 & -1\ 1
\end{pmatrix}.
$$
(48)

Assume we want to compute $W_{3,2,4}$. We denote s_i the representative of the ith symmetry class of E_2^4 and s'_i the representative of the ith symmetry class of E_3^4. Using proposition 4 we can build W' with K_4, where:

For instance, as $s'_8 \bmod 2 = (2,1,1,0) \bmod 2 = (0,1,1,0) \in P_{2,4}((1,1,0,0))$ and $(1,1,0,0) = s_3$, the 8th column of W' is equal to the 3rd column of K_4.

Using proposition 5 we can build $W_{3,2,4}$ with W'. For instance, the 8th row of $W_{3,2,4}$ is equal to 2 ($= 12/6$) times the 3rd row of W'. And so,

$$
W_{3,2,4} = \begin{pmatrix}
1 & 1 & 1 & 1 & 1 & 1 & 1 & 1 & 1 & 1 & 1 & 1 & 1 & 1 & 1 \\
4 & 2 & 4 & 0 & 2 & 4 & -2 & 0 & 2 & 4 & -4 & -2 & 0 & 2 & 4 \\
4 & 4 & 4 & 4 & 4 & 4 & 4 & 4 & 4 & 4 & 4 & 4 & 4 & 4 & 4 \\
6 & 0 & 6 & -2 & 0 & 6 & 0 & -2 & 0 & 6 & 6 & 0 & -2 & 0 & 6 \\
12 & 6 & 12 & 0 & 6 & 12 & -6 & 0 & 6 & 12 & -12 & -6 & 0 & 6 & 12 \\
6 & 6 & 6 & 6 & 6 & 6 & 6 & 6 & 6 & 6 & 6 & 6 & 6 & 6 & 6 \\
4 & -2 & 4 & 0 & -2 & 4 & 2 & 0 & -2 & 4 & -4 & 2 & 0 & -2 & 4 \\
12 & 0 & 12 & -4 & 0 & 12 & 0 & -4 & 0 & 12 & 12 & 0 & -4 & 0 & 12 \\
12 & 6 & 12 & 0 & 6 & 12 & -6 & 0 & 6 & 12 & -12 & -6 & 0 & 6 & 12 \\
4 & 4 & 4 & 4 & 4 & 4 & 4 & 4 & 4 & 4 & 4 & 4 & 4 & 4 & 4 \\
1 & -1 & 1 & 1 & -1 & 1 & -1 & 1 & -1 & 1 & 1 & -1 & 1 & -1 & 1 \\
4 & -2 & 4 & 0 & -2 & 4 & 2 & 0 & -2 & 4 & -4 & 2 & 0 & -2 & 4 \\
6 & 0 & 6 & -2 & 0 & 6 & 0 & -2 & 0 & 6 & 6 & 0 & -2 & 0 & 6 \\
4 & 2 & 4 & 0 & 2 & 4 & -2 & 0 & 2 & 4 & -4 & -2 & 0 & 2 & 4 \\
1 & 1 & 1 & 1 & 1 & 1 & 1 & 1 & 1 & 1 & 1 & 1 & 1 & 1 & 1
\end{pmatrix}.
$$
(49)

C   Complexities Comparison

Here, we focus on the complexity of computing the Walsh transform of all symmetric functions of $\mathcal{SM}_n(q, 2)$. We compare complexity using Gray codes, described in this paper, $\left(\mathcal{O}\left(r^{\binom{n+q-1}{q-1}}\binom{n+q-1}{q-1}\right)\right)$ to the one using Fourier transform $\mathcal{O}\left(r^{\binom{n+q-1}{q-1}}nq^n\log(q)\right)$.

Example 6.

q	2					3			4		
n	10	20	30	40	100	5	10	20	5	10	
Our complexity	2^{14}	2^{25}	2^{36}	2^{46}	2^{108}	2^{19}	2^{61}	2^{218}	2^{62}	2^{294}	
Other complexity	2^{24}	2^{45}	2^{66}	2^{86}	2^{208}	2^{25}	2^{74}	2^{246}	2^{68}	2^{309}	

$$(50)$$

A first look on those results shows that our method is very efficient in the Boolean case.

When q grows, the number of symmetric functions grows exponentially, thus the complexity of the search is for both methods close to the number of symmetric functions. For instance, $\text{Card}(\mathcal{SM}_{10}(4, 2)) = 2^{286}$. In this case, with our method we need 2^8 operations to compute the Walsh spectrum of a symmetric function, whereas we need 2^{23} operations if we use the Fast Fourier Transform.

So, our method is also efficient when q grows.

Verification of Restricted EA-Equivalence for Vectorial Boolean Functions

Lilya Budaghyan and Oleksandr Kazymyrov

Department of Informatics, University of Bergen,
P.O.Box 7803, N-5020 Bergen, Norway
{Lilya.Budaghyan,Oleksandr.Kazymyrov}@uib.no

Abstract. We present algorithms for solving the restricted extended
affine equivalence (REA-equivalence) problem for any m-dimensional
vectorial Boolean functions in n variables. The best of them has com-
plexity $O(2^{2n+1})$ for REA-equivalence $F(x) = M_1 \cdot G(x \oplus V_2) \oplus M_3 \cdot x \oplus V_1$.
The algorithms are compared with previous effective algorithms for solv-
ing the linear and the affine equivalence problem for permutations by
Biryukov et. al [1].

Keywords: EA-equivalence, Matrix Representation, S-box, Vectorial
Boolean Function.

1 Introduction

Vectorial Boolean functions play very important role in providing high-level security for modern ciphers. They are used in cryptography as nonlinear combining
or filtering functions in the pseudo-random generators (stream ciphers) and as
substitution boxes (S-boxes) providing confusion in block ciphers. Up to date
an important question of generation of vectorial Boolean functions with optimal
characteristics to prevent all known types of attacks remains open. Sometimes
equivalence (i.e. EA or CCZ) is used for achieving necessary properties without
losing other ones (i.e. δ-uniformity, nonlinearity) [2].

But very often inverse problem occurs: it is needed to check several functions
for equivalence. For instance, when finding a new vectorial Boolean function it is
necessary to verify whether it is equivalent to already known ones as it happens
with some of block ciphers, where several substitutions are used, (i.e. ARIA [3] or
Kalyna [4,5]). The complexity of exhaustive search for checking EA-equivalence
for functions from $\mathbb{F}_2^n$ to itself equals $O\left(2^{3n^2+2n}\right)$. When $n = 6$ the complexity
is already 2^{120} that makes it impossible to perform exhaustive computing.

In the paper [1] Alex Biryukov et al. have shown that in case when given functions are permutations of $\mathbb{F}_2^n$, the complexity of determining REA-equivalence
equals $O\left(n^2 \cdot 2^n\right)$ for the case of linear equivalence and $O\left(n \cdot 2^{2n}\right)$ for affine
equivalence. In this paper we consider more general cases of REA-equivalence
for functions from $\mathbb{F}_2^n$ to $\mathbb{F}_2^m$ and specify results, when complexity can be reduced
to polynomial. The complexities of our algorithms and the best previous known
ones are given in Table 1.

F. Özbudak and F. Rodríguez-Henríquez (Eds.): WAIFI 2012, LNCS 7369, pp. 108–118, 2012.
© Springer-Verlag Berlin Heidelberg 2012

Table 1. Best Complexities for Solving REA-equivalence Problem

Restricted EA-equivalence	Complexity	m	$G(x)$	Source
$F(x) = M_1 \cdot G(M_2 \cdot x)$	$O\left(n^2 \cdot 2^n\right)$	$m = n$	Permutation	[1]
$F(x) = M_1 \cdot G(M_2 \cdot x \oplus V_2) \oplus V_1$	$O\left(n \cdot 2^{2n}\right)$	$m = n$	Permutation	[1]
$F(x) = M_1 \cdot G(x \oplus V_2) \oplus V_1$	$O\left(2^{2n+1}\right)$	$m \geq 1$	†	Sec. 3
$F(x) = M_1 \cdot G(x \oplus V_2) \oplus V_1$	$O\left(m \cdot 2^{3n}\right)$	$m \geq 1$	Arbitrary	Sec. 3
$F(x) = G(M_2 \cdot x \oplus V_2) \oplus V_1$	$O\left(n \cdot 2^m\right)$	$m \geq 1$	Permutation	Sec. 3
$F(x) = G(x \oplus V_2) \oplus M_3 \cdot x \oplus V_1$	$O\left(n \cdot 2^n\right)$	$m \geq 1$	Arbitrary	Sec. 3
$F(x) = M_1 \cdot G(x \oplus V_2) \oplus M_3 \cdot x \oplus V_1$	$O\left(2^{2n+1}\right)$	$m \geq 1$	‡	Sec. 3
$F(x) = M_1 \cdot G(x \oplus V_2) \oplus M_3 \cdot x \oplus V_1$	$O\left(m \cdot 2^{3n}\right)$	$m \geq 1$	Arbitrary	Sec. 3

† - G is under condition $\{2^i \mid 0 \leq i \leq m - 1\} \subset \text{img}(G')$ where $G'(x) = G(x) + G(0)$.
‡ - G is under condition $\{2^i \mid 0 \leq i \leq m - 1\} \subset \text{img}(G')$ where G' is defined as (4).

2   Preliminaries

For any positive integers n and m, a function F from $\mathbb{F}_2^n$ to $\mathbb{F}_2^m$ is called *differentially δ-uniform* if for every $a \in \mathbb{F}_2^n \setminus \{0\}$ and every $b \in \mathbb{F}_2^m$, the equation $F(x) + F(x + a) = b$ admits at most δ solutions [6]. Vectorial Boolean functions used as S-boxes in block ciphers must have low differential uniformity to allow high resistance to differential cryptanalysis (see [7]). In the important case when $n = m$, differentially 2-uniform functions, called *almost perfect nonlinear* (APN), are optimal (since for any function $\delta \geq 2$). The notion of APN function is closely connected to the notion of *almost bent* (AB) function [8] which can be described in terms of the *Walsh transform* of a function $F : \mathbb{F}_2^n \mapsto \mathbb{F}_2^m$

$$\lambda(u, v) = \sum_{x \in \mathbb{F}_2^n} (-1)^{v \cdot F(x) + u \cdot x},$$

where "$\cdot$" denotes inner products in $\mathbb{F}_2^n$ and $\mathbb{F}_2^m$, respectively. The set $\{\lambda(u, v) \mid (u, v) \in \mathbb{F}_2^n \times \mathbb{F}_2^m, v \neq 0\}$ is called the *Walsh spectrum* of F and the set $\Lambda_F = \{|\lambda(u, v)| \mid (u, v) \in \mathbb{F}_2^n \times \mathbb{F}_2^m, v \neq 0\}$ the *extended Walsh spectrum* of F. If $n = m$ and the Walsh spectrum of F equals $\{0, \pm 2^{\frac{n+1}{2}}\}$ then the function F is called AB [8]. AB functions exist for n odd only and oppose an optimum resistance to linear cryptanalysis (see [9]). Every AB function is APN but the converse is not true in general (see [10] for comprehensive survey on APN and AB functions).

The natural way of representing F as a function from $\mathbb{F}_2^n$ to $\mathbb{F}_2^m$ is by its algebraic normal form (ANF):

$$\sum_{I \subseteq \{1, \ldots, n\}} a_I \left(\prod_{i \in I} x_i\right), \qquad a_I \in \mathbb{F}_2^m,$$

(the sum being calculated in $\mathbb{F}_2^m$). The algebraic degree $deg(F)$ of F is the degree of its ANF. F is called affine if it has algebraic degree at most 1 and it is called linear if it is affine and $F(0) = 0$.

Any affine function $A : \mathbb{F}_2^n \mapsto \mathbb{F}_2^m$ can be represented in matrix form

$$A(x) = M \cdot x \oplus C, \tag{1}$$

where M is an $m \times n$ matrix and $C \in \mathbb{F}_2^m$. All operations are performed in $\mathbb{F}_2$, thus (1) can be rewritten as

$$\begin{pmatrix} a_0 \\ a_1 \\ \cdots \\ a_{m-1} \end{pmatrix}_x = \begin{pmatrix} k_{0,0} & \cdots & k_{0,n-1} \\ k_{1,0} & \cdots & k_{1,n-1} \\ \vdots & \ddots & \vdots \\ k_{m-1,0} & \cdots & k_{m-1,n-1} \end{pmatrix} \cdot \begin{pmatrix} x_0 \\ x_1 \\ \cdots \\ x_{n-1} \end{pmatrix} \oplus \begin{pmatrix} c_0 \\ c_1 \\ \cdots \\ c_{m-1} \end{pmatrix}$$

with $a_i, x_i, c_i, k_{j,s} \in \mathbb{F}_2$.

Two functions $F, G : \mathbb{F}_2^n \mapsto \mathbb{F}_2^m$ are called *extended affine equivalent* (EA-equivalent) if there exist an affine permutation A_1 of $\mathbb{F}_2^m$, an affine permutation A_2 of $\mathbb{F}_2^n$ and a linear function L_3 from $\mathbb{F}_2^n$ to $\mathbb{F}_2^m$ such that

$$F(x) = A_1 \circ G \circ A_2(x) + L_3(x).$$

Clearly A_1 and A_2 can be presented as $A_1(x) = L_1(x) + c_1$ and $A_2(x) = L_2(x) + c_2$ for some linear permutations L_1 and L_2 and some $c_1 \in \mathbb{F}_2^m$, $c_2 \in \mathbb{F}_2^n$.

Definition 1. *Functions F and G are called restricted EA-equivalent (REA-equivalent) if some elements of the set $\{L_1(x), L_2(x), L_3(x), c_1, c_2\}$ are in $\{0, x\}$.*

There are two special cases

- linear equivalence when $\{L_3(x), c_1, c_2\} = \{0, 0, 0\}$;
- affine equivalence when $L_3(x) = 0$.

In matrix form EA-equivalence is represented as follows

$$F(x) = M_1 \cdot G(M_2 \cdot x \oplus V_2) \oplus M_3 \cdot x \oplus V_1$$

where elements of $\{M_1, M_2, M_3, V_1, V_2\}$ have dimensions $\{m \times m, n \times n, m \times n, m, n\}$.

We say that functions F and F' from $\mathbb{F}_2^n$ to $\mathbb{F}_2^m$ are *CCZ-equivalent* if there exists an affine permutation $\mathcal{L}$ of $\mathbb{F}_2^n \times \mathbb{F}_2^m$ such that $G_F = \mathcal{L}(G_{F'})$, where $G_H = \{(x, H(x)) \mid x \in \mathbb{F}_2^n\}, H \in \{F, F'\}$. CCZ-equivalence is the most general known equivalence of functions for which differential uniformity and extended Walsh spectrum are invariants. In particular every function CCZ-equivalent to an APN (respectively, AB) function is also APN (respectively, AB). EA-equivalence is a particular case of CCZ-equivalence [11]. The algebraic degree of a function is invariant under EA-equivalence but, in general, it is not preserved by CCZ-equivalence.

3   Verification of Restricted EA-Equivalence

Special types of REA-equivalence, which are considered in this paper, are shown in Table 2.

Table 2. Special types of REA-equivalence

REA-equivalence	Type
$F(x) = M_1 \cdot G(x) \oplus V_1$	I
$F(x) = G(M_2 \cdot x \oplus V_2)$	II
$F(x) = G(x) \oplus M_3 \cdot x \oplus V_1$	III
$F(x) = M_1 \cdot G(x) \oplus M_3 \cdot x \oplus V_1$	IV

Hereinafter assume that obtaining the value $F(x)$ for any x takes one step. Pre-computed values of function $F(x), F^{-1}(x)$ and corresponding substitutions are used as input for the algorithms. Thereafter, complexity of representing functions in needed form is not taken into account, as well as memory needed for data storage. This assumptions are introduced to be able to compare complexities of algorithms of the present paper with those of [1] where the same assumptions were made.

There are $2^{m \cdot n}$ choices of linear mappings. The complexity of obtaining the $m \times n$ matrix M satisfying the equation

$$F(x) = M \cdot G(x)$$

using exhaustive search method is $O(2^n \cdot 2^{m \cdot n})$, where $O(2^{m \cdot n})$ and $O(2^n)$ are complexities of checking all matrices for all possible $x \in \mathbb{F}_2^n$. Another natural method is based on system of equations. The complexity in this case depends only on the largest of the parameters n and m. Indeed, for square matrices we can benefit from the asymptotically faster Williams method based on system of equations with complexity $O(n^{2.3727})$ [12]. Besides, for $n \leq 64$ we can use 64-bit processor instructions to bring the complexity to $O(n^2)$ because two rows (columns) can be added in 1 step. Since any system of m equations with n variables can be considered as a system of k equations with k variables where $k = \max\{n, m\}$ then the complexity of solving such a system is

$$\mu = O(k^2) \,, \tag{2}$$

which gives the complexity of finding M by this method.

Proposition 1. *Any linear function $L : \mathbb{F}_2^n \mapsto \mathbb{F}_2^m$ can be converted to a matrix with the complexity $O(n)$.*

Proof. We need to find an $m \times n$ matrix M satisfying

$$L(x) = M \cdot x$$

Suppose $\text{rows}_M(i) = (m_{ij})$, $\forall j \in \{0, 1, \ldots, n-1\}$ and $\text{cols}_M(j) = (m_{ij})$, $\forall i \in \{0, 1, \ldots, m-1\}$ are the i-th row and the j-th column of matrix M, respectively. Each value of $x \in \{2^i \mid 0 \leq i \leq n-1\}$ is equivalent to a vector with 1 at the i-th row

$$2^0 = \begin{pmatrix} 1 \\ 0 \\ \cdots \\ 0 \end{pmatrix} \quad 2^1 = \begin{pmatrix} 0 \\ 1 \\ \cdots \\ 0 \end{pmatrix} \quad 2^{n-1} = \begin{pmatrix} 0 \\ 0 \\ \cdots \\ 1 \end{pmatrix}.$$

Clearly, every column, except the i-th, becomes zero when multiplying the matrix M to $x = 2^i$. Hence, each column of matrix M can be computed from

$$L(2^i) = \text{cols}_M(i), \ i \in \{0, 1, \ldots, n-1\}.$$

For finding all columns of M it is necessary to compute n values of $L(2^i)$, $0 \le i \le n-1$. Consequently the complexity of transformation is $O(n)$. $\square$

Proposition 2. *Let* $F, G : \mathbb{F}_2^n \mapsto \mathbb{F}_2^m$ *and* $G'(x) = G(x) \oplus G(0)$. *Then the complexity of checking* F *and* G *for REA-equivalence of type I equals*

- $O(2^{n+1})$ *in case when for any* $i \in \{0, \ldots, m-1\}$ *there exists* $x \in \mathbb{F}_2^n$ *such that* $G'(x) = 2^i$;
- $O(m \cdot 2^{2n})$ *in case* G *is arbitrary.*

Proof. Let $F'(x) = F(x) \oplus F(0)$. Then REA-equivalent of type I

$$F'(x) \oplus F(0) = M_1 \cdot G'(x) \oplus M_1 \cdot G(0) \oplus V_1$$

can be rewritten in the following form

$$\begin{cases} F(0) = M_1 \cdot G(0) \oplus V_1 \\ F'(x) = M_1 \cdot G'(x) \end{cases} . \tag{3}$$

In case of $G(0) = 0$ we get $V_1 = F(0)$, but in general it's necessary first to find M_1 from equation $F'(x) = M_1 \cdot G'(x)$. If the set $\{2^i \mid 0 \le i \le m-1\}$ is the subset of the image set of G', then the problem of finding $m \times m$ matrix M_1 is equivalent to the problem of converting linear function to matrix form with additional testing for all x in $\mathbb{F}_2^n$. It is possible to find M_1 with the complexity $O(m)$ as was shown in Proposition 1. The complexity of finding the pre-images of G' of elements 2^i, $\forall i \in \{0, \ldots, m-1\}$ equals $O(2^n)$ as well as the complexity of checking $F'(x) = M_1 \cdot G'(x)$ for given M_1. In cryptography, in most cases $2^n \gg m$, so the complexity $O(m)$ can be neglected. Therefore the total complexity of verification for equivalence of F and G equals $O(2^n + 2^n + m) \approx O(2^{n+1})$.

Let now G be arbitrary and $F'(x)_i$ be the i-th bit of $F'(x)$. Denote $\text{img}(G')$ the image set of G' and $u_{G'} = |\text{img}(G')|$ the number of elements of $\text{img}(G')$. Let also $N_{G'}$ be any subset of $\mathbb{F}_2^n$ such that $|N_{G'}| = u_{G'}$ and $|\{G'(a)|a \in N_{G'}\}| = u_{G'}$. Then to find M_1 it is necessary to solve a system below for all $i \in \{0, \ldots, m-1\}$

$$F'(x_j)_i = \text{rows}_{M_1}(i) \cdot G'(x_j), \ \forall x_j \in N_{G'}, \ 0 \le j \le u_{G'} - 1 \Leftrightarrow$$

$$\Leftrightarrow \begin{cases} F'(x_0)_i = \text{rows}_{M_1}(i) \cdot G'(x_0) \\ F'(x_1)_i = \text{rows}_{M_1}(i) \cdot G'(x_1) \\ \cdots \\ F'(x_{u_{G'}-1})_i = \text{rows}_{M_1}(i) \cdot G'(x_{u_{G'}-1}) \end{cases} .$$

For every i, $i \in \{0, \ldots, m-1\}$, the complexity of solving the system highly depends on $u_{G'}$ and m and equals $O(\max\{u_{G'}, m\}^2)$ according to (2). Then the total complexity of finding M_1 for all m bits is $O(m \cdot \max\{u_{G'}, m\}^2)$. If value $u_{G'} \approx 2^n$, then $O(m \cdot 2^{2n})$. $\qquad\square$

Remark 1. If it is known in advance that functions F and G in Proposition 2 are REA-equivalent of type I, then the complexity of verification $F'(x) = M_1 \cdot G'(x)$ can be ignored and the total complexity for the case $\{2^i \mid 0 \le i \le m-1\} \subset \operatorname{img}(G')$ becomes $O(2^n)$.

Proposition 3. *Let $F, G : F_2^n \mapsto F_2^n$ and G be a permutation. Then the complexity of checking F and G for REA-equivalence of type II is $O(n)$.*

Proof. Denote $H(x) = G^{-1}(F(x))$. Then the equality $F(x) = G(M_2 \cdot x \oplus V_2)$ becomes

$$H(x) = M_2 \cdot x \oplus V_2 .$$

Taking $x = 0$ we get $V_2 = H(0)$ and the equivalence can be represented as $H'(x) = M_2 \cdot x$, where $H'(x) = H(x) \oplus H(0)$. The method and the complexity of finding n by n matrix M_2 are similar to finding the matrix corresponding to the linear function. Therefore, the complexity equals $O(n)$. $\qquad\square$

Proposition 4. *Let $F, G : F_2^n \mapsto F_2^m$. Then the complexity of checking F and G for REA-equivalence of type III equals $O(n)$.*

Proof. Denote $H(x) = F(x) \oplus G(x)$, then REA-equivalence

$$F(x) = G(x) \oplus M_3 \cdot x \oplus V_1$$

takes the form

$$H(x) = M_3 \cdot x \oplus V_1 .$$

And we have the same situation as in Proposition 3, but with $m \times n$ matrix. Thus the complexity of finding M_3 and V_1 (or showing its non-existence) equals $O(n)$. $\qquad\square$

Every vectorial Boolean function admits the form

$$H(x) = H'(x) \oplus L_H(x) \oplus H(0) , \tag{4}$$

where L_H is a linear function and H' has terms of algebraic degree at least 2.

Proposition 5. *Let $F, G : F_2^n \mapsto F_2^m$ and G' be defined by (4) for G. Then the complexity of checking F and G for REA-equivalence of type IV equals*

- $O(2^{n+1})$ *in case $\{2^i \mid 0 \le i \le m-1\} \subset \operatorname{img}(G')$,*
- $O(m \cdot 2^{2n})$ *in case G is arbitrary.*

Algorithm 1. Checking Functions for REA-equivalence of Type IV
___
Input: $F'(x), L_F(x), F(0), G'(x), L_G(x), G(0)$
Output: True if F is EA-equivalent to G
for $V_2 = 0$ **to** 2^n **do**
   $H'(x) \leftarrow G'(x \oplus V_2)$;
   $L_H(x) \leftarrow L_G(x \oplus V_2)$;
   $H(0) \leftarrow G(V_2)$;
   for $i = 0$ **to** $m - 1$ **do**
      x $\leftarrow 2^i$;
      $find(2^i == G(y))$;
      SetColumn(M_1,i,$H(y)$);
   end for
   $V_1 \leftarrow M_1 \cdot H(0) \oplus F(0)$;
   for $i = 0$ **to** $n - 1$ **do**
      x $\leftarrow 2^i$;
      SetColumn(M_3,i,$L_F(x) \oplus M_1 \cdot L_H(x)$);
   end for
   for $i = 0$ **to** $2^n - 1$ **do**
      if $F(x) \mathrel{!=} M_1 \cdot H(x \oplus V_2) \oplus M_3 \cdot x \oplus V_1$ **then**
         goto next V_2;
      end if
   end for
   return True
end for
return False
___

Proof. Using (4) REA-equivalence of type IV can be rewritten as

$$F'(x) \oplus L_F(x) \oplus F(0) = M_1 \cdot G'(x) \oplus M_1 \cdot L_G(x) \oplus M_3 \cdot x \oplus M_1 \cdot G(0) \oplus V_1$$

and gives the system of equations

$$\begin{cases} F'(x) = M_1 \cdot G'(x) \\ L_F(x) = M_1 \cdot L_G(x) \oplus M_3 \cdot x \\ F(0) = M_1 \cdot G(0) \oplus V_1 \end{cases}$$

It's easy to see that for a given M_1 one can easily compute M_3 and V_1 from the second and the third equations of the system. The first equation of the system leads to the two different cases for the function G' considered in Proposition 2. Hence, according to Proposition 2, the total complexity for finding G' equals $O(2^{n+1})$ and $O(m \cdot 2^{2n})$, respectively. It should be noted that the complexity of finding the matrix M_3 is not taken into account since $2^{n+1} \gg n$. □

If we add one of V_1, V_2 values to REA-equivalence, then the complexity will increase in 2^m or 2^n times respectively. REA-equivalance with V_1, V_2 and corresponding complexities are shown in Table 1. It should be mentioned that types I and III of REA-equivalence are particular cases of type IV. But taking into account different restrictions for the function G it is necessary to check all these types of EA-equivalence.

The presented methods of verification of REA-equivalence were checked using the free open source mathematical software system Sage [13]. An example of a program for the most general case (type IV) of REA-equivalence in case $\{2^i \mid 0 \leq i \leq m - 1\} \subset \operatorname{img}(G')$ is shown in Appendix A. The corresponding algorithm is presented in Algorithm 1.

4   Conclusions

The present paper studies complexities of checking functions for special cases of EA-equivalence and it is shown that for some of this cases the complexity of checking takes polynomial time. Obtained results give a practical method for checking functions on equivalence. The best result is with the complexity $O(2^{2n+1})$ for checking REA-equivalence of the form $F(x) = M_1 \cdot G(x \oplus V_2) \oplus M_3 \cdot x \oplus V_1$ under some condition on G.

References

1. Biryukov, A., De Canniere, C., Braeken, A., Preneel, B.: A Toolbox for Cryptanalysis: Linear and Affine Equivalence Algorithms. In: Biham, E. (ed.) EUROCRYPT 2003. LNCS, vol. 2656, pp. 33–50. Springer, Heidelberg (2003)
2. Daemen, J., Rijmen, V.: The Design of Rijndael. Springer, Heidelberg (2002)
3. Kwon, D.: New Block Cipher: ARIA. In: Lim, J.-I., Lee, D.-H. (eds.) ICISC 2003. LNCS, vol. 2971, pp. 432–445. Springer, Heidelberg (2004)
4. Oliynykov, R., Gorbenko, I., Dolgov, V., Ruzhentsev, V.: Symmetric block cipher "Kalyna". Applied Radio Electronics 6, 46–63 (2007) (in Ukrainian)
5. Oliynykov, R., Gorbenko, I., Dolgov, V., Ruzhentsev, V.: Results of Ukrainian National Public Cryptographic Competition. Tatra Mt. Math. Publ. 47, 99–113 (2010), http://www.sav.sk/journals/uploads/0317154006ogdr.pdf
6. Nyberg, K.: Differentially Uniform Mappings for Cryptography. In: Helleseth, T. (ed.) EUROCRYPT 1993. LNCS, vol. 765, pp. 55–64. Springer, Heidelberg (1994)
7. Biham, E., Shamir, A.: Differential Cryptanalysis of DES-like Cryptosystems. Journal of Cryptology 4(1), 3–72 (1991)
8. Chabaud, F., Vaudenay, S.: Links between Differential and Linear Cryptanalysis. In: De Santis, A. (ed.) EUROCRYPT 1994. LNCS, vol. 950, pp. 356–365. Springer, Heidelberg (1995)
9. Matsui, M.: Linear Cryptanalysis Method for DES Cipher. In: Helleseth, T. (ed.) EUROCRYPT 1993. LNCS, vol. 765, pp. 386–397. Springer, Heidelberg (1994)
10. Carlet, C.: Vectorial Boolean Functions for Cryptography. In: Crama, Y., Hammer, P. (eds.) Chapter of the Monography Boolean Models and Methods in Mathematics, Computer Science, and Engineering, pp. 398–469. Cambridge University Press (2010)
11. Carlet, C., Charpin, P., Zinoviev, V.: Codes, bent functions and permutations suitable for DES-like cryptosystems. Designs, Codes and Cryptography 15(2), 125–156 (1998)
12. Williams, V.V.: Breaking the Coppersmith-Winograd barrier (November 2011), http://www.cs.berkeley.edu/~virgi/matrixmult.pdf
13. Stein, W.A., et al.: Sage Mathematics Software (Version 4.8.2), The Sage Development Team (2012), http://www.sagemath.org

A Source Code for Verification of REA-equivalence of Type IV

```
1  #!/usr/bin/env sage

3  # Global variables
   bits=0
5  length=0
   k=0
7  P=0

9  def check_rEA4(F,G):
     r ' ' '
11     Return True if
        - F(x) = M1 * G(x) + M3 * x + V
13      - G'(x) is permutation, where G(x) = G'(x) + L_G(x) + G(0)
     ' ' '
15   M1 = matrix(GF(2),nrows=bits,ncols=bits)
     M3 = matrix(GF(2),nrows=bits,ncols=bits)
17
     polF = F
19   polG = G

21   V1 = polF.constant_coefficient()
     V2 = polG.constant_coefficient()
23
     polF += V1
25   polG += V2

27   V1 = V1.integer_representation()
     V2 = V2.integer_representation()
29
     polFc=polF.coeffs()
31   polFc += [P("0") for i in xrange(length-len(polFc))]
     polGc=polG.coeffs()
33   polGc += [P("0") for i in xrange(length-len(polGc))]

35   L1 = zero_vector(length).list()
     L2 = zero_vector(length).list()
37
     for i in xrange(bits):
39     if polFc[1<<i] != 0:
         L1[1<<i] = polFc[1<<i]
41       polFc[1<<i] = 0

43     if polGc[1<<i] != 0:
         L2[1<<i] = polGc[1<<i]
45       polGc[1<<i] = 0

47   L1 = P(L1)
     L2 = P(L2)
49   polF = P(polFc)
     polG = P(polGc)
51
     sboxF = range(length)
53   sboxG = range(length)

55   sboxL1 = [L1.subs(k(ZZ(i).digits(2))).integer_representation()
         for i in xrange(length)]
     sboxL2 = [L2.subs(k(ZZ(i).digits(2))).integer_representation()
         for i in xrange(length)]
57   sboxF  = [polF.subs(k(ZZ(i).digits(2))).integer_representation
         () for i in xrange(length)]
     sboxG  = [polG.subs(k(ZZ(i).digits(2))).integer_representation
         () for i in xrange(length)]
59
```

```
      sboxFt=sboxF [:]
61    sboxGt=sboxG [:]

63    if len(set(sboxG).intersection(set([2^g for g in xrange(bits)])
         )) != bits:
         #print ">>> sboxG hasn't all values of {0} <<<".format([2^g
            for g in xrange(bits)])
65       return None

67    for i in xrange(bits):
         x=sboxGt.index(1<<i)
69       M1.set_column(i,ZZ(sboxFt[x]).digits(base=2,padto=bits))

71    sboxM = range(length)

73    V = ZZ((M1*vector(GF(2),ZZ(V2).digits(base=2,padto=bits))).list
         (),2) ^^ V1
      for i in xrange(length):
75       sboxM[i] = sboxL1[i] ^^ ZZ((M1*vector(GF(2),ZZ(sboxL2[i]).
            digits(base=2,padto=bits))).list(),2) ^^ V

77    sboxT=sboxM [:]

79    V = vector(GF(2),ZZ(sboxT[0]).digits(base=2,padto=bits))

81    if sboxT[0] != 0:
         sboxT = [ g^^sboxT[0] for g in sboxT ]
83
      for i in xrange(bits):
85       x=1<<i
         M3.set_column(i,ZZ(sboxT[x]).digits(base=2,padto=bits))
87
      sbox = range(length)
89
      sF   = [F.subs(k(ZZ(i).digits(2))).integer_representation() for
            i in xrange(length)]
91    sG   = [G.subs(k(ZZ(i).digits(2))).integer_representation() for
            i in xrange(length)]

93    for i in xrange(length):
         sbox[i]=vector(GF(2),ZZ(sG[i]).digits(base=2,padto=bits))
95
         sbox[i]=M1*sbox[i]
97
         tx=M3*vector(GF(2),ZZ(i).digits(base=2,padto=bits))
99
         sbox[i]=vector(GF(2),[ZZ(sbox[i].get(j)) ^^ ZZ(tx.get(j)) ^^
            ZZ(V.get(j)) for j in xrange(len(sbox[i]))])
101
         sbox[i]=ZZ(sbox[i].list(),2)
103
      if sbox == sF:
105      return [M1,M3,V]
      else:
107      return None

109 def is_EA_equivalent(F,G,functions):

111   for v2 in xrange(length):
         polG=G.subs(P("x+{0}".format(k(ZZ(v2).digits(2))))).mod(P("x
            ^{0}+x".format(length)))
113
         ret=check_rEA4(F,polG)
115
         if ret != None:
117         M1=ret[0]
            M3=ret[1]
119         V1=ret[2]
```

```
121          V2=vector(GF(2),ZZ(v2).digits(base=2,padto=bits))
             if functions == True:
                 return [M1,V1,V2,M3]
123          else:
                 return True
125
         return False
127
     def main(argv=None):
129      global bits,length,k,P

131      bits=6
         length=1<<bits
133      k=GF(2^bits,'a')
         P=PolynomialRing(k,'x')
135
         F=P.random_element(length-1)
137      G=P.random_element(length-1)

139      # Test polynomials for bits=6
         #G=P("a^63*x^0 + a^61*x^1 + a^23*x^2 + a^39*x^3 + a^15*x^4 + a
             ^21*x^5 + a^57*x^6 + a^37*x^7 + a^3*x^8 + a^23*x^9 + a^26*x
             ^10 + a^40*x^11 + a^48*x^12 + a^26*x^13 + a^51*x^14 + a^43*
             x^15 + a^32*x^16 + a^13*x^17 + a^33*x^18 + a^48*x^19 + a
             ^36*x^20 + a^1*x^21 + a^11*x^22 + a^40*x^23 + a^42*x^24 + a
             ^62*x^25 + a^11*x^26 + a^22*x^27 + a^5*x^28 + a^6*x^29 + a
             ^59*x^30 + a^10*x^31 + a^51*x^32 + a^4*x^33 + a^13*x^34 + a
             ^63*x^35 + a^54*x^36 + a^26*x^37 + a^58*x^38 + a^39*x^39 +
             a^53*x^40 + a^34*x^41 + a^28*x^42 + a^27*x^43 + a^40*x^44 +
             a^25*x^45 + a^10*x^46 + a^58*x^47 + a^30*x^48 + a^34*x^49
             + a^35*x^50 + a^49*x^51 + a^53*x^52 + a^35*x^53 + a^49*x^54
             + a^7*x^55 + a^55*x^56 + a^39*x^57 + a^53*x^58 + a^29*x^59
             + a^52*x^60 + a^45*x^61 + a^9*x^62 + a^26*x^63")
141      #F=P("a^44*x^0 + a^34*x^1 + a^7*x^2 + a^5*x^3 + a^51*x^4 + a
             ^40*x^5 + a^27*x^6 + a^23*x^7 + a^28*x^8 + a^63*x^9 + a^20*
             x^10 + a^38*x^11 + a^12*x^12 + a^16*x^13 + a^18*x^14 + a
             ^39*x^16 + a^53*x^17 + a^62*x^18 + a^17*x^19 + a^50*x^20 +
             a^13*x^21 + a^15*x^22 + a^29*x^23 + a^33*x^24 + a^12*x^25 +
             a^22*x^26 + a^49*x^27 + a^7*x^28 + a^43*x^29 + a^28*x^30 +
             a^53*x^31 + a^5*x^32 + a^59*x^33 + a^22*x^34 + a^26*x^35 +
             a^45*x^36 + a^39*x^37 + a^49*x^38 + a^9*x^39 + a^58*x^40 +
             a^13*x^41 + a^14*x^42 + a^43*x^43 + a^61*x^44 + a^38*x^45
             + a^10*x^46 + a^9*x^47 + a^25*x^48 + a^44*x^49 + a^30*x^50
             + a^12*x^51 + a^16*x^52 + a^24*x^53 + a^56*x^54 + a^3*x^55
             + a^40*x^56 + a^23*x^57 + a^49*x^58 + a^39*x^59 + a^58*x^60
             + a^11*x^61 + a^55*x^62 + a^29*x^63")

143      print "F\t= {0}".format(F)
         print "G\t= {0}".format(G)
145
         ret=is_EA_equivalent(F,G,functions=True)
147
         if ret != False:
149          [M1,V1,V2,M3]=ret
             print "EA\t\t\t\t= {0}".format(True)
151          print "V1:\n{0}".format(V1)
             print "V2:\n{0}".format(V2)
153          print "M1:\n{0}".format(M1)
             print "M3:\n{0}".format(M3)
155      else:
             print "EA\t\t\t\t= {0}".format(False)
157
     if __name__ == "__main__":
159      sys.exit(main())
```

Software Implementation of Modular Exponentiation, Using Advanced Vector Instructions Architectures

Shay Gueron[1, 2] and Vlad Krasnov[2]

[1] Department of Mathematics, University of Haifa, Israel
[2] Intel Corporation, Israel Development Center, Haifa, Israel

Abstract. This paper describes an algorithm for computing modular exponentiation using vector (SIMD) instructions. It demonstrates, for the first time, how such a software approach can outperform the classical scalar (ALU) implementations, on the high end x86_64 platforms, if they have a wide SIMD architecture. Here, we target speeding up RSA2048 on Intel's soon-to-arrive platforms that support the AVX2 instruction set. To this end, we applied our algorithm and generated an optimized AVX2-based software implementation of 1024-bit modular exponentiation. This implementation is seamlessly integrated into OpenSSL, by patching over OpenSSL 1.0.1. Our results show that our implementation requires 51% less instructions than the current OpenSSL 1.0.1 implementation. This illustrates the potential significant speedup in the RSA2048 performance, which is expected in the coming (2013) Intel processors. The impact of such speedup on servers is noticeable, especially since migration to RSA2048 is recommended by NIST, starting from 2013.

Keywords: modular arithmetic, modular exponentiation, Montgomery multiplication, RSA, SIMD, AVX, AVX2.

1   Introduction

The cryptographic algorithms that underlie SSL/TLS connections are a critical computational load for the supporting servers. As the major ingredient in the SSL/TLS handshake, RSA is an important factor, and NIST's recommendation for key-lengths [1], makes RSA2048 an important optimization target.

Currently, software implementations of RSA are "scalar code" that use ALU instructions (e.g., *ADD/ADC/MUL*). Improvements in the performance of *ADD/ADC/MUL* have made the scalar implementations very efficient on the modern x86_64 processors (see [7], [5] for details).

In this paper, we show that RSA software can gain significant performance by "vectorized" implementations that utilize the modern SIMD (vector) architectures. This is done by implementing a version of the RSAZ algorithm (short for RSA ZARIZ - Hebrew for "quick") [7]. SIMD (aka vector architecture), an acronym for "Single Instruction Multiple Data", is an architecture where a single instruction computes a function of several inputs, simultaneously (see e.g., [9]). These inputs are called "elements" and reside in registers that hold a few of them together. Early SIMD

F. Özbudak and F. Rodríguez-Henríquez (Eds.): WAIFI 2012, LNCS 7369, pp. 119–135, 2012.
© Springer-Verlag Berlin Heidelberg 2012

architectures used the MMX instructions that operate on *64*-bit SIMD registers. It was followed by the SSE architecture that introduced *128*-bit registers. The Advanced Vector Extensions (AVX) extends the SSE architecture in several respects, e.g., by introducing non-destructive destination and floating point operations over *256*-bit registers [11]. AVX2 is the latest SIMD architecture. It has been recently disclosed, and will be first introduced in the next architecture (Codename "Haswell") in 2013 [12]. AVX2 includes sixteen *256*-bit registers (called YMM's), each one is capable of holding eight *32*-bit elements, or four *64*-bit elements. It also offers new integer instructions that operate on these wide registers.

Many algorithms (often in media processing, e.g., DCT) operate on multiple independent elements, and are therefore inherently suitable for SIMD architectures. However, big-numbers (multi-digit) arithmetic, which is in the heart of RSA computations, is not naturally suitable for vector architectures: the digits of the multi-digit numbers are not independent, due to carry propagation during arithmetic operations such as addition and multiplication.

We offer here an efficient method for using SIMD architecture for big-numbers arithmetic, in particular for modular exponentiation. Some attempts to use SIMD for big-numbers arithmetic (and RSA) have been made. For example, Page and Smart [19] suggested using SIMD architectures to calculate several exponentiations in parallel, and using a "redundant Montgomery representation" (which we call here Non Reduced) to avoid conditional final subtractions in Montgomery Multiplications. Lin [16] implemented a *128*-by-*128* bits integer multiplication function using SIMD instructions on Freescale's e600 *32*-bit processor, and used it as a building block for larger multiplications. Reference [10] suggests converting the big-numbers to numbers that have only *29*-bit digits, and use SIMD operations for multiplying them. Such a method is also used in [2] for prime field ECC. This is and underlying idea in this paper as well, although we do not use it (directly) for integer multiplications.

We use our algorithms for efficiently computing (a variant of) Montgomery Multiplications (and Squaring). The novelty in our approach includes the balancing of the computational workload between the SIMD and the ALU units, in an efficient manner. This resolves the bottlenecks that exist in a purely SIMD or purely ALU implementation.

2   Preliminaries

2.1   The RSA Context

In this paper, we discuss RSA cryptosystem with a *2n*-bit modulus, $N = P \times Q$, where P and Q are *n*-bit primes. Let the *2n*-bit private exponent be d. Decryption of a *2n*-bit message C requires one *2n*-bit modular exponentiation $C^d \bmod N$. To use the Chinese Remainder Theorem (CRT), the following quantities are pre-computed: $d_1 = d \bmod (P\text{-}1)$, $d_2 = d \bmod (Q\text{-}1)$, and $Qinv = Q^{-1} \bmod P$. Then, two *n*-bit modular exponentiations, namely $M_1 = C^{d1} \bmod P$ and $M_2 = C^{d2} \bmod Q$, are computed (M_1, M_2, d_1, d_2 are *n*-bit integers). The results are recombined by using $C^d \bmod$

$N = M_2 + (Qinv \times (M_1 - M_2) \bmod P) \times Q$. Using the CRT, the computational cost of a $2n$-bit RSA decryption is well approximated by twice the computational cost of one n-bit modular exponentiations. In our context, we can assume that by construction (of the RSA keys), $2^{n-1} < P, Q < 2^n$.

2.2   The Non Reduced Montgomery Multiplication

For a detailed description of software implementation of modular exponentiation, and the resulting performance, we refer the readers to [7]. In general, the critical building block in modular exponentiation computations is modular multiplication, or an equivalent. Here, we use the Non Reduced Montgomery Multiplication (*NRMM*), as defined in [4] (see also [20], [22]). *NRMM* is a variation of the well known Montgomery Multiplication (MM hereafter; see also [3], [14], [15], [21])

Definition 1. [Montgomery Multiplication] Let M be an odd integer (modulus), a, b be two integers such that $0 \le a$, $b < M$, and t be a positive integer (hereafter, all the variables are non-negative integers). The Montgomery Multiplication of a by b, modulo M, with respect to t, is defined by $MM(a, b) = a \times b \times 2^{-t} \bmod M$.

Definition 2. [Non Reduced Montgomery Multiplication] Let M be an odd integer (modulus), a, b be two integers such that $0 \le a$, $b < 2M$, and t be *a* positive integer such that $2^t > 4M$. The Non Reduced Montgomery Multiplication of a by b, modulo M, with respect to t, is defined by

$$NRMM\ (a,b) = \frac{a \times b + \left(\left(-M^{-1} \times a \times b\right) \bmod 2^t\right) \times M}{2^t} \qquad (1)$$

We say that 2^t is the Montgomery parameter. For the Non Reduced Montgomery Square, we denote $NRMM\ (a, a) = NRMSQR\ (a)$.

The following lemma shows how *NRMM* can be used, similarly to *MM*, for efficient computations of modular exponentiation.

Lemma 1. Let M be an odd modulus a, b be two integers such that $0 \le a$, $b < 2M$, and t be a positive integer such that $2^t > 4M$. Then, a) $NRMM(a, b) < 2M$; b) $NRMM(a, 1) < M$.

Proof. To prove part a, we write

$$NRMM\ (a,b) < \frac{a \times b + 2^t \times M}{2^t} < \frac{2M \times 2M}{2^t} + M < \frac{4M^2}{4M} + M = 2M \qquad (2)$$

To prove part b, we write

$$NRMM\ (a,1) < \frac{a + 2^t \times M}{2^t} < \frac{2M}{2^t} + M < \frac{1}{2} + M \qquad (3)$$

The last inequality follows from $2^t > 4M$ and $a < 2M$. Therefore, $NRMM$ $(a, 1)$ is fully reduced modulo M.

Remark 1. Lemma 1 (part a) shows the "stability" of $NRMM$: the output of one $NRMM$ can be used as an input to a subsequent $NRMM$.

Remark 2. Since $NRMM(a, b)$ mod $M = MM(a, b)$, it follows, from the bound in Lemma 1, that $NRMM(a, b)$ is either $MM(a, b)$ or $MM(a, b)+M$.

Remark 3. Suppose that $0 \le a, b < 2M$, $c2 = 2^{2t}$ mod M, $a` = NRMM(a, c2)$, $b` = NRMM(b, c2)$, $u` = NRMM(a`, b`)$, and $u = NRMM(u`, 1)$. Then, Lemma 1 implies that $a`, b`, u`$ are smaller than $2M$, and $u = a \times b$ mod M.

These remarks indicate how $NRMM$ can be used for computing modular exponentiation in a way that is similar to the way in which MM is used. For a given modulus M, the constant $c2 = 2^{2t}$ mod M can be pre-computed. Then, a^x mod M (for $0 \le a < M$ and some integer x) can be computed by: a) Mapping the base (a) to the (non reduced) Montgomery domain, $a' = NRMM$ $(a, c2)$ b) Using an exponentiation algorithm while replacing modular multiplications with $NRMM$'s c) Mapping the result back to the residues domain, $u = NRMM$ $(u', 1)$.

2.3   The Relevant AVX2 Instructions

The AVX2 vector operations we use in our context are (see [12] for details):

- *VPADDQ* – addition of four *64*-bit integer values, from one YMM register and four *64*-bit values from another YMM register, and storing the result in a third YMM register.
- *VPMULUDQ* – multiplication of four *32*-bit unsigned integer values, from one YMM register, by four *32*-bit values from another YMM register, producing four 64-bit products into a third YMM register.
- *VPBROADCASTQ* – copying a given *64*-bit value, four times, to produce a YMM register with four equal elements (with that value).
- *VPERMQ* – Permutes *64*-bit values inside a YMM register, according to an *8*-bit immediate value.
- *VPSRLQ/VPSLLQ* – Shift *64*-bit values inside a YMM register, by an amount specified by an *8*-bit immediate value.

3   Modular Exponentiation with Vector Instructions

SIMD instructions are designed to repeat the same operation on *independent* elements stored in a register. Therefore, it has an inherent difficulty with efficiently handling carry propagation associated with big-numbers arithmetic. As an example, the carry propagation in a (*2* digit) × (*3* digit) multiplication, is illustrated in Fig. 1.

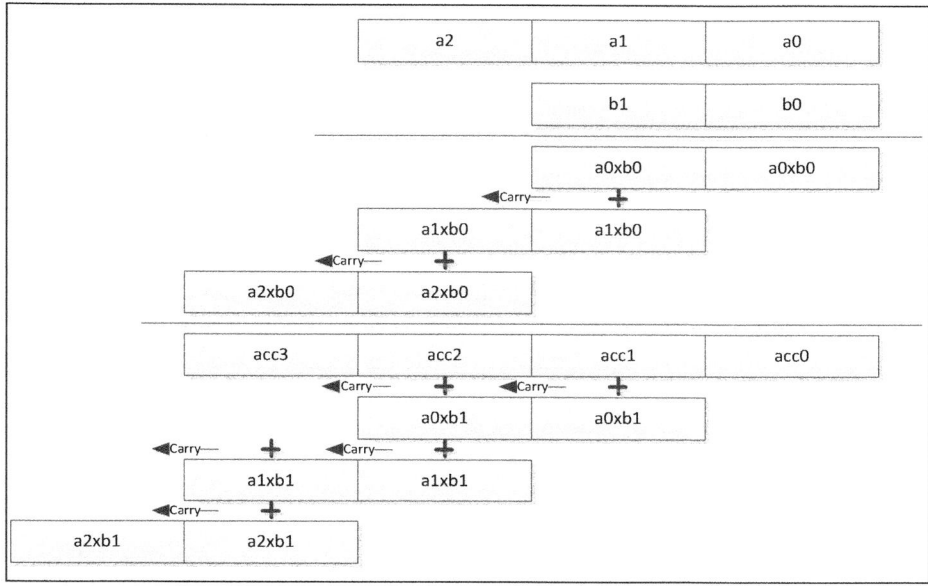

Fig. 1. Illustration of the carry propagation during a multiplication of the two integers A (3 digits) and B (2 digits). The schoolbook method is used. Each digit of A is multiplied by each digit of B, and the appropriate sub-products are aligned and summed accordingly. The digits of the partial sums are not independent of each other, due to the carry propagation, and this is why SIMD architectures would require some cumbersome manipulations to handle such a flow.

3.1   Redundant Representation

Modular exponentiation can be translated to a sequence of *NRMM/NRMSQR*'s. We show here how to optimize these operations, using vector instructions. The underlying idea is to always operate on "small" elements (i.e., less than 2^{32}). This allows two big-numbers products to be summed up, without causing an overflow inside the *64*-bit "container" that holds the digits of the accumulator. The cumbersome handling of the carry propagation can therefore be avoided. To this end, we define here an alternative representation of long (multi-digits) integers.

Let A be an *n*-bit integer, written in a radix 2^{64} as an *l*-digits integer, where $l = \lceil n/64 \rceil$, and where each *64*-bit digit a_i satisfies $0 \le a_i < 2^{64}$. This representation is unique. Consider a positive m such that $1 < m < 64$. We can write A in radix 2^m as

$$A = \sum_{i=0}^{k-1} x_i \times 2^{m \times i}$$. This representation is unique, and requires $k = \lceil n/m \rceil > l$ digits, x_i,

satisfying $0 \le x_i < 2^m$, for $i = 0, ..., k\text{-}1$. See Fig. 2 for an example.

If we relax the requirement $0 \le x_i < 2^m$, and allow the digits to satisfy only the inequality $0 \le x_i < 2^{64}$, we say that A is written in a Redundant-radix-2^m Representation (redundant representation for short). This representation is not unique. Figuratively speaking, the redundant representation is simply "embedding the digits of a number in a big container".

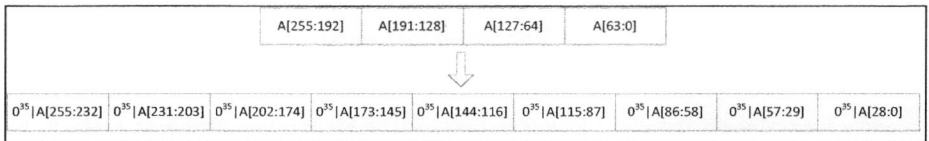

Fig. 2. A 256-bit integer in radix 2^{64} ($n=256$, $l=4$) written in radix 2^{29} ($m=29$) using 9 digits ($k = ceil(n/m) = 9$). Each digit can be stored as a 64-bit "element" in a vector of $k=9$ elements.

An integer A, written with k digits in redundant representation, satisfying $A < 2^{m \times k}$, can be converted to a radix 2^m representation with the *same* number of digits (k), as shown in Algorithm 1 (Fig. 3).

```
Algorithm 1: Redundant-to-2ᵐ
Input: U in redundant representation using k digits(assumption: U < 2ᵐˣᵏ)
Output: U in radix 2ᵐ representation
Flow:
   1.  temp = 0
   2.  For i = 0 to k-1
            a.  temp = temp + uᵢ
            b.  vᵢ = temp mod 2ᵐ
            c.  temp = temp/2ᵐ
   End for
   Return V
```

Fig. 3. Redundant-to-2^m conversion (see explanation in the text)

Example 1: take $n=1024$ and the 1024-bit number $A = 2^{1024}-105$. It has $l = 16$ digits in radix 2^{64}. The least significant digit is $0xFFFFFFFFFFFFFF97$, and the other 15 digits are $0xFFFFFFFFFFFFFFFF$. With $m = 28$, A becomes a number with $k = 37$ digits in radix 2^{28} the least significant digit is $0xFFFFF97$, the most significant digit is $0xFFFF$, and the rest are $0xFFFFFFF$. For $m=29$, A becomes a number with $k=36$ digits in radix 2^{29} the least significant digit is $0x1FFFFF97$ the most significant digit is $0x1FF$ and the rest are $0x1FFFFFFF$.

Operations on integers that are given in redundant representation can be "vectorized", as follows:

Let X and Y be two numbers given in redundant representation, such that $X = \sum_{i=0}^{k-1} x_i \times 2^{m \times i}$, $Y = \sum_{i=0}^{k-1} y_i \times 2^{m \times i}$, with $0 \le x_i, y_i < 2^{64}$. Let $t > 0$ be an integer.

Addition: If $x_i + y_i < 2^{64}$ for $i = 0, 1, \ldots, (k-1)$, then the sum $X + Y$ is given, in redundant representation, by $X + Y = \sum_{i=0}^{k-1} z_i \times 2^{m \times i}$ with $z_i = x_i + y_i; 0 \le z_i < 2^{64}$.

Multiplication by constant: If $x_i \times t < 2^{64}$ for $i = 0, 1, \ldots, (k-1)$, then the product $t \times X$ is given by $t \times X = \sum_{i=0}^{k-1} z_i \times 2^{m \times i}$ with $z_i = t \times x_i; 0 \le z_i < 2^{64}$.

To illustrate, we provide the following example.

Example 2: *n=256, m=29.*

```
A = fa5401a8593c981b||fd42a2802a750928||e930850d63bc2c5f||da8d4ca9655091ad
B = eb9a100d6e586233||50608103451895a2||5572dfe2de045f13||132ba675e3adb497
A in radix 2^29 =
0000000000fa5401||00000000150b2793||00000000006ff50a||000000001140153a||0000000010
928e93||00000000010a1ac7||000000000f0b17f6||00000000146a654b||00000000055091ad
B in radix 2^29 =
0000000000eb9a10||0000000001adcb0c||0000000008cd4182||000000000081a28c||0000000009
5a2557||0000000005bfc5bc||000000000117c4c4||00000000195d33af||0000000003adb497
A + B in redundant representation =
0000000001e5ee11||0000000016b8f29f||00000000093d368c||0000000011c1b7c6||0000000019
ecb3ea||0000000006c9e083||000000001022dcba||000000002dc798fa||0000000008fe4644
A + B in radix 2^29 =
0000000001e5ee11||0000000016b8f29f||00000000093d368c||0000000011c1b7c6||0000000019
ecb3ea||0000000006c9e083||000000001022dcbb||000000000dc798fa||0000000008fe4644
t = 0x00000001fedcba98
t × A in redundant representation =
01f38b30a4869a98||29fe5dc375b44d48||00df6ab21b1ac1f0||226c89ee9750be70||2112420fb2
ef7548||021306c96a787c28||1e05123ba966f610||28bd901be338a288||0a9b1753885a30b8
t × A in radix 2^29 =
000000000f9c598f||00000000147988b3||000000001cafa2e1||000000000e7f116c||000000001f
e2ceee||000000000387ab9a||000000001aa10e0f||000000000f5376f1||0000000018115d24||00
000000085a30b8
```

In Fig. 5, we illustrate how redundant representation helps delaying the carry propagation to the last stage of the multiplication.

3.2   NRMM

NRMM can be computed in a "word-by-word" style, as shown in Algorithm 2, Fig. 4.

```
Algorithm 2: Word-by-word computation of NRMM (WW-NRMM)
Input:
M, an odd modulus such that 2^(n-1) < M < 2^n (M has n bits in binary form)
Integer 1 < m < 64, such that if k = ⌈n/m⌉, then k×m>n+2.
A, B < 2M < 2^(n+1) given in radix 2^m (a_i, b_i denote their radix 2^m digits)
Pre-computed:k_0 = -M^(-1) mod 2^m
Output: X = NRMM (A, B)
Flow:
   1. X = 0
   2. For i = 0 to k-1
      2.1. X = X + A×b_i
      2.2. x_0 = X mod 2^m
      2.3. y = x_0×k_0 mod 2^m
      2.4. X = X + M×y
      2.5. X = X/2^m
   3. Output X
```

Fig. 4. Word-by-word computation of NRMM

Proof of correctness: note that $X = (A \times B + ((-M^{-1} \times A \times B) \mod 2^{k \times m}) \times M)/2^{k \times m}$. Therefore, (a) follows immediately and (b) follows from Lemma 1. Note also that if *B* = *1*, then *X mod M* = $A \times B \times 2^{-l \times m}$ exactly. In step 2.5, *X* is divisible by 2^m due to steps 2.4-2.4 and the definition of k_0.The number of digits in the final result remains unchanged (*k*), because the result $X<2^{n+1}$, and $k \times m>n+2$.

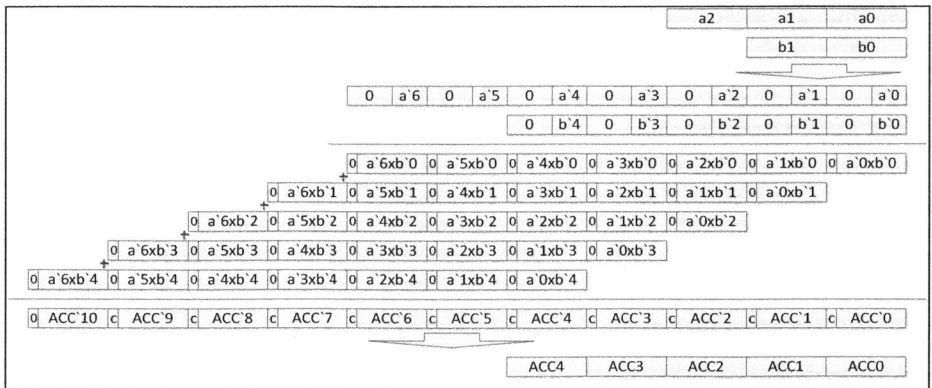

Fig. 5. Illustration of carry accumulation during a multiplication of two integers, A and B, using vectorized computations (compare to Fig. 1). Suppose that $m=29$. A (3 digits) and B (2 digits) are first converted into redundant representation in radix 2^{29}. In this representation, they have 7 and 5 digits, where each digit is smaller than 2^{29}, and is stored in a 64-bit "container" (SIMD element). Then, the sub-products are accumulated, while the carry bits spill into the "empty space" inside the 64-bit container (which is initially 0). In the end, the result is "normalized" back to a standard 2^m representation, according to Algorithm 1. This result can be fed to a consecutive multiplication, or transformed back to radix 2^{64}.

Remark 4. In the redundant representation, steps 2.1 and 2.4 can be computed efficiently using vector instructions (*VPMULUDQ*). Steps 2.2 and 2.3 can be computed efficiently using scalar instructions, because they operate on a single digit.

The above remark indicates that Algorithm 2 is useful for computing *NRMM* via a mix of scalar and vector instructions. Indeed, our implementation uses scalar instructions to compute the low digits of *X*, and vector instructions for the other computations, described in Algorithm 3.

Algorithm 3, for computing *NRMM* (with the parameter *m*) assumes the inputs *A*, *B* and *M* are represented in redundant form, with each digit is strictly smaller than 2^m (we call it "normalized redundant representation").

Since we want to integrate such an *NRMM* implementation into a "standard" implementation, we need to transform our input/output from/to a regular (radix 2^{64}) representation. The cost of such transformation is only few tens of cycles, but if done for every *NRMM*, it can add up to a noticeable overhead. For efficiency, we transform the inputs to a redundant form in the beginning of the exponentiation; carry out all the operations in the redundant form, during the entire exponentiation; in the end, transform the result back to the standard representation. This makes the overhead of the to-and-from transformation negligible, while keeping a standard interface for the exponentiation function (transparent to the user).

```
Algorithm 3 [VNRMM]: Vectorized implementation of NRMM(A,B)
Input: A, B and M, in radix 2^m
Pre-computed: k_0 = -M^{-1} mod M
Output: X = NRMM (A, B)
Flow:
1.  x_0 = 0, X_q, ...,X_0 = 0
2.  a_0 = A mod 2^m (i.e., digit 0 of A)
3.  m_0 = M mod 2^m (i.e., digit 0 of M)
4.  load digits 1, 2, ..., (k-1) of A into SIMD registers A_1..A_q (q as
    required)
5.  load digits 1, 2, ..., (k-1) of M into SIMD registers M_1..M_q (q as
    required)
6.  addCounter = 0
7.  for i = 0 to k-1
    7.1. x_0 = x_0 + a_0 × b_i
    7.2. T = Broadcast b_i
    7.3. for j = 1 to q
            7.3.1. X_j = X_j + A_j × T
    7.4. y_0 = x_0 × k_0 mod 2^m
    7.5. x_0 = x_0 + m_0 × y_0
    7.6. T = Broadcast y_0
    7.7. for j = 1 to q
            7.7.1. X_j = X_j + M_j × T
    7.8. x_0 = x_0 >> m
    7.9. x_0 = x_0 + X_1[0]
    7.10. X_q,...,X_1 = X_q,...,X_1 >> 64
    7.11. addCounter = addCounter + 2
    7.12. if addCounter ≥ (2^{64-2m})
            7.12.1. perform X "cleanup"
            7.12.2. addCounter = 0
End for
8.  Convert X_q,..,X_1, x_0: from redundant representation to radix 2^m
                            (using Algorithm 1)
```

Fig. 6. Computing NRMM using combination of vector and scalar operations. In order to avoid carry overflow, a "cleanup" procedure may be initiated, converting X to normalized 2^m representation.

3.3   Modular Exponentiation Using VNRMM

We note that a windowed method modular exponentiation requires judicious preparation and use of tables (see [7] for details). In the vectorized implementations, these tables are larger than the tables that are used for the scalar implementation (that uses *64*-bit digits), because the redundant representation requires more than twice as many digits. However, since the top *64-m* bits of each digit are zeroed in the end of each NRMM call, and therefore do not need to be stored, we can decrease the size of the required tables, to some extent.

We briefly mention that in order to be side-channel protected, our implementation operates in "constant time": the memory access patterns (and timing) do not depend on the secret exponent. This is achieved (among other factors) by holding the tables in a way that a portion of each entry lies in a portion of each cache-line that is used by the table. Details on choosing the optimal table size and on implementing side-channel protected table access, is provided in [7].

```
Algorithm 4: w-ary modular exponentiation using VNRMM
Input: A, X and M - n-bit integers, in radix 2⁶⁴ representation
Pre-computed: k₀ = -M⁻¹ mod 2⁶⁴, RR = 2²ⁿ mod M, w - window size
Output: C = Aˣ mod M
Flow:
1.  Let m be the largest integer such that 2⁶⁴⁻²ᵐ > 2×⌈n/m⌉
2.  Let A`, RR` and M` be A, RR and M converted to normalized radix 2ᵐ
3.  Let X be x0 + x1×2ʷ + … + xj×2ʲʷ, where 0≤x0,x1…xj<2ʷ
4.  Let k₀` = k₀ mod 2ᵐ = -M⁻¹ mod 2ᵐ
5.  C2 = VNRMM(RR`, RR`) (congruent to 2⁴ⁿ⁻ᵏˣᵐ mod M)
6.  C2 = VNRMM(C2, 4k×m - 4n) (congruent to 2²ᵏˣᵐ mod M)
7.  Table[0] = VNRMM(C2, 1)
8.  Table[1] = VNRMM(C2, A`)
9.  For i = 2,..,2ʷ⁻¹-1 do
        9.1.    Table[i×2]=VNRMM(M[i], M[i])
        9.2.    Table[i×2]=VNRMM(M[i×2], M[1])
    End for
10. h = m[xj]
11. For i=j-1,…0 do
        11.1.   For l = 1,…,w
                    11.1.1. h = VNRMM(h,h)
                End for
        11.2.   h = VNRMM(h, xi)
    End for
12. h = VNRMM(h, 1)
13. hh = radix-2ᵐ-to-radix-2⁶⁴(h)
14. Return hh

 •  The access to the Table is side channel protected
```

Fig. 7. The w-ary modular exponentiation, using VNRMM

4   Implementation, Choice of Parameters and Optimizations

4.1   Choice of Parameters

For our usages, we are mainly interested in $n=512$, 1024, 2048, (for RSA1024, RSA2048, RSA4096, respectively). We choose $m=29$ if $n=512$ or $n=1024$, and $m=28$ when $n=2048$.

To explain this choice of parameters, we first point out that $m = 29$ for $n = 1024$ is larger than the value of m that is specified in Step 1 of Algorithm 4 (namely $m=28$). Indeed, the correctness of Algorithm 4 can be maintained with different choices of m, as long as the "cleanup" steps are properly applied, to prevent any overflows beyond the range allowed by the 64 bits container. The tradeoff is clear: a large m decreases the number of digits of the operands – which improves the efficiency of the computations. On the other hand, it requires a more frequent "cleanup", because a fewer "spare" bits are left for accumulation.

In our case, where $n=1024$, choosing $m=29$ leads to 36 digits operands, which results in 58-bit products, and leaves only 6 "spare" bits for carry-accumulation. Therefore, cleanup is required after $2^6=64$ accumulations, that is, every 32 iterations of the loop (Step 7) in Algorithm 3. For $n=1024$, this loop repeats 36 iterations, so the cleanup is required only once. With $m=28$, we have 37 digits operands, and 8 "spare" bits, therefore the cleanup is required every *128* iterations of the loop, allowing exponents of up to *3584* bits without any cleanup.

In our implementation, we optimize the cleanup step (shaving off only the necessary number of bits), and make $m=29$ the preferable parameter choice.

4.2   Why Is the AVX2 Architecture Sufficient for an Efficient Vectorized Implementation?

We explain here why AVX2 is the first SIMD architecture that can support vectorized *NRMM* implementation that can outperform the scalar implementation.

For simplicity, we consider a schoolbook scenario, and count the operations and the tradeoffs. Computing *NRMM* in redundant representation requires $2 \times \lceil n/m \rceil^2$ single precision multiplications. Similarly, the scalar implementation (in radix 2^{64}) requires $2 \times \lceil n/64 \rceil^2$ single precision multiplications. However, *NRMM* in redundant representation requires only one single precision addition per multiplication, whereas the scalar implementation requires three single precision add-with-carry operations.

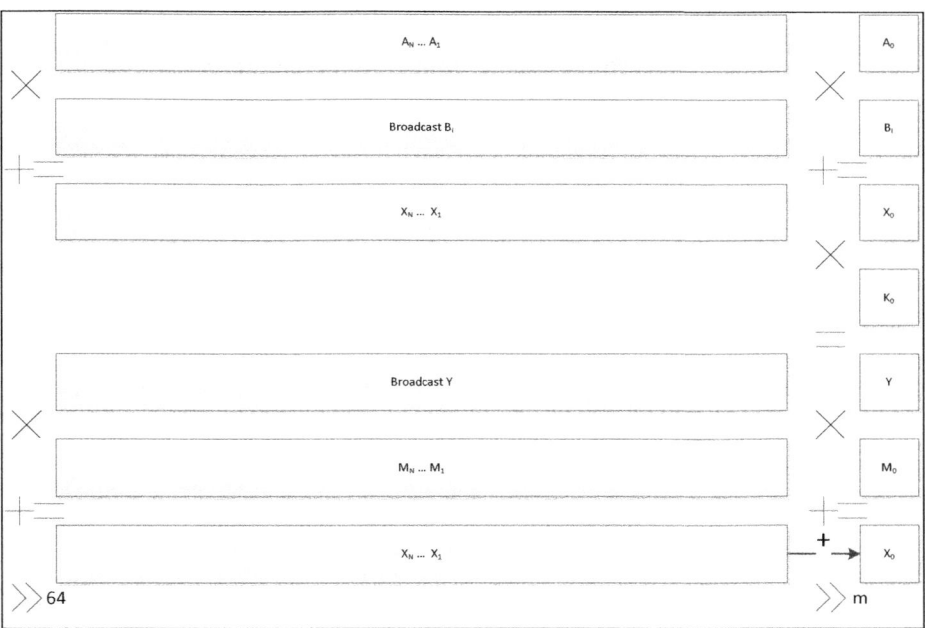

Fig. 8. Illustration of VNRMM as described in Fig. 6, as a flow of vector instructions (on the left) and scalar (ALU) instructions (on the right)

For example, the *1024*-bit *NRMM* using scalar implementation, requires *~512* multiplications and *~1536* additions. The redundant implementation with $m=29$ requires *~2592* multiplications and *~2592* additions. The total number of multiplications and additions for the scalar implementation with radix 2^{64} is *~2048* and for the *NRMM* the instruction total is *~5184* instructions.

Consequently, to make the vectorized code outperform the scalar implementation, it has to run on a SIMD architecture that can execute a factor of *2.53* more single precision operations than the scalar implementation. In other words, we need a SIMD architecture that accommodates *2.53* digits (of *64* bits), implying registers of at least *162* bits. The soon-to-appear AVX2 architecture has registers of *256* bits (*4* digits) with the appropriate integer instructions, and is therefore the first SIMD generation with the potential to support fast vectorized implementations. We demonstrate here how this potential can be realized.

By plain counting, one might suspect that the existing (narrower) SIMD architectures, could also support a vectorized implementation (of modular exponentiation) that outperforms the scalar implementation. We argue here that this is not the case on platforms with an efficient *64*-bit *mul* and *adc* performance, because the overhead associated with the vectorized implementation is too high. To illustrate, we consider the SSE3 instructions set, with SSSE3 extensions. We point out that:

- The SSE architectures have only *16* xmm registers. This allows for storing only *2048* bit of data at a time, which is not enough to keep the accumulators in registers. This adds overhead for memory traffic.
- The SSE architectures operate only on *16*-byte aligned memory operands, which hinders the "shift-save" optimization.
- The lack of a "broadcast" instruction adds an overhead for loading the digits of the operands.

4.3   Optimizing the Implementation

Implementing *NRMM* as in Algorithm 3 is rather straightforward, as illustrated in Fig. 8. However, we identify two bottlenecks in that implementation:

- The expensive right shifting of a vector (of digits) across several registers.
- The latency between the computation of y_0 (in step 7.4 of Algorithm 3), followed by broadcasting to a SIMD register, and the point in time where the multiplications in step 7.7 can start.

To address these bottlenecks, we use the following optimizations. We first note that instead of right shifting X, we can keep the values $A_q.A_1$ and $M_q.M_1$ in memory, and use "unaligned" *VPMULUDQ* operations with the proper address offset. To do this correctly, the operands A and M are padded with zeroes. For the second bottleneck, we preemptively calculate a few digits, using ALU instructions, to achieve a better pipelining of the ALU and SIMD units.

These optimizations are illustrated, schematically, in Fig. 10. Furthermore, they are implemented in real code, in the form of an OpenSSL patch, which the readers can download from [8] and examine.

4.4   Vectorized Redundant Montgomery Square

Modular exponentiation involves *NRMM*, majority of which are of the form *NRMM(A,A)*. We therefore add dedicated optimization for this case and call it

NRMSQR. Unlike *NRMM*, where we interleave the operations, the function *NRMSQR(A)* starts with calculating A^2, followed by a Montgomery Reduction. Our implementation employs the big-numbers squaring method published in [6]: first creating a copy of *A*, left shifted by *1*. In the redundant representation, this operation is simple: merely left shifting by *1*, of each element. Subsequently, the elements of *A*, and "*A<<1*" are multiplied, as described in Fig. 9.

```
Algorithm 5 [VNRMSQR]: Vectorized implementation of NRMSQR(A)
Input: A and M, in radix 2^m
Pre-computed: k_0 = -M^-1 mod M
Output: X, such that X mod M = A^2×2^-k×m mod M and X<2^n+1
Flow:
1.  Let A` = A×2 (i.e. a`_i = a_i << 1)
2.  Let s be the number of 64-bit elements in a SIMD register
3.  X_2q+1, …, X_0 = 0
//First stage - perform the square
4.  load digits 0, 2, …, (k-1) of A into SIMD registers A_0..A_q load
    digits 0, 2, …, (k-1) of A` into SIMD registers A`_0..A`_q
5.  for i = 0 to ⌈k-1/s⌉
  5.1. for j = 0 to s-1
          5.1.1. T = Broadcast a_ixs+j
          5.1.2. X_2q+1,…,X_i = X_2q+1,…,X_i + (A`_q,…,A`_i+1,A_i * T << 2^64×(ixs+j))
//Second stage - perform word-by-word reduction
6.  m_0 = M mod 2^m (i.e., digit 0 of M)
7.  x_0 = X_0[0]
8.  load digits 1, 2, …, (k-1) of M into SIMD registers M_0..M_q
9.  X_q,…,X_0 = X_q,…,X_0 >> 64
10. for i = 0 to k-1
  10.1. y_0 = x_0 × k_0 mod 2^m
  10.2. x_0 = x_0 + m_0 × y_0
  10.3. T = Broadcast y_0
  10.4. X_q,…,X_0 = X_q,…,X_0 + (M_0,…,M_0 * T)
  10.5. x_0 = x_0 >> m
  10.6. x_0 = x_0 + X_0[0]
  10.7. X_q,…,X_0 = X_q,…,X_0 >> 64
  End for
11. for j = 1 to q
  11.1. X_j = X_j + X_j+q+1
12. Convert X_q,..,X_1, x_0 from redundant representation to radix 2^m
    according to Algorithm 1.
```

Fig. 9. Optimized NRMSQR: using combination of vector and scalar instructions

5   Results

To assess our algorithm, we implemented an optimized *1024*-bit modular exponentiation code, using the described mix of vector (AVX2) and scalar instructions. We integrated this code into OpenSSL, in the form of a fully functional OpenSSL patch, which accelerates RSA2048. We call this implementation "RSAZ-AVX2" (the "Z" is for "Zariz" - "fast" in Hebrew). The patch is available at [8].

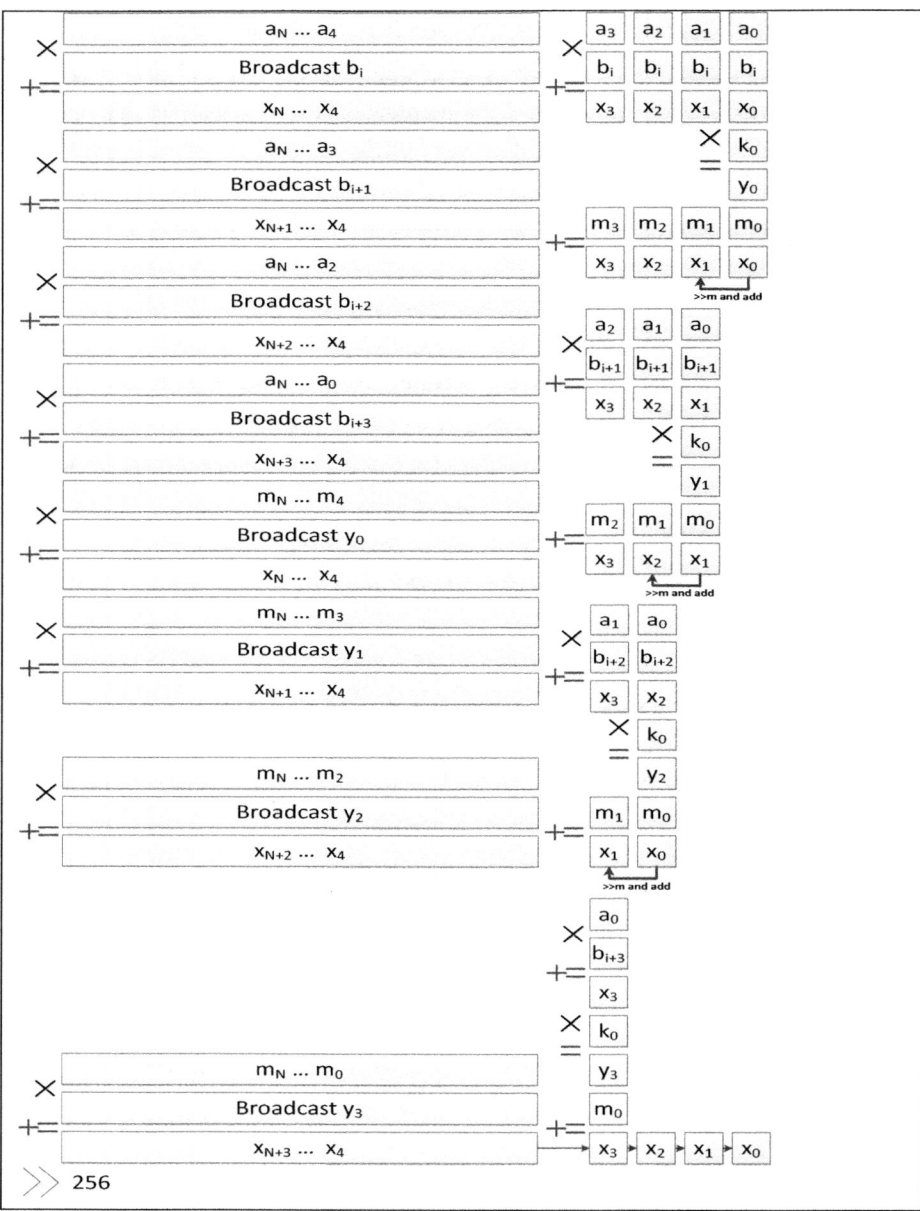

Fig. 10. Optimized VNRMM computation, via a flow of vector instructions (on the left) and scalar instructions (on the right). To reduce bottlenecks, four redundant digits are loaded into GPRs, and handled using ALU instructions, while the rest is handled via SIMD instructions. Vector shifting is performed only once per four digits.

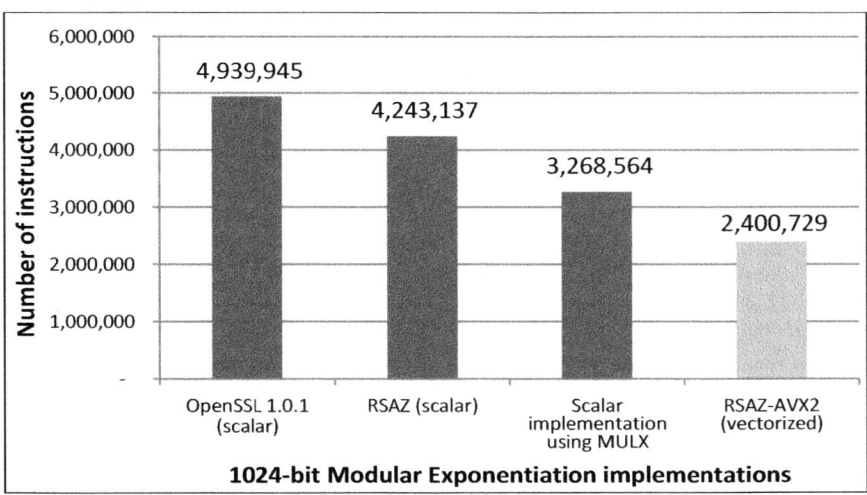

Fig. 11. The number of instructions in scalar and vector implementations of 1024-bit modular exponentiation (see explanation in the text)

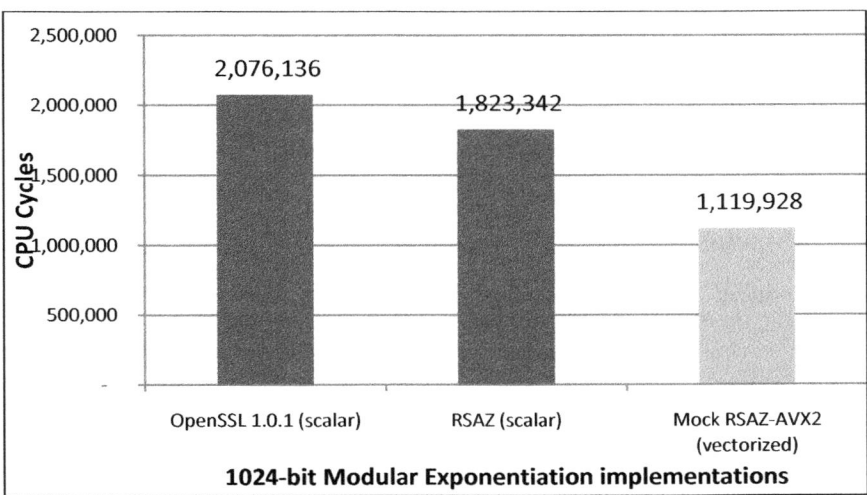

Fig. 12. Cycles count for a *1024*-bit modular exponentiation: Mock RSAZ-AVX2 (see explanation in the text) compared to other (real) scalar codes. The performance was measured on an Intel® Core™ i7-2600K processor.

This code can be compiled and tested for correctness using the existing public tools [23], [13]. However, we cannot report real performance figures, in cycles, at the time this paper is published: the Intel processor (Codename "Haswell") that supports AVX2 will be available only in 2013 [12].

To demonstrate the power of our method, we overcome this difficulty by approximating the speedup via counting the number of instructions in different modular exponentiation implementations. The instruction can be counted using the Intel SDE tool, with the "-mix" flag, as described in [13]. As the baseline, we used three implementations. The first two are public and can be run on the existing processors: the current (official) OpenSSL 1.0.1 [18], and the best known implementation (which we call RSAZ) [5]. In addition, to make sure that we compare to the best-future-implementation, we also generated a new scalar implementation that uses the new scalar instruction *MULX*, which is expected to be faster than the existing scalar implementations (the *MULX* instruction will appear, together with the AVX2 instructions, in the coming "Haswell" processor [12]). The instructions count for each implementation is provided in Fig. 11.

Fig. 11 shows that our method requires less than half of the number required by the current OpenSSL 1.0.1 implementation, ~*43%* fewer instructions than the currently best known scalar implementation, and ~*26%* fewer instructions than a future scalar code that uses the (coming) *MULX* instruction. This clearly indicates that our implementation outperforms the alternatives by a significant margin.

We point out that the instructions count is only an approximation, and cannot be translated directly to CPU cycles, in an out-of-order architecture. To this end, we add another comparative approximation via a "Mock implementation". We took our new modular exponentiation code and replaced all of the AVX2 instructions with AVX1 instructions (the rationale is that most of these instructions are merely a wider version of their AVX1 counterparts). This allowed us to measure the performance (of the mock modular exponentiation) on an existing processor. Performance wise, this gives us yet another *hint* to the expected performance (although the output is functionally incorrect). The numbers are presented in Fig. 12, showing that the mock RSAZ-AVX2 implementation is *1.85* times faster than OpenSSL 1.0.1, and *1.63* times faster than the best known scalar implementation (RSAZ). This improvement is quite consistent with the improvement shown in Fig. 11

6   Conclusion

We introduced here a new method and implementation for computing modular exponentiation, thus accelerating software performance of RSA on modern high end processors. Our algorithm utilized the coming AVX2 architecture, and a special balance between scalar and vector operations. The demonstrated results show that a significant performance gain (up to *40%* over current implementations) will be available together with the release of the new Intel® Architecture Codename Haswell.

Our vectorization method is scalable, and can gain performance from any wide (and wider) SIMD architectures. Finally, we point out that our method can also be used on AVX/SSE architectures. This achieves a significant performance gain for (low end) processors that have AVX/SSE instructions, and only a *32*-bit ALU unit, such as the current Atom processors.

References

1. Barker, E., Roginsky, A.: Transitions: Recommendation for Transitioning the Use of Cryptographic Algorithms and Key Lengths, p. 5. NIST Special Publication 800-131A (2011) http://csrc.nist.gov/publications/nistpubs/800-131A/sp800-131A.pdf

2. Bernstein, J.D.: Curve25519: New Diffie-Hellman speed records (2006)
3. Brent, R., Zimmermann, P.: Modern Computer Arithmetic. Cambridge University Press (2010), http://www.loria.fr/~zimmerma/mca/pub226.html (retrieved)
4. Gueron, S.: Enhanced Montgomery Multiplication. In: Kaliski Jr., B.S., Koç, Ç.K., Paar, C. (eds.) CHES 2002. LNCS, vol. 2523, pp. 46–56. Springer, Heidelberg (2003)
5. Gueron, S., Krasnov, V.: Efficient and side channel analysis resistant 512-bit and 1024-bit modular exponentiation for optimizing RSA1024 and RSA2048 on x86_64 platforms, OpenSSL #2582 patch (posted August 2011), http://rt.openssl.org/Ticket/Display.html?id=2582&user=guest&pass=guest
6. Gueron, S., Krasnov, V.: Speeding up Big-numbers Squaring. In: IEEE Proceedings of 9th International Conference on Information Technology: New Generations (ITNG 2012), pp. 821–823 (2012)
7. Gueron, S.: Efficient Software Implementations of Modular Exponentiation. Journal of Cryptographic Engineering 2, 31–43 (2012),
8. Gueron, S., Krasnov, V.: Efficient, and side channel analysis resistant 1024-bit modular exponentiation, for optimizing RSA2048 on AVX2 capable x86_64 platforms, OpenSSL patch (posted June 2012), http://rt.openssl.org/
9. Hassaballah, M., Omran, S., Mahdy, Y.B.: A Review of SIMD Multimedia Extensions and their Usage in Scientific and Engineering Applications. The Computer Journal 51(6), 630–649 (2007)
10. Intel: Using Streaming SIMD Extensions (SSE2) to Perform Big Multiplications (2006)
11. Intel: Intel Advanced Vector Extensions Programming Reference, http://software.intel.com/file/36945
12. Buxton, M. (Intel): Haswell New Instruction Descriptions Now Available!, http://software.intel.com/en-us/blogs/2011/06/13/haswell-new-instruction-descriptions-now-available/
13. Intel: Software Development Emulator (SDE), http://software.intel.com/en-us/articles/intel-software-development-emulator/
14. Koc, Ç.K., Kaliski, B.S.: Analyzing and Comparing Montgomery Multiplication Algorithms. Micro 16(3), 26–33 (1996), http://islab.oregonstate.edu/papers/j37acmon.pdf
15. Koç, Ç.K., Walter, C.D.: Montgomery Arithmetic. In: van Tilborg, H. (ed.) Encyclopedia of Cryptography and Security, pp. 394–398. Springer (2005)
16. Lin, B.: Solving Sequential Problems in Parallel. Application Note, Freescale Semiconductor (2006)
17. Menezes, A.J., van Oorschot, P.C., Vanstone, S.A.: Handbook of Applied Cryptography, 5th printing. CRC Press (2001)
18. OpenSSL: The Open Source toolkit for SSL/TLS, http://www.openssl.org/
19. Page, D., Smart, P.: Parallel Cryptographic Arithmetic Using a Redundant Montgomery Representation. IEEE Transactions on Computers 53(11), 1474–1482 (2004)
20. Walter, C.D.: Montgomery exponentiation needs no final subtractions. Electron. Lett. 35, 1831–1832 (1999)
21. Walter, C.D.: Montgomery's Multiplication Technique: How to Make It Smaller and Faster. In: Koç, Ç.K., Paar, C. (eds.) CHES 1999. LNCS, vol. 1717, pp. 80–93. Springer, Heidelberg (1999)
22. Walter, C.D.: Precise Bounds for Montgomery Modular Multiplication and Some Potentially Insecure RSA Moduli. In: Preneel, B. (ed.) CT-RSA 2002. LNCS, vol. 2271, pp. 30–39. Springer, Heidelberg (2002)
23. YASM: The YASM Modular Assembler Project, http://yasm.tortall.net/

Efficient Multiplication over Extension Fields

Nadia El Mrabet[1] and Nicolas Gama[2]

[1] LIASD - Université Paris 8
[2] Université de Versailles - PRISM - CNRS

Abstract. The efficiency of cryptographic protocols rely on the speed of the underlying arithmetic and finite field computation. In the literature, several methods on how to improve the multiplication over extensions fields $\mathbb{F}_{q^m}$, for prime q were developped. These optimisations are often related to the Karatsuba and Toom Cook methods. However, the speeding-up is only interesting when m is a product of powers of 2 and 3. In general cases, a fast multiplication over $\mathbb{F}_{q^m}$ is implemented through the use of the naive school-book method. In this paper, we propose a new efficient multiplication over $\mathbb{F}_{q^m}$ for any power m. The multiplication relies on the notion of Adapted Modular Number System (AMNS), introduced in 2004 by [3]. We improve the construction of an AMNS basis and we provide a fast implementation of the multiplication over $\mathbb{F}_{q^m}$, which is faster than GMP and NTL.

1   Introduction

The efficiency of an algebraic cryptosystem is directly related to the speed of the underlying arithmetic computation over finite fields $\mathbb{F}_q$ and their extensions $\mathbb{F}_{q^m}$, wherein q is a prime number and m an integer. Depending on the mathematics required for the cryptosystem, the prime q is either 2, 3 or a large prime number. For cryptographic applications, the NIST provides recommendations for the appropriate size of q and m [2,1], for elliptic curves cryptography [16,7], for pairing based cryptography [5,6] or torus based cryptography [21,26].

For a large prime number q the arithmetic over $\mathbb{F}_{q^m}$ is rather expensive and several methods to improve it are described in the literature [27]. In characteristic 2, different optimizations of the multiplication are described in [14,25], when the field $\mathbb{F}_{2^m}$ is defined modulo a cyclotomic polynomial, or as the evaluations of polynomials over a root of unity. Usually, the finite field is embedded in a larger field using the cyclotomic polynomials, and the multiplication is then optimized via the Fast Fourier Transform. When the characteristic of the finite field is different from 2, classical optimizations depend on the value of m. For extensions of degree 2, the Karatsuba method is the most efficient [27], while the Toom-Cook method is recommended for extensions of degree 3 [27]. Consequently, when m is the product of a power of 2 with a power of 3 ($m = 2^i 3^j$), the multiplication is done by recursive applications of the Karatsuba and Toom Cook methods [4,9]. When m is a very large power of 2, the Fast Fourier Transformation (FFT) can be used to improve the multiplication over $\mathbb{F}_{q^{2^n}}$, with q being a prime and

F. Özbudak and F. Rodríguez-Henríquez (Eds.): WAIFI 2012, LNCS 7369, pp. 136–151, 2012.
© Springer-Verlag Berlin Heidelberg 2012

n being an integer such that the number q^{2^n} has several thousand digits [27]. In [18], Montgomery proposed efficient multiplications for extensions of small degrees $m = 5$, 6 and 7. The case of degree 5 has been recently improved in [19].

For other extension degrees m, very few improvements have been proposed.

In [10], El Mrabet and Nègre proposed an efficient multiplication for several values of m. This multiplication is based on the Discrete Fourier Transform (DFT) scheme [27]. One multiplication using the DFT scheme costs $(2m - 1)$ multiplications in $\mathbb{F}_q$ and $\mathcal{O}(m^2)$ multiplications by a root of unity in $\mathbb{F}_q$. In a classical representation of the field $\mathbb{F}_q$, multiplication by a root of unity or by a random element of $\mathbb{F}_q$ would both have the same complexity. Consequently, the $\mathcal{O}(m^2)$ multiplications by a root of unity in $\mathbb{F}_q$ would be very expensive. In [10], the authors use an original representation of finite fields: the Adapted Modular Number System (AMNS) introduced by Bajard et al. in [3]. The main advantage of the AMNS representation is that all multiplications by a root of unity correspond to cyclic shifts, and become very cheap. This novel representation was designed in order to improve the multiplication over $\mathbb{F}_q$ when q is a pseudo Mersenne prime. In [8], a new class of specific moduli for cryptography was introduced, the more generalized Mersenne numbers. In [3], Bajard et all extend the work of [8] with the definition of the AMNS representation. The AMNS can in fact be applied for every prime q, and as well when q is a power of prime. Furthermore, the AMNS representation can be used to improve the arithmetic for the protocol RSA or more generally for a set $\mathbb{Z}/q\mathbb{Z}$.

We improve the result of [10] and we propose an efficient implementation of the multiplication for the extensions of field of prime degree. Our article is organized as follows. In Section 2 we recall the definition of the AMNS representation. In Section 2.2 we review the arithmetic and the multiplication in $\mathbb{F}_{q^m}$ with the AMNS representation, and we propose a simplified multiplication. In Section 3 we improve the multiplication in $\mathbb{F}_{q^m}$ by an efficient construction of an AMNS basis. In Section 4 we describe the first implementation of an AMNS representation. We use it in order to implement an efficient multiplication over $\mathbb{F}_{q^m}$, then we compare our results with the GMP implementation. We conclude in Section 5.

2   AMNS

The efficiency of the arithmetic in a field $\mathbb{F}_q$ relies on the representation of its elements. Usually the representation of a finite field is based on a positional number system representation. An element a of a field $\mathbb{F}_q$ is decomposed in the base β as $a = \sum_{i=0}^{\ell-1} a_i \beta^i$, with $a_i \in [0; \beta - 1]$. The element β is such that $\beta^\ell \geq q$. As an example, we can cite the binary decomposition ($\beta = 2$), or the hexadecimal ($\beta = 8$) and decimal ($\beta = 10$) decomposition. The Adapted Modular Number System (AMNS) was introduced by Bajard et al. in [3]. Initially, this representation was described for pseudo Mersenne primes, but it was extended to any group $\mathbb{Z}/q\mathbb{Z}$.

2.1 Definition of an AMNS Representation

The AMNS representation relies on the decomposition of numbers as a combination $\sum a_i \gamma^i$ of powers of a base γ, but with more drastic constraints on γ and on the size of the coefficients a_i. Namely, given an integer n, the base γ of an AMNS representation verifies that γ^n equals a very small number λ modulo q, for example $\lambda = \pm 1, \pm 2 \mod q$. The coefficients a_i in an AMNS representation are all bounded by $|a_i| \leq \rho$ for some parameter $\rho \approx q^{1/n}$. The AMNS representation is defined as follow.

Definition 1. *A basis $\mathcal{B}$ of an AMNS representation is a tuple $(q, n, \gamma, \rho, \lambda)$ such that: $\gamma^n = \lambda \mod q$; ρ is a covering radius such that for each element $a \in \mathbb{F}_q$,*

$$\textit{it exists a polynomial } \mathfrak{a}(X) = \sum_{i=0}^{n-1} a_i X^i \textit{ of degree } < n \textit{ and } \|\mathfrak{a}\|_\infty \leq \rho \textit{ such that }$$

$\mathfrak{a}(\gamma) = a \mod q$.

Then for each polynomial $\mathfrak{p} \in \mathbb{Z}_n[X]$, we say that $\mathfrak{p}$ is an AMNS representation of $\mathfrak{p}(\gamma) \mod q$ when $\|\mathfrak{p}\|_\infty = \max_{i=1\ldots n} |p_i| \leq 2n\lambda\rho^2$ and that it is a reduced AMNS representation when $\|\mathfrak{p}\|_\infty \leq \rho$.

Note that in general, the smallest covering radius ρ will be a small multiple of $q^{1/n}$. Also, a good basis for an AMNS representation should for efficiency be designed so that the largest entries fits into raw integer types (namely $2n\lambda\rho^2$ should be smaller than 2^{63} on a classical 64-bit processors).

Example 1. Let $q = 19$, an AMNS basis for $\mathbb{F}_q$ is $\mathcal{B} = (q = 19, n = 3, \gamma = 7, \rho = 2, \lambda = 1)$. Each element of $\mathbb{F}_q$ in signed representation is a polynomial in $\gamma = 7$, of degree at the most 2, and with coefficients in $[-1, 0, 1]$. We can check that $\gamma^3 \equiv 1 \mod q$. The decomposition of $\mathbb{F}_q$ is given in Table 1

We could note that the evaluation of the polynomial $-1 - X + X^2$ in γ is 3 mod q and that $\left\| -1 - X + X^2 \right\|_\infty = 1 < 2$. Since $\|a_i\|_\infty < \rho$, it follows that the multiplication of two coefficients is efficient. This consideration leads to an efficient multiplication in AMNS systems where the prime q is a pseudo Mersenne number. For general primes, we will use a Montgomery's like multiplication.

Remark 1. When q is not a prime but is a prime power, the construction of an AMNS basis presented in [3] is not efficient. Indeed, we have to find an $n^{th}-$root

Table 1. Decomposition of $\mathbb{F}_q$

1	2	3	4	5	6
1	$-X^2 - X + 1$	$X^2 - X - 1$	$X^2 - X$	$-X^2 - X + 1$	$X - 1$

7	8	9	10	11	12
X	$X + 1$	$-X^2 + 1$	$X^2 - 1$	X^2	$X^2 + 1$

13	14	15	16	17	18
$-X + 1$	$-X^2 + X - 1$	$-X^2 + X$	$-X^2 + X + 1$	$X^2 + X - 1$	-1

of λ. Or, in a field $\mathbb{F}_{p^t}$ for p a prime and t an integer greater than 2; finding an nth-root of λ is difficult because it is equivalent to find the factorisation of $q - 1$, which is a difficult issue for a large value of q.

Remark 2. We could note that the AMNS representation of an integer is redundant. For example, 12 can be seen as X^2+1 or $-X$. If the protocol needs to compare two elements we can either go back to classical representation (we can use the Horner scheme to efficiently evaluate the polynomial.), or we can use Babai nearest plane algorithm to test whether the difference is in the lattice.

Both are indeed costly, but this cost can be amortized if the scheme requires a great number of aritmetic operations before being evaluated.

2.2   Arithmetic in AMNS Representation

As the AMNS representation is quite original, we could wonder if the arithmetic in AMNS representation is more or less efficient than with a a classical representation. In [20], Nègre and Plantard present an efficient multiplication in the AMNS representation. This multiplication is constructed over Montgomery's scheme, and works for a prime q or for a prime power.

In order to give a simple explanation, we can say that the arithmetic in AMNS representation is based on these three primitives:

1. Addition and substraction of two elements $\mathfrak{a}$ and $\mathfrak{b}$ (denoted respectively by $\mathfrak{a} + \mathfrak{b}$ and $\mathfrak{a} - \mathfrak{b}$) is performed term by term on the AMNS representation, and the norm of the result is bounded by $\|\mathfrak{a}\|_\infty + \|\mathfrak{b}\|_\infty$.
2. Multiplication of $\mathfrak{a}$ and $\mathfrak{b}$ (denoted by $\mathfrak{a} \times \mathfrak{b}$) is performed by multiplying the two polynomials and reducing modulo $(X^n - \lambda)$. This leads to an element of norm at the most $n(\lambda + 1) \|\mathfrak{a}\|_\infty \|\mathfrak{b}\|_\infty$.
3. Division by the constant parameter ϕ is used to reduce the size of coefficients. Given a representation $\mathfrak{a}$ of norm $|\mathfrak{a}|_\infty \leq 2n\lambda\rho^2$, Algorithm 1 outputs a representation of $\mathfrak{a}(\delta) \times \phi^{-1}$ of norm $\leq \rho$.

Algorithm 1. Division by ϕ in AMNS

Require: An AMNS basis $\mathcal{B}$, a polynomial $\mathfrak{m}$ and its inverse $\mathfrak{m}' \bmod \phi$, and an element $\mathfrak{a}$ of norm $\leq 2n\lambda\rho^2$.
Ensure: A reduced AMNS representation of $\mathfrak{a}(\gamma) \times \phi^{-1}$ of norm $\leq \rho$.
  1: Compute $\mathfrak{a}' = \mathfrak{a} \times \mathfrak{m}' \bmod \phi$ having all its coefficients in $[-\phi/2, \phi/2]$.
  2: **Return** $\frac{1}{\phi}(\mathfrak{a} - \mathfrak{a}' \times \mathfrak{m})$ ▷ the division is exact

---

By applying successively the last two primitives, one obtains an efficient Montgomery's like multiplication between any two reduced AMNS representations $\mathfrak{a}$ and $\mathfrak{b}$. Like in the classical Montgomery's multiplication, the result is not exactly $\mathfrak{a} \times \mathfrak{b} \bmod q$ but $\mathfrak{a} \times \mathfrak{b} \times \phi^{-1} \bmod q$. Note that the whole process performs only three multiplications of polynomials with small coefficients modulo $(X^n - \lambda)$, which makes the AMNS arithmetic efficient. The Montgomery's like multiplication is resume in Algorithm 2.

Algorithm 2. Montgomery's like multiplication in AMNS

Require: $\mathfrak{a}$ and $\mathfrak{b}$ in an AMNS basis $\mathcal{B} = (q, n, \gamma, \rho, \lambda)$, $\phi \geq 2n\lambda\rho$, a polynomial $\mathfrak{m}$
and its inverse $\mathfrak{m}'$ mod $(X^n - \lambda)$ mod ϕ.
Ensure: The element $\mathfrak{t} \in \mathcal{B}$ such that $\mathfrak{t}(\gamma) = \mathfrak{a}(\gamma) \times \mathfrak{b}(\gamma) \times \phi^{-1}$ mod q.
1: Compute $\mathfrak{c} = \mathfrak{a} \times \mathfrak{b}$ mod $(X^n - \lambda)$
2: Compute $\mathfrak{q} = \mathfrak{c} \times \mathfrak{m}'$ mod $(X^n - \lambda)$ mod ϕ
3: **Return** $\mathfrak{t} = \frac{1}{\phi}(\mathfrak{c} + \mathfrak{m}')$ mod $(X^n - \lambda)$ ▷ the division is exact

---

The third primitive requires the preprocessing of a valid parameters ϕ and two polynomials $\mathfrak{m}$ and $\mathfrak{m}'$ which are inverse to each other modulo $(X^n - \lambda)$ mod ϕ.

Lemma 1. *If ϕ is larger than $4n\lambda\rho^2$ and the polynomials $\mathfrak{m}$ and $\mathfrak{m}'$ satisfy the conditions*

$$\mathfrak{m}(\gamma) \equiv 0 \quad \mathrm{mod} \ q,$$
$$\|\mathfrak{m}\|_\infty \leq \rho/n,$$
$$\mathfrak{m} \times \mathfrak{m}' \equiv 1 \quad \mathrm{mod} \ (X^n - \lambda) \quad \mathrm{mod} \ \phi,$$

then for any input AMNS representation $\mathfrak{a}$ of norm $\leq 2n\lambda\rho^2$, Algorithm 1 outputs a reduced AMNS representation (of norm $\leq \rho$) of $\mathfrak{a}(\gamma)/\phi$.

Proof. The division by ϕ in the last line of Algorithm 1 is exact, because $\mathfrak{a}' \times \mathfrak{m} = \mathfrak{a} \times \mathfrak{m} \times \mathfrak{m}'$ mod ϕ which is by definition equal to $\mathfrak{a}$ mod ϕ. When taken modulo q, the evaluation in γ of $\phi^{-1} \times (\mathfrak{a} - \mathfrak{a}' \times \mathfrak{m})$ is $\mathfrak{a}(\gamma)/\phi$ because $\mathfrak{m}(\gamma) \equiv 0$ mod q. Finally, the norm of $\mathfrak{a}'$ is $\|\mathfrak{a}'\|_\infty = \phi/2$, so the norm of $\|\mathfrak{a}' \times \mathfrak{m}\|_\infty$ is smaller than $\phi\rho/2$, i.e. $(\|\mathfrak{a}' \times m\|_\infty \leq \phi\rho/2)$ and $\left\|\frac{1}{\phi}(\mathfrak{a} - \mathfrak{a}' \times \mathfrak{m})\right\|_\infty \leq (2n\lambda\rho^2 + \phi\rho/2)/\phi \leq \rho.$

The coefficients of the polynomials are smaller than ρ and ϕ. The execution of the algorithm needs to perform two different reductions. One reduction is performed modulo the polynomial $(X^n - \lambda)$, and one is an integer reduction modulo ϕ. The polynomial $(X^n - \lambda)$ is sparse, thus the polynomial reduction consists essentially in additions and shifts.

In the same way as in [20], the integer ϕ must be larger than $2n\lambda\rho$, and $\mathfrak{m}$ must be a polynomial with very small coefficients, which admits γ as a root modulo $\mathbb{F}_q$ and which is invertible   mod ϕ. Unfortunately, the algorithm proposed in [20] to generate $\mathfrak{m}$ requires to approximate the shortest vector problem in a n-dimensional lattice within a constant factor, which can be difficult in high dimension. Furthermore, since a lot of integer divisions by ϕ occur in the Montgomery's like multiplication algorithm, in practice it would be ideal to have ϕ equal to a power of 2 in order to speed-up those divisions. But once again, the construction of [20] does not guarantee that ϕ can be even.

In the next sections, we propose a polynomial time construction of all parameters which is inspired from the key generation algorithm used by Gentry et al. for the fully Homomorphic Encryption scheme challenges in [13].

2.3   Efficient Multiplication in $\mathbb{F}_{q^m}$ Using AMNS

In [10], El Mrabet and Nègre use the AMNS representation of $\mathbb{F}_q$ to improve the multiplication in an extension $\mathbb{F}_{q^m}$. They combine the Discrete Fourier Transform (DFT) multiplication with the AMNS representation. The DFT multiplication needs several multiplications by roots of unity. In a classical representation, a root of unity can be any element of $\mathbb{F}_q$ and consequently a multiplication by roots of unity is equivalent to a random multiplication in $\mathbb{F}_q$. The advantage of the AMNS representation is that the element γ chosen to construct the base can be a root of unity.

The extension $\mathbb{F}_{q^m}$ of $\mathbb{F}_q$ is defined as the quotient $\mathbb{F}_q[Y]/(P(Y)\mathbb{F}_q[Y])$, where $P(Y)$ is an irreducible polynomial of degree m over $\mathbb{F}_q$ and $\mathbb{F}_q[Y]$ represente the polynomial in Y and with coefficients in $\mathbb{F}_q$. An element of $\mathbb{F}_{q^m}$ is a polynomial in Y of degree smaller than m and with coefficients in $\mathbb{F}_q$,

$$\mathbb{F}_{q^m} = \{R(Y) \in \mathbb{F}_q[Y] \text{ such that } \deg(R(Y)) < m\}.$$

We resume the combination of the DFT multiplication of two polynomials $U(Y)$ and $V(Y)$ with the AMNS representation. We denote $W(Y) = U(Y) \times V(Y)$.

Let l be an integer such that we can define α to be a $2l^{th}$−root of unity in $\mathbb{F}_q$, and let $\alpha_i = \alpha^i$ for $i = 0, \ldots, 2l - 1$.

The DFT multiplication is the composition of three steps:

1. The evaluation of the polynomials U and V in the α_is,
2. The $2l$ multiplications $U(\alpha_i) \times V(\alpha_i)$, in order to find the evaluation in the α_is of $W(Y)$,
3. The interpolation of $W(Y) = U(Y) \times V(Y)$.

The evaluation and interpolation steps can be performed with a matrix vector product. The evaluation corresponds to the product $\Omega \times^t U$ and $\Omega \times^t V$, where $U = [u_0, u_1, \ldots, u_{l-1}]$ and $V = [v_0, v_1, \ldots, v_{l-1}]$ are the vectors of the coefficients of $U(Y)$ and $V(Y)$. The interpolation step is composed by the product $\Omega^{-1} \times^t W$, where Ω^{-1} is the matrix inverse of Ω.

The matrix Ω and Ω^{-1} are the following

$$\Omega = \begin{bmatrix} 1 & 1 & 1 & \ldots & 1 \\ 1 & \alpha & \alpha^2 & \ldots & \alpha^{l-1} \\ \vdots & \vdots & \vdots & \vdots & \vdots \\ 1 & \alpha^i & (\alpha^i)^j & \ldots & (\alpha^i)^{(l-1)} \\ \vdots & \vdots & \vdots & \vdots & \vdots \\ 1 & \alpha^{(l-1)} & \alpha^{2(l-1)} & \ldots & \alpha^{(l-1)(l-1)} \end{bmatrix}, \Omega^{-1} = \frac{1}{l} \begin{bmatrix} 1 & 1 & 1 & \ldots & 1 \\ 1 & \alpha^{-1} & (\alpha^{-1})^2 & \ldots & (\alpha^{-1})^{(l-1)} \\ \vdots & \vdots & \vdots & \ldots & \vdots \\ 1 & (\alpha^{-i}) & (\alpha^{-i})^2 & \ldots & (\alpha^{-i})^{(l-1)} \\ \vdots & \vdots & \vdots & \vdots & \vdots \\ 1 & \alpha^{-(l-1)} & \alpha^{-2(l-1)} & \ldots & \alpha^{-(l-1)(l-1)} \end{bmatrix}.$$

The complexity of the DFT multiplication is the sum of the complexity of 3 matrix vector products and the $2l$ products in $\mathbb{F}_q$. In an AMNS representation defined by $\mathcal{B} = (q, n, \gamma, \rho, \lambda)$, we can choose the parameters such that the matrix-vector products are composed only with shifts and additions in $\mathbb{F}_q$. Indeed the matrix-vector products are only composed with multiplications of the α_is with the coefficients of $U(X)$ and $V(X)$. If we choose $l = 2n$, γ and λ in the AMNS base $\mathcal{B}$ to be such that $\lambda = -1$ and then $\gamma^n \equiv -1 \mod q$. With this choice, γ is an $l-$root of unity in $\mathbb{F}_q$ and can be use to define the matrix Ω. As a consequence, the multiplications by powers of γ consist only in shifts and additions in $\mathbb{F}_q$, as explained in the following lemma.

Lemma 2. *Let u_i be the ith coefficient of $U(Y)$, $u_i \in \mathbb{F}_q$ and u_i is decomposed in the AMNS basis $\mathcal{B} = (q, n, \gamma, \rho, \lambda)$. Let $\mathfrak{u}_i$ be a representation of u_i in $\mathcal{B}$. Then the product $u_i \times \gamma^j = \mathfrak{u}_i \times X^j$ and*

$$\mathfrak{u}_i \times \gamma^j = (\sum_{k=0}^{n-1} u_i^k Y^k) \times Y^j \mod (Y^n + 1) = \sum_{k=0}^{j-1} u_i^{n-j+k} X^k + \sum_{k=j}^{n-1} u_i^{k-j} X^k.$$

Proof. The proof consists in writing down the equation and use the fact that we work modulo the polynomial $(X^n + 1)$. □

Remark 3. We can notice that the Lemma 2 is writen for $\lambda = -1$, but it is very easy to obtain the same result for a more generic value of λ. The important point is that λ must be choosen such that the multiplication by λ are for free. For example, λ can be ± 1, or $\pm 2^d$ with d being a small integer.

An important consequence of Lemma 2 is the fact that a multiplication by a power of γ is a very easy operation, which preserves the norm. Indeed, multiplying a vector $\mathfrak{u}$ by a power of γ consists only in a permutation of the coefficients of $\mathfrak{u}$, which is only linear in the bit-length.

Lemma 3. *The use of the DFT multiplication on powers of γ in an AMNS representation leads to an efficient multiplication over an extension field.*

Indeed, the evaluation of a polynomial in AMNS representation in a power of γ consists in multiplications by power of γ and additions. A multiplication by a power of γ in the AMNS representation is a mere shift in the representation. As a consequence, the evaluation of a polynomial in power of γ costs only $O(n)$ additions instead of $O(n)$ multiplications. The dual operation of interpolating a function from its values on powers of γ is also easy. Indeed, both operations can be viewed as a multiplication by a Lagrange matrix containing only powers of γ.

Multiplication of two polynomials $\mathfrak{f}, \mathfrak{g}$ can therefore be performed in a DFT manner by evaluating $\mathfrak{f}$ and $\mathfrak{g}$, pairwise multiplying the evaluations, interpolating the result, and reducing modulo P. Overall, this requires $O(n)$ multiplications in $\mathbb{F}_q$ instead of $O(n \log n)$ or $O(n^2)$ for classical algorithms. □

The above multiplication algorithm has the drawback that a multiplication requires $4n$ evaluations of the input polynomial on the roots of unity. Instead, it is preferable to directly represent a polynomial $P(Y)$ in a Lagrange representation, by its evaluations $(P(\gamma^0), \ldots, P(\gamma^{2m-1}))$ rather than by its coefficients. Once again, the multiplication we propose is a Montgomery like algorithm, because the result of the product of $U(Y)$ times $V(Y)$ is $(m/\phi^3) \times UV$. The multiplication in Lagrange representation is described in Algorithm 3.

Algorithm 3. AMNS-Multiplication in $\mathbb{F}_{q^m}$

Require: Two reduced Lagrange-AMNS representations $\left(U(\gamma^0), \ldots, U(\gamma^{2m-1})\right)$ and $\left(V(\gamma^0), \ldots, V(\gamma^{2m-1})\right)$ of polynomials $U, V \in \mathbb{F}_q[Y]$ of degree $\leq m - 1$.
Ensure: The reduced Lagrange-AMNS representation $(W(\gamma^0), \ldots, W(\gamma^{2m-1}))$ of the product $W \equiv (m/\phi^3).UV \mod Y^m - \alpha$ of degree $\leq m - 1$
1: Compute $A = UV \times \phi^{-1} = (U(\gamma^i) \times V(\gamma^i) \times \phi^{-1})_{i=0,\ldots,2m-1}$ using the AMNS Montgomery's product
2: Compute the coefficients of $(\mathfrak{a}_0, \ldots, \mathfrak{a}_{2m-1})$ of $m \cdot A[Y] = \sum_{i=0}^{2m-1} \mathfrak{a}_i Y^i$ in AMNS representation using the inverse DFT
3: Reduce the polynomial modulo $Y^m - \alpha$ and divide it by ϕ to obtain $B = \displaystyle\sum_{i=0}^{m-1} \mathfrak{b}_i Y^i$

with $\mathfrak{b}_i = (\mathfrak{a}_i + \alpha\mathfrak{a}_{i+m}) \times \phi^{-1}$.
4: **Return** $B \times \phi^{-1}$ as $(B(\gamma^0) \times \phi^{-1}, \ldots, B(\gamma^{2m-1}) \times \phi^{-1})$ using DFT

If the integer n is a power of 2, the multiplication by the Fast Fourier Transform (FFT [27]) method can be an improvement of the DFT method. The FFT method is very interesting because it consists in factoring the computation in order to not to compute twice the same operation. In our case, this is very efficient, and the operations in the FFT method are only composed with multiplications by powers of γ. We recall here the major steps of the FFT multiplication.

Let $U(Y)$ be a polynomial that we want to evaluate for the DFT multiplication. The FFT method is a divide and conquer scheme. Let γ be a root of unity.

The FFT method consists in dividing the polynomial $U(Y)$ in two parts:

$$U_1 = \sum_{k=0}^{n/2-1} u_{2k} Y^{2k},$$

$$U_2 = \sum_{k=0}^{n/2-1} u_{2k+1} Y^{2k}, \text{ such that } U = U_1 + YU_2.$$

We denote $\widehat{U} = [U(1), U(\gamma), \ldots, U(\gamma^{n-1})]$ and $\widehat{U_j} = [U_j(1), U_j(\gamma), \ldots, U_j(\gamma^{n-1})]$ for $j = 1, 2$. The element $\widehat{U}[i]$ is the ith coefficient of $\widehat{U}$. The evaluation of U in a power of γ can be expressed as follow

$$\text{for } i \in [0; n/2[$$
$$\widehat{U}[i] = \widehat{U_1}[i] + \gamma^i \widehat{U_2}[i],$$
$$\widehat{U}[i + n/2] = \widehat{U_1}[i] - \gamma^i \widehat{U_2}[i].$$

We use the fact that $\gamma^{i+n/2} = -\gamma^i$, which is evident since γ is a root of unity. Since n is a power of 2, when we apply the FFT method, we can recursively use this formula.

In [3], the AMNS representation was proposed randomly and the polynomial $\mathfrak{m}$ was not inversible for each construction. We propose below an efficient way to construct an AMNS representation and to assure the fact that the polynomial $\mathfrak{m}$ is invertible. We split the analyse in two parts. First, we analysis in Section 3.2, the simultaneous generation of $\mathfrak{m}$ and q, which can be done very efficiently even for large dimensions. Then in Section 3.3, we consider cases where q is fixed in advance, and the goal is to generate $\mathfrak{m}$ accordingsly. This second case, which occurs in pairing based cryptography, is much harder than the previous one, but can still be achieved in practice for extensions of degree ≈ 100. these cases occur for example in pairing based cryptography.

3   Theory

We propose a new construction of the AMNS parameters by adapting the key generation algorithm used by Gentry [13] in the fully homomorphic encryption scheme. Given as input an extension of degree n, the procedure generates a polynomial $\mathfrak{m}$ and a prime number q. This is for instance the case of almost all Elliptic Curve Cryptography scheme based on Diffie Hellman. However, this does not work for pairings, which have strong external constraints on q.

3.1   Some Theory about Lattices

The approach of [20] to generate AMNS parameters can be viewed as follow: given q a prime number, and γ such that $\gamma^n = \lambda \mod q$ is small, one construct the lattice $\mathcal{L}$ of all polynomials having the root γ modulo q.

$$\mathcal{L} = \{\mathfrak{a}(X) \in \mathbb{Z}[X], \text{ such that } \deg(\mathfrak{a}) < n \text{ and and } \mathfrak{a}(\gamma) = 0\}.$$

The polynomial $\mathfrak{m}$ must simply be a short vector of $\mathcal{L}$. However, it remains an open problem to efficiently construct this short vector for a large n. Once we identify the coefficients vectors $(a_0, \ldots, a_{n-1}) \in \mathbb{Z}^n$ with the corresponding polynomial $\mathfrak{a}(X) = \sum_{i=0}^{n-1} a_i X^i \in \mathbb{Z}[X]$, the lattice $\mathcal{L}$ is the set of all linear combinations of rows of its Hermite normal form basis $\mathcal{M}$.

$$\mathcal{M} = \begin{pmatrix} q & 0\ 0\ 0\ \ldots 0 \\ -\gamma & 1\ 0\ 0\ \ldots 0 \\ -\gamma^2 & 0\ 1\ 0\ \ldots 0 \\ \vdots & 0\ 0\ \ddots\ 0\ \vdots \\ -\gamma^{n-2} & 0\ 0\ \vdots\ \ 1\ 0 \\ -\gamma^{n-1} & 0\ 0\ \ldots\ 0\ 1 \end{pmatrix} \begin{array}{l} \leftarrow q \\ \leftarrow X - \gamma \\ \leftarrow X^2 - \gamma^2 \\ \vdots \\ \leftarrow X^{n-2} - \gamma^{n-2} \\ \leftarrow X^{n-1} - \gamma^{n-1} \end{array}$$

The parameter $\mathfrak{m}$ which is used in Algorithm 1, and which is necessary for the Montgomery's multiplication in [20], must be a vector of $\mathcal{L}$ of norm $\|\mathfrak{u}\|_\infty \approx q^{1/n}$. On one hand, the existence of such short vector is guaranteed by a variant of Minkowski's theorem [17] for the $\|\cdot\|_\infty$. In [20], the authors experimentally used the LLL algorithm [15] in small dimension to reduce the lattice and produce a good $\mathfrak{m}$. In medium dimensions ($n \leq 115$), one could still use extreme pruning as described in [12] to find $\mathfrak{m}$. But for larger dimensions, no polynomial algorithm is known to produce such short vector: All known lattice reduction algorithm (either polynomial or practical) outputs vectors exponentially larger than the optimum in practice (namely, $\|\mathfrak{m}\|_2$ would be $\geq 1.01^n q^{1/n}$ see [11]).

Now suppose that we overcome this hardness and obtain a very short vector $\mathfrak{m}$ of $\mathcal{L}$, we still need to ensure that it is invertible   mod ϕ. Note that when $\mathfrak{m}$ has a non-constant GCD with $X^n - \lambda$, this fails for all values of ϕ. Therefore, whenever the resultant of $\mathfrak{m}$ and $X^n - \lambda$ is even, ϕ cannot be chosen as a power of 2. Since Algorithm 1 involves a lot of exact divisions by ϕ, being able to take ϕ equal to a power of two would have a strong impact on the efficiency of the AMNS multiplication.

In the next subsection, we propose several solutions to overcome these problems, depending on how much freeness we have for the choice of the prime number q.

3.2   Construction When We Can Choose q

In ECC cryptography, for example the Diffie Helman protocol, we can freely choose the prime q, the only condition is that q must be large enough considering the security level we want to reach. We propose in that case an efficient way to construct an AMNS representation of the finite field $\mathbb{F}_q$. We adapt the key generation of the last challenges of Gentry's fully homomorphic cryptosystem, published in [13].

The method is a reversal construction of an AMNS representation of a finite field, in a sense that the lattice is built around a chosen short vector $\mathfrak{m}$, which satisfies the most favorable properties. The expensive lattice reduction step is not needed any more. Furthermore, we can ensure that the resultant of $\mathfrak{m}$ and $X^n - \lambda$ is odd, which enables us to set ϕ as a power of 2, and ensures that the AMNS multiplication in $\mathbb{F}_q$ is efficient.

The generation algorithm we propose in Algorithm 4 takes as input the dimension n and a boundary s on the expected norm of $\mathfrak{m}$. First, we choose $\lambda = -1$. This ensures the existence of $2n$-th roots of unity, and most of all, this removes the negative impact of all $|\lambda|$ in the bounds of Lemmas 2 and 1. Then we pick a short vector $\mathfrak{m}$ at random, with coefficients of s bits, and test whether its resultant with $X^n + 1$ is either a prime number, or contains a large prime factor q if one just needs to represent a ring $\mathbb{Z}/q\mathbb{Z}$ instead of a field. Note that in Gentry's challenges [13], the condition was more restrictive, since the resultant itself had to be prime. However, in all cases, it is only necessary to repeat a polynomial number of times the process in order to get a valid polynomial $\mathfrak{m}$ and its associated modulus q. All other parameters (ρ, ϕ, $\mathfrak{m}', \gamma$) are easy to deduce:

Theorem 1. *Given a polynomial $\mathfrak{m}$ of odd resultant r with $X^n - \lambda$ and q a divisor of r, one deduces a valid AMNS representation basis $\mathcal{B} = (q, n, \gamma, \rho, \lambda)$ and additional parameters $\mathfrak{m}, \mathfrak{m}'$ and ϕ needed by Algorithm 1 as follows:*

- $\rho = n\lambda \|m\|_\infty$ *is a valid covering radius,*
- ϕ *is set to the smallest power of* 2 *larger than* $4n\lambda\rho$, *and* $\mathbf{m}'$ *the inverse of* $\mathbf{m} \mod X^n - \lambda \mod \phi$ *exists,*
- γ *is a n-th root of* λ, *which can be extracted from the Hermite Normal Form of the circulant lattice basis*

$$HNF \begin{pmatrix} \mathbf{m}_0 & \mathbf{m}_1 & \mathbf{m}_2 & \cdots & \mathbf{m}_{n-1} \\ \lambda\mathbf{m}_{n-1} & \mathbf{m}_0 & \mathbf{m}_1 & \ddots & \mathbf{m}_{n-2} \\ \lambda\mathbf{m}_{n-2} & \lambda\mathbf{m}_{n-1} & \mathbf{m}_0 & \ddots & \vdots \\ \vdots & \ddots & \ddots & \ddots & \mathbf{m}_1 \\ \lambda\mathbf{m}_1 & \cdots & \lambda\mathbf{m}_{n-2} & \lambda\mathbf{m}_{n-1} & \mathbf{m}_0 \end{pmatrix} = \begin{pmatrix} r & 0 \cdots 0 \\ -\gamma & 1 \ddots \vdots \\ \vdots & 0 \ddots 0 \\ -\gamma^{n-1} & 0 \; 0 \; 1 \end{pmatrix}$$

Proof. The rows of the above circulant matrix generate the lattice formed by coefficient vectors of all algebraic multiples of $\mathbf{m}$ modulo $X^n - \lambda$. By definition, the determinant of this lattice is exactly r, which is also the resultant of $\mathbf{m}$ and $X^n - \lambda$. This means that they share a common root γ such that $\mathbf{m}(\gamma) = 0 \mod r$ and $X^n = \lambda \mod r$. The right side HNF basis codes all polynomials which zero on γ modulo r: it contains the previous lattice, and has the same determinant r, so the generated lattices are equal. Therefore the right side HNF is the hermite normal form of the circulant basis, and allows to compute a root γ of λ even when the factorization of q is unknown. From the circulant basis, we see that a sublattice contains n independent vectors of euclidean norm $\leq \lambda \|\mathbf{m}\|_2$, so the covering radius of the lattice is $\leq \sqrt{n}\lambda \|\mathbf{m}\|_2 \leq n\lambda \|\mathbf{m}\|_\infty$. Therefore we can choose $\rho = n\lambda \|m\|_\infty$.

The algorithm is sumarized in Algorithm 4.

Algorithm 4. Generate AMNS parameters

Require: A dimension n, a size s, and a small integer λ
Ensure: An AMNS parameter set $\mathcal{B} = (q, n, \lambda, \gamma, \rho)$ and $(\mathbf{m}, \mathbf{m}', \phi)$
1: **repeat**
2:     Choose a vector $(m_0, \ldots, m_{n-1})$ with all coefficients $\in [-2^s, 2^s]$
3:     Compute the resultant q' of $m = \sum_{i=0}^n m_i X^i$ and $P_\mathcal{B}(X) = X^n - \lambda$.
4: **until** q' is prime or has a large prime factor q
5: Chose $\rho = \lambda n \|\mathbf{m}\|_\infty$ and $\phi = 2^{\lceil 4n\lambda\rho \rceil}$
6: Compute the common root γ of $X^n - \lambda$ and $\mathbf{m}$ modulo q (using a gcd or an HNF algorithm)
7: Compute the inverse $\mathbf{m}'$ of $\mathbf{m}$ modulo ϕ
8: **Return** the parameter set $\mathcal{B} = (q, n, \lambda, \gamma, \rho)$ and $(\mathbf{m}, \mathbf{m}', \phi)$

The advantage of this construction is that we do not have to proceed to an LLL reduction in order to generate the parameters for the AMNS basis. Furthermore the vector m is invertible $\mod (P_\mathcal{B}, \phi)$ by construction. This algorithm returns the parameters of an AMNS basis, also the vector m and its inverse mI modulo $P_\mathcal{B}$ and ϕ.

In cases where Gentry's construction cannot be used. Typically, in pairing based cryptography, where the prime q is imposed by the choices of the parameters. We have to construct an efficient base AMNS dealing with a fixed value of the prime q.

3.3   Construction When We Cannot Choose q

In pairing based cryptography, the prime q, the fields $\mathbb{F}_q$ and the extensions $\mathbb{F}_{q^m}$ are fixed during the construction of the elliptic curve. We do not have enough freedom to use the Gentry-like algorithm of Section 3.2 to generate the AMNS basis.

Of course, selecting another elliptic curve would produce another prime q. This fact can be used to tune q until $2m$ divides $q-1$. This ensures the existence of $2m$ roots of unity, and allows to set $\lambda = -1$. However, once q is chosen, the only way to generate $\mathfrak{m}$ is to find a short vector of the lattice spanned by

$$
\mathcal{B} = \begin{pmatrix}
q & 0 & 0 & 0 & \dots & 0 \\
-\gamma & 1 & 0 & 0 & \dots & 0 \\
-\gamma^2 & 0 & 1 & 0 & \dots & 0 \\
\vdots & & 0 & 0 & \ddots & 0 & \vdots \\
-\gamma^{n-2} & 0 & 0 & \vdots & & 1 & 0 \\
-\gamma^{n-1} & 0 & 0 & \dots & & 0 & 1
\end{pmatrix}.
$$

This lattice reduction phase already existed in [3]. It can be performed using the LLL algorithm. Here, we prove that after these steps, we can choose the small vector $\mathfrak{m}$ having an odd resultant with $X^m + 1$. This allows ϕ to be a power of 2.

Lemma 4. *Let $\mathcal{L}$ be the lattice of the AMNS representation, generated by the basis $\mathcal{B}$ above. Let $\mathcal{M}$ be another basis of $\mathcal{L}$. Then, for any $\phi = 2^k$ a power of 2, there exist at least one vector $\mathfrak{m}$ of $\mathcal{M}$ such that $\mathfrak{m}$ is invertible modulo $\mathrm{mod}\ (P_\mathcal{B}, \phi)$*

Proof. First, note that a polynomial $\mathfrak{m}$ is invertible $\mathrm{mod}\ (P_\mathcal{B}, \phi)$ as soon as the evaluation of $\mathfrak{m}$ over all integers are odd. By interpolation arguments, it is enough to verify that $\mathfrak{m}(x)$ is odd on at least n integers $\in [0, \phi - 1]$. In the public basis $\mathcal{B}$, this is the case of the constant polynomial $\mathfrak{b}_0 = q$. All the other rows can be put on the form $\mathfrak{b}_k = X^k - \alpha_k$ where α_k is odd (by adding q if necessary). Of course, $\mathfrak{b}_1, \dots \mathfrak{b}_{k-1}$ are not invertible since their evaluations on 1 are always even.

Let $\mathcal{M} = [\mathfrak{m}_0, \mathfrak{m}_1, \dots, \mathfrak{m}_{n-1}]$ be another basis of $\mathcal{L}$, it is obtained by left multiplication of $\mathcal{B} = [\mathfrak{b}_0, \mathfrak{b}_1, \dots, \mathfrak{b}_{n-1}]$ by a unimodular matrix. Consequently, there exists an index i such that $\mathfrak{m}_i = u_0\mathfrak{b}_0 + \cdots + u_{n-1}\mathfrak{b}_{n-1}$ where u_0 is odd. Then by construction, $\mathfrak{m}_i$ evaluates to an odd number on every odd integer, and therefore, it is invertible $\mathrm{mod}\ (P_\mathcal{B}, \phi)$. $\square$

By this lemma, once a lattice reduction algorithm (like LLL [15], BKZ [23], or better, HKZ [22]) has been run on $\mathcal{B}$, this property implies that we always have a vector in $\mathcal{M}$ invertible   mod $(P_{\mathcal{B}}, \phi)$. We can choose this vector for the Montgomery's like multiplication, together with $\phi = 2^k$. Even if this is not the first vector of $\mathcal{M}$, the norm of $\mathfrak{m}$ remains short.

4   Implementation and Results

In order to illustrate our approach, we implemented the AMNS arithmetic for $\mathbb{F}_{q^m}$ where q had between 150 bits and 1500 bits, and for various extensions $m = 8, 16, 32$ and 64 with a FFT. Other small non-power-of-two extensions, like $m = 17$ use the DFT. Both algorithms are given in Section 2.

Our implementation is written in C++ language, and is tested on an Intel Core i5 laptop with a 32-bit Linux platform[1]. The AMNS parameters are chosen such that the representation can fit on 32-bit integers, and during the computation, temporary variables are stored on 64-bit integers, so that they can store products without overflow. Note that on 64-bit platforms, the size of the parameters can be doubled, so that AMNS vectors fit on 64-bit integers and temporaries on 128-bit long integers.

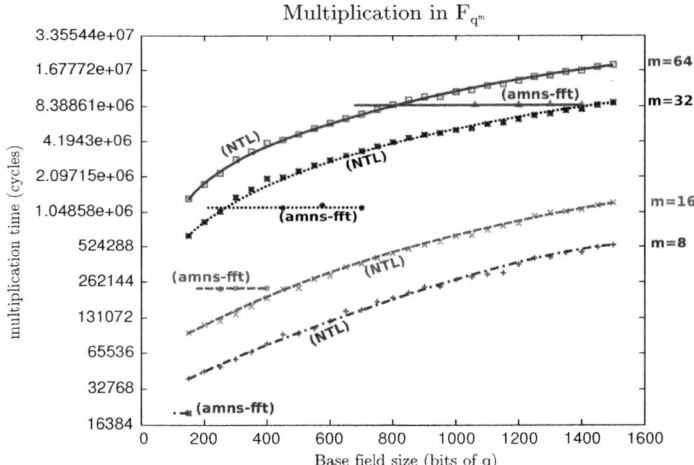

Fig. 1. Running time in cycles of multiplication in $\mathbb{F}_{q^m}$, AMNS with FFT versus NTL ZZ_pE

By construction, the complexity of the arithmetic in AMNS representation only depends on the dimension n and the extension degree $m \leq n$. When these two dimensions are fixed, the running time does not depend on the bit-size of the corresponding base field q, as soon as there exist AMNS parameters for $\mathbb{F}_q$ (basically, $2n|q-1$ and and $q = O(\rho^n)$).

---

[1] Ubuntu precise 12.04, with kernel 3.2.0-21-generic-pae, 4Gb RAM, and compiler g++ version 4.6.3 (Ubuntu/Linaro 4.6.3-1ubuntu3)

Note that the AMNS arithmetic requires some precomputations, for instance of the polynomial $\mathfrak{m}$. In order to make a fair comparison, we chose to compare our results with the NTL 5.4.2 [24][2] version of arithmetic on the same field extensions using the ZZ_pE module. This is the fastest module of NTL to perform arithmetic in a fixed field extension $\mathbb{F}_q[X]/P$, and allows for instance NTL to preprocess some data, based on q and P, in order to speed-up all operations. Of course, the arithmetic in the base field of NTL is performed by GMP, which is also highly optimized in assembly.

Figure 1 provides the reader with a comparison between the computational time required by our AMNS library and NTL for multiplication over $\mathbb{F}_{q^m}$.

The running time of a multiplication in cycles of AMNS and NTL based multiplications. We restricted the AMNS parameters to the case $m = n$, and the results are obtained with the -O9 -static -funroll-all-loops -masm=intel -ftree-vectorize -msse3 optimization flags of g++.

For $m = 8$, our AMNS implementation is more efficient than NTL for q smaller than 180 bit. For $m = 16$, our AMNS implementation less efficient than NTL. However, for larger extension degrees, according to the results reported in Figure 1, our implementation of AMNS multiplication is faster than NTL for large q, like extension degree $m = 32$ and $q \geq 300$ bits, or $m = 64$ and $q \geq 800$ bits.

5   Conclusion

We revised the study of AMNS bases in order to improve the arithmetic over finite field extensions. We propose new efficient routines to efficiently construct AMNS bases. An easy one, when the characteristic can be freely chosen, is derived from Gentry's algorithm. Else, when the prime number q is constrained by other parameters of the protocol, we develop an explicit construction of efficient AMNS bases. We also propose the first software implementation of the AMNS arithmetic over fields extensions, which runs faster than GMP over large fields. It could be interesting to compare an implementation of a pairing based protocole in AMNS representation with an implementation in a classical representation.

Open problems:

– As mentionned in Remark 2 the equality test in the AMNS representation is not straightforward. It would be interesting to compare the different methods to compare a number in the AMNS representation.
– According to the protocol, it could be interesting to find an efficient method to implement the division in the AMNS representation.
– We compare the multiplication in an extension field in the AMNS representation with the classical multiplication implemented in NTL. This work can be completed with the comparaison of the implementation of a cryptographic protocol over finite field in classical representation and in the AMNS representation.

---

[2] libntl-dev, as packaged by default on ubuntu 11.10. It is statically linked with GMP 4.3.2 for large integer operations.

Acknowledgments. The authors would like to thank Jean Luc Beuchat, Thomas Plantard, and Peter Schwabe for their invaluable comments and helpful suggestions on our preliminary manuscript, and the anonymous reviewers for their numerous suggestions and remarks which have enables us to substantially improve the paper.

References

1. NIST Key Length Recommendations, http://www.keylength.com/
2. Recommendations for Key Management. Special Publication 800-57 Part 1 (2007)
3. Bajard, J.-C., Imbert, L., Plantard, T.: Modular Number Systems: Beyond the Mersenne Family. In: Handschuh, H., Hasan, M.A. (eds.) SAC 2004. LNCS, vol. 3357, pp. 159–169. Springer, Heidelberg (2004)
4. Bodrato, M.: Towards Optimal Toom-Cook Multiplication for Univariate and Multivariate Polynomials in Characteristic 2 and 0. In: Carlet, C., Sunar, B. (eds.) WAIFI 2007. LNCS, vol. 4547, pp. 116–133. Springer, Heidelberg (2007)
5. Boneh, D., Franklin, M.: Identity-Based Encryption from the Weil Pairing. In: Kilian, J. (ed.) CRYPTO 2001. LNCS, vol. 2139, pp. 213–229. Springer, Heidelberg (2001)
6. Boneh, D., Lynn, B., Shacham, H.: Short Signatures from the Weil Pairing. J. Cryptology 17(4), 297–319 (2004)
7. Brier, E., Joye, M.: Fast Point Multiplication on Elliptic Curves Through Isogenies. In: Fossorier, M.P.C., Høholdt, T., Poli, A. (eds.) AAECC 2003. LNCS, vol. 2643, pp. 43–50. Springer, Heidelberg (2003)
8. Chung, J., Hasan, A.: More Generalized Mersenne Numbers (Extended Abstract). In: Matsui, M., Zuccherato, R.J. (eds.) SAC 2003. LNCS, vol. 3006, pp. 335–347. Springer, Heidelberg (2004)
9. Devegili, A.J., Ó hÉigeartaigh, C., Scott, M., Dahab, R.: Multiplication and squaring on pairing-friendly fields. Cryptology ePrint Archive, Report 2006/471 (2006), http://eprint.iacr.org/
10. El Mrabet, N., Negre, C.: Finite Field Multiplication Combining AMNS and DFT Approach for Pairing Cryptography. In: Boyd, C., González Nieto, J. (eds.) ACISP 2009. LNCS, vol. 5594, pp. 422–436. Springer, Heidelberg (2009)
11. Gama, N., Nguyen, P.Q.: Predicting Lattice Reduction. In: Smart, N.P. (ed.) EUROCRYPT 2008. LNCS, vol. 4965, pp. 31–51. Springer, Heidelberg (2008)
12. Gama, N., Nguyen, P.Q., Regev, O.: Lattice Enumeration Using Extreme Pruning. In: Gilbert, H. (ed.) EUROCRYPT 2010. LNCS, vol. 6110, pp. 257–278. Springer, Heidelberg (2010)
13. Gentry, C., Halevi, S.: Implementing Gentry's Fully-Homomorphic Encryption Scheme. In: Paterson, K.G. (ed.) EUROCRYPT 2011. LNCS, vol. 6632, pp. 129–148. Springer, Heidelberg (2011)
14. Katti, R., Brennan, J.: Low Complexity Multiplication in a Finite Field Using Ring Representation. IEEE Transactions on Computers 52(4), 418–427 (2003)
15. Lenstra, A.K., Lenstra Jr., H.W., Lovász, L.: Factoring polynomials with rational coefficients. Mathematische Ann. 261, 513–534 (1982)
16. Miller, V.S.: Use of Elliptic Curves in Cryptography. In: Williams, H.C. (ed.) CRYPTO 1985. LNCS, vol. 218, pp. 417–426. Springer, Heidelberg (1986)
17. Minkowski, H.: Geometrie der Zahlen. Druck und Verlag von B.G. Teubner, Leipzig und Berlin (1910)

18. Montgomery, P.L.: Five, six, and seven-term Karatsuba-like formulae. IEEE Transactions on Computers 54(3), 362–369 (2005)
19. El Mrabet, N., Guillevic, A., Ionica, S.: Efficient Multiplication in Finite Field Extensions of Degree 5. In: Nitaj, A., Pointcheval, D. (eds.) AFRICACRYPT 2011. LNCS, vol. 6737, pp. 188–205. Springer, Heidelberg (2011)
20. Negre, C., Plantard, T.: Efficient Modular Arithmetic in Adapted Modular Number System Using Lagrange Representation. In: Mu, Y., Susilo, W., Seberry, J. (eds.) ACISP 2008. LNCS, vol. 5107, pp. 463–477. Springer, Heidelberg (2008)
21. Rubin, K., Silverberg, A.: Torus-Based Cryptography. In: Boneh, D. (ed.) CRYPTO 2003. LNCS, vol. 2729, pp. 349–365. Springer, Heidelberg (2003)
22. Schnorr, C.-P.: A hierarchy of polynomial lattice basis reduction algorithms. Theoretical Computer Science 53, 201–224 (1987)
23. Schnorr, C.-P., Euchner, M.: Lattice basis reduction: improved practical algorithms and solving subset sum problems. Math. Programming 66, 181–199 (1994)
24. Shoup, V.: Number Theory Library (1996), http://www.shoup.net/ntl
25. Silverman, J.H.: Rings of low multiplicative complexity. In: Finite Fields and Their Applications, vol. 6, pp. 175–191. Academic Press (2000)
26. van Dijk, M., Granger, R., Page, D., Rubin, K., Silverberg, A., Stam, M., Woodruff, D.: Practical Cryptography in High Dimensional Tori. In: Cramer, R. (ed.) EUROCRYPT 2005. LNCS, vol. 3494, pp. 234–250. Springer, Heidelberg (2005)
27. Von ZurGathen, J., Gerhard, J.: Modern Computer Algebra. Cambridge University Press, New York (2003)

GF(2^m) Finite-Field Multipliers with Reduced Activity Variations

Danuta Pamula[1,2] and Arnaud Tisserand[1]

[1] IRISA, CNRS, INRIA, Univ. Rennes 1, Lannion, France.
arnaud.tisserand@irisa.fr
[2] Silesian University of Technology, Gliwice, Poland
danuta.pamula@polsl.pl

Abstract. Electrical activity variations in a circuit are one of the information leakage used in side channel attacks. In this work, we present GF(2^m) multipliers with reduced activity variations for asymmetric cryptography. Useful activity of typical multiplication algorithms is evaluated. The results show strong shapes, which can be used as a small source of information leakage. We propose modified multiplication algorithms and multiplier architectures to reduce useful activity variations during an operation.

Keywords: finite field arithmetic, cryptography, side channel attacks

1 Introduction

Side channel attacks (SCA) are nowadays a major threat for embedded cryptographic systems. Power analysis based SCAs exploit correlations between internal secret values (e.g. keys) and the current, which flows in the circuit [14]. Similar information leakage can be exploited from electromagnetic radiations [7].

In elliptic curve cryptography (ECC) [9], many SCAs [19] have been proposed. To protect circuits against those attacks researchers propose various countermeasures, or protections, for ECC see [10]. Moreover, specific protections at the arithmetic level have been proposed for ECC (and other asymmetric cryptosystems). For instance, addition chains allow performing only one type of operation, point addition, during scalar multiplications [2]. In [3] randomized and very redundant representations of the scalar are used. But these protections are at the curve level not the finite field one. Efficient and secure computation units for finite-field arithmetic are important elements of ECC processors. For instance, see [18] and [8] as examples of GF(2^m) ECC processors.

In this work, we investigate protection elements at the field level in GF(2^m) multiplication algorithms and their architectures for ECC. The proposed solutions are not autonomous countermeasures but an additional protection element, which should enhance higher-level countermeasures. In some SCAs, power variations provide sources of information leakage about the executed operations.

Instantaneous power is linked with the number of useful transitions in the operator. Useful transitions are the theoretical changes during the operation

F. Özbudak and F. Rodríguez-Henríquez (Eds.): WAIFI 2012, LNCS 7369, pp. 152–167, 2012.

(from one clock cycle to the next one). This is also called useful circuit activity. To estimate information leakage in typical GF(2^m) multipliers, we first accurately measured their useful activity. The results show very strong shapes for useful activity variations. These shapes reveal multiplications time boundaries. Next, we propose modifications of the multiplication algorithms and architectures for reducing useful activity variations during the computation.

The paper presents background on power consumption in digital circuits and activity evaluation methods in Section 2. In Section 3, GF(2^m) multiplication algorithms used for our experiments are described. In Section 4, we present the results and an analysis of useful activity measurements as an estimation of information leakage source. Section 5 presents our modifications for reducing the activity variations in the multipliers and the new implementation results.

2   Activity in Hardware Arithmetic Operators

This section gives a short introduction to electrical activity in digital integrated circuits and its links to SCAs. Power analysis based SCAs use possible correlations between internal secret values (e.g. keys) and information leakage related to instantaneous power of the executed operations (see [14] for details).

Instantaneous power at time t is $P_{\mathrm{DD}}(t) = i_{\mathrm{DD}}(t) \times V_{\mathrm{DD}}$ where $i_{\mathrm{DD}}(t)$ is the instantaneous current and V_{DD} is the power supply. Power consumption components are: *static* power and *dynamic* power. See [26, Sec. 4.4] for circuit-level details in CMOS circuits. Static power does not depend on circuit activity and is not used in this work. Dynamic power is due to circuit activity: charging and discharging load/parasitic capacitances and short-circuit currents. It strongly depends on the executed operations and data values. Dynamic power variations are used as a source of information leakage for power attacks.

Dynamic power components are: *useful* activity and *parasitic* activity as illustrated on Figure 1. Useful (or theoretical) activity is due to complete and stable transitions required by computations from one clock cycle to the next one (i.e. $0 \rightarrow 1$ and $1 \rightarrow 0$ for each bit). Parasitic (or glitching) activity is due to non-useful transitions. For instance, in case of non-equal arrival times for a gate inputs, the output may have multiple transitions before reaching a steady state.

Parasitic activity in GF(2^m) multipliers is small. This is not the case for all arithmetic operators (e.g. operators in high-performance CPUs [24]). In GF(2^m) arithmetic units, the logical depth is small. Power consumption of memory elements (e.g. flip-flops) used in GF(2^m) multipliers is important compared to

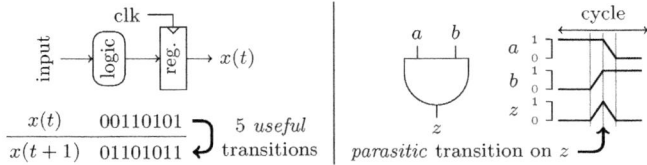

Fig. 1. Useful (left) and parasitic (right) transitions

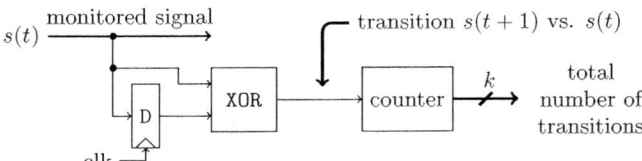

Fig. 2. Activity counter architecture for a 1-bit signal $s(t)$ (control not represented)

power consumption in logic gates. In this work, we only focus on useful activity as a large contribution to $i_{DD}(t)$.

Several methods can be used to evaluate useful activity: cycle accurate and bit accurate (CABA) simulation of a low-level architecture description, electrical simulation or FPGA emulation. Fast high-level behavioral simulation is not sufficient to catch cycle accurate and bit-level coding aspects. As the target operators have large operands (e.g. 160 to 500 bits for ECC) and long computations CABA simulation would be to slow. This is even more critical with electrical simulation. So, we use FPGA emulation for evaluating useful activity. An activity counter is added on each monitored signal [25]. It counts the number of useful transitions as illustrated on Figure 2. The D flip-flop and the XOR gate produce a 1 for each useful transition between $s(t+1)$ and $s(t)$. The k-bit counter accumulates transitions numbers (k depends on test vector length).

We insert activity counters at the output of each internal register and for each bit of the multiplier. Comparisons with electrical simulations in [25] show that this is reasonable assumption for small parasitic activity. Outputs of all XOR gates (radix-1 representation of transitions number) are compressed into a binary value as the total transitions number for cycle t. This value is stored in a memory as an estimation of $i_{DD}(t)$. At the end of the evaluation period, the memory content is sent to the host computer for the analysis step. Using FPGA emulation, it is possible to quickly and accurately evaluate useful activity in $GF(2^m)$ finite-field multipliers for large and relevant test vectors (this cannot be done using "slow" software simulations). Activity counters do not change the multiplier mathematical behavior. FPGA emulation is obviously CABA.

In Section 4, we present the implementation results of $GF(2^m)$ multipliers without and with activity counters. Table 1 reports huge area overhead and about a $\div3$ frequency decrease due to the counters. These overheads are very important, but they only appear during evaluation not in final circuit. FPGA emulation leads to activity evaluation running at more than 100 MHz (cf. Table 1) which would not be possible using software simulations.

3 $GF(2^m)$ Finite-Field Multiplication Operators

A complete introduction to finite fields and their applications can be found in book [13]. Algorithms for finite-field arithmetic are presented in book [4]. Binary-field extension algorithms are presented in [21, chapter 6]. A summary on $GF(2^m)$ multiplication algorithms is given in paper [22].

Our goal was to design efficient finite-field multipliers in such a way that their architectures can be easily modified to add protection against SCAs. We have analyzed many algorithms, with different variants, to be able to take and combine those parts, which will allow us to create the most efficient algorithm fulfilling, assumed requirements. We have considered GF(2^m) elements represented by polynomial basis of the form $\{1, x, x^2, ..., x^{m-2}, x^{m-1}\}$. As a result of our study, we have prepared three GF(2^m) multipliers based on three different algorithms: classical two-step, Montgomery and Mastrovito.

All proposed hardware solutions are analyzed for standard size $m = 233$ (similar results are obtained for other field extension sizes). See [9, annexes A and B] for ECC standards and parameters. The multipliers are described using VHDL, synthesized, place and routed using Xilinx ISE 12.2 environment and implemented in a Xilinx Virtex-6 LX240T FPGA.

3.1   Classic Two-Step Multiplication Algorithm

Classic multiplication comprises two steps: multiplication $d(x) = a(x)b(x)$ and reduction $c(x) = d(x) \bmod f(x)$. There exist many versions of two-step multipliers, which combine different methods for multiplication and reduction. In our hardware solution we have combined Karatsuba-Ofman multiplication principle with features of matrix-vector approach. Moreover to perform reduction we have used classical method optimized for a specific irreducible polynomial.

Multiplication part: In order to avoid managing large vectors a, b we have partitioned them into smaller vectors according to divide-and-conquer method optimized with Karatsuba-Ofman trick [11]. Karatsuba-Ofman partitioning of polynomials is assumed for polynomials of even sizes. Because field sizes used for cryptographic purposes are usually odd, we add redundant zeroes on missing most significant positions of A_H, B_H chunks. Utilizing Karatsuba-Ofman optimization aiming at reducing number of the most complex operation in the equation, we denote multiplication as follows:

$$d(x) = a(x)b(x) = (x^{\frac{m}{2}} A_H + A_L)(x^{\frac{m}{2}} B_H + B_L)$$
$$= x^m A_H B_H + x^{\frac{m}{2}}((A_H + A_L)(B_H + B_L) - A_H B_H - A_L B_L) + A_L B_L.$$

After vast analysis we have found that the best results in terms of speed and area, for most input sizes are for halved inputs. For some partitions sizes, designed multipliers were significantly faster but also significantly bigger and otherwise. We had to find some trade-off. Thus having $m = 233$ and utilizing Karatsuba-Ofman trick to perform multiplication we have decided to use three 117-bit multipliers.

Our 117-bit multipliers are based on matrix-vector approach. There, polynomial $a(x)$ is represented by a specific matrix A of size $2m - 1 \times m$, in which each column represents consecutive left shifts of $a(x)$, element $b(x)$ is represented in form of m-bit vector and product $d(x)$ is also a vector but of size $(2m - 1)$.

Our idea for a multiplier based on this approach is to store only two columns of the matrix A at a time, which in fact means working on two registers representing matrix A columns, exchanging in the first values of consecutive columns of A and accumulating partial results in the other. We may actually regard one vector as a column of matrix A and the other as a vector storing product d.

Reduction part: Reduction module uses reduction method optimized for irreducible polynomial $f(x)$. In classical method (see [9], [4]) one looks for bits equal 1 in the upper part of $d(x)$, from ranks $(2m - 2)$ down to m, and step by step reduces vector $d(x)$, XORs vector $d(x)$ by appropriate shift left of irreducible polynomial $f(x)$. The drawback of this method is that it is very time consuming. Observing properties of special irreducible polynomials (e.g. trinomials, pentanomials) one may optimize classical algorithm. In case of field size $m = 233$, the irreducible trinomial $f(x) = x^{233} + x^{74} + 1$ is used. With such polynomial we were able to significantly decrease number of steps needed to reduce the product as depicted in Figure 3. Similar algorithms can be found in [9].

CONSTANTS: $f(x) = x^{233} + x^{74} + 1$,
INPUT: $d(x)$,
OUTPUT: $c(x) = d(x) \bmod f(x)$

1. $e = d[2m - 2, ..., m]$ // assign part of vector d to e
2. $e1 = e \times f$
3. $d1 = d$ XOR $e1$ // first reduction step
4. $e = d1[74 + (m - 1), ..., m]$ // assign part of new vector d to e
5. $e2 = e \times f$
6. $c = d1$ XOR $e2$ // second reduction step
7. **return** c

Fig. 3. Classical reduction algorithm

Multiplication in lines 2. and 5. is a short chain of XOR operations. Combining described multiplication and reduction block we have obtained a multiplier with size of 3638 LUTs, frequency of 302 MHz and operation time of 264 clock cycles.

3.2   Montgomery Multiplication Multiplier

The Montgomery algorithm is constructed in a specific way to avoid most costly operations. Instead of performing $c(x) = a(x)b(x) \bmod f(x)$ it performs $c(x) = a(x)b(x)r^{-1}(x) \bmod f(x)$. To perform complete finite-field multiplication $c(x) = a(x)b(x) \bmod f(x)$ one must run the algorithm twice, at first for $a(x)$ and $b(x)$ and then for the obtained result $d(x)$ and $r^2(x) \bmod f(x)$. Operation $c(x) = a(x)b(x) \bmod f(x)$ comprises in fact two steps: 1) $d = MontMult(a, b)$ and 2) $c = MontMult(d, r^2 \bmod f)$ where $MontMult(a, b)$ symbolizes Montgomery multiplication. The algorithm we have used based on Montgomery method [16] calculating the modular product $c(x)$ is as presented in Figure 4.

CONSTANTS: $r(x) = x^m, f(x), f'(x), r^2(x) \bmod f(x)$
INPUT: $a(x), b(x)$
OUTPUT: $c(x) = a(x)b(x) \bmod f(x)$
// $MontMult(a, b)$
1. $t(x) = a(x)b(x)$
2. $u(x) = t(x)f'(x) \bmod r(x)$
3. $d(x) = [t(x) \text{ XOR } u(x)f(x)]/r(x)$
// $MontMult(d, r^2 \bmod f)$
4. $t(x) = d(x)(r^2(x) \bmod f(x))$
5. $u(x) = t(x)f'(x) \bmod r(x)$
6. $c(x) = [t(x) \text{ XOR } u(x)f(x)]/r(x)$
7. **return** c

Fig. 4. Modular multiplication algorithm based on Montgomery method

To be able to utilize the algorithm we need three values, $r(x)$, $r^2(x) \bmod f(x)$, $f'(x)$, which for known irreducible polynomial can be pre-calculated. Element $r(x)$ for field GF(2^m) is chosen to be a simple polynomial x^m (see [12]).

The most complicated in the algorithm is the first step where we need to perform multiplication of two large binary vectors. To perform it we have used multiplier based on shift-and-add method. For $m = 233$ we divide vector b into 16-bit chunks (we add bits equal to 0 on MSB positions of chunk if necessary), multiply sequentially a by all parts of vector b (we need to perform 15 multiplications) and sequentially accumulate partial results. To construct full finite-field multiplier based on Montgomery method, we may use different types of multipliers but we have to remember that they strongly influence final solution.

All other multiplications needed during execution of the algorithm, multiplication by $r(x), r^2(x) \bmod f(x), f'(x), f(x)$ are simpler due to the fact that we know the values of those operands and the number of bits equal to 1 in those polynomials is very low. Thus we may substitute those multiplications with short chains of XOR operations. In modulo operation in line 2. and 5., we just cut out all elements of order higher or equal to m. The division operation in line 3. is just simple right shift by m positions. The resulting multiplier uses 2178 LUTs, runs at 323 MHz and needs 270 clock cycles to compute the complete product.

3.3   Mastrovito Multiplication Algorithm

Mastrovito matrix method [15] is an interleaved version of basic matrix-vector approach. In standard matrix-vector approach, we perform two steps: multiplication and reduction with use of special reduction matrix R. Reduction matrix R is a matrix, which coefficients are defined in terms of irreducible polynomial $f(x)$, generating the field. In Mastrovito approach we perform only one step $c = Mb = (A^L + A^H R)b$, where M is so called Mastrovito matrix. Mastrovito matrix M is a combination of A^L, A^H matrices (parts of matrix A representing polynomial a) and special reduction matrix R, see [5].

Matrix M construction and storage is very problematic. Our idea is to partition it into sub-matrices, to save area and ease optimization and synchronization of operators. The chosen size of sub-matrices is 16×16 bits. In order to save space, coefficients of sub-matrices are not stored in the operator. They are calculated on-the-fly during computations, from parts of matrices A^L, A^H and R (a, b, f operands), by dedicated sub-multipliers.

Sub-multipliers schedule is controlled using a finite-state machine. Results of sub-multiplications are independent of each other and can be calculated in arbitrary order. We can group sub-multipliers in different manners and that way easily change computation time or somehow the design area. We can group sub-matrices into rows, and try to use each sub-multiplier as efficiently as possible; we may change the order of sub-multiplications to adapt the circuit to our needs.

Our basic algorithm for computing the result of $a(x)b(x) \bmod f(x)$ works as follows: the 16×16-bit sub-matrices of M are grouped into rows, spanning one row of matrix M, and for each row 16-bit part of ab product is calculated:

$$c_0 = M_{(0,0)}b_0 + M_{(0,1)}b_1 + \quad \cdots \quad + M_{(0,m/16)}b_{m/16}$$
$$c_1 = M_{(1,0)}b_0 + M_{(1,1)}b_1 + \quad \cdots \quad + M_{(1,m/16)}b_{m/16}$$
$$\vdots$$
$$c_{m/16} = M_{(m/16,0)}b_0 + M_{(m/16,1)}b_1 + \quad \cdots \quad + M_{(m/16,m/16)}b_{m/16},$$

where c_i denotes 16-bit chunk of final result $c(x)$, M denotes Mastrovito matrix resulting from $A^L + A^H R$.

It is easy to observe that there exist many variations of sub-multiplications schedule. The most efficient solution, obtained after many experiments, has the area of 3760 LUTs, frequency of 297 MHz and needs 75 clock cycles to compute the multiplication.

4   Useful Activity Analysis for Multiplication Algorithms

Useful activity of $GF(2^m)$ multipliers has been evaluated using FPGA emulation and activity counters from Section 2. The activity counters are monitoring transitions of each register used in an operator. Corresponding implementation results without and with activity counters are reported in Table 1. The $3\times$ area overhead is due to the numerous counters inserted. Frequency of monitored multipliers is divided by 3. But measurements still can be performed at 100 MHz or more. Such an evaluation speed would not be possible using software simulation.

For all experiments, random operands have been used with uniform and equiprobable distribution for all bits. We performed numerous experiments (corresponding to hundreds of thousands clock cycles for each tested solution). The traces reported below correspond to typical traces. Using average trace is not possible since this may flatten the activity variations and mask information leakage. Thus we have been evaluating our modifications by running modified multipliers for several various sets of experimental data.

Figure 5 (left) presents useful activity measurement results for a typical sequence of $GF(2^m)$ multiplications of random operands using classical algorithm

Table 1. FPGA implementation results of GF(2^m) multipliers without (original operators) and with (monitored operators) activity counters

Algorithms	*without* activity counters			*with* activity counters		
	area LUT	freq. MHz	clock cycles	area LUT	freq. MHz	clock cycles
Classical	3638	302	264	11383	133	264
Montgomery (full)	2178	323	270	6100	121	270
Mastrovito	3760	297	75	5956	110	75

from Section 3.1. There is a high peak at the beginning of each multiplication due to the initialization phase. Figure 5 (right) presents an extract for a single representative multiplication (all random operands lead to the similar overall shape). We have noticed that dependency of the shape of activity variation curves on input data is rather low.

Measurement results for a sequence of random GF(2^m) multiplications using Montgomery algorithm from Section 3.2 are presented in Figure 6 (left) with an extract of a single representative multiplication (right). The reported measurements are shown for complete multiplications with final reduction (for conversion from Montgomery "representation"). We will provide comments on that point at the end of this section. There is a large activity drop at the end of each multiplication due to the reduction step.

Figure 7 (left) presents useful activity measurement results for a typical sequence of random GF(2^m) multiplications using Mastrovito algorithm from Section 3.3 with an extract for a single representative multiplication (right). The variations of the useful activity during a multiplication have a very specific decreasing "step-wave" shape.

Measurements for all three multiplication algorithms show very specific shapes for useful activity variations, which may lead to some information leakage. Those specific shapes provide the attacker with strong temporal references of the operations time location. Based on these references about field-level operations, higher-level operations (e.g. point addition and doubling) can be guessed.

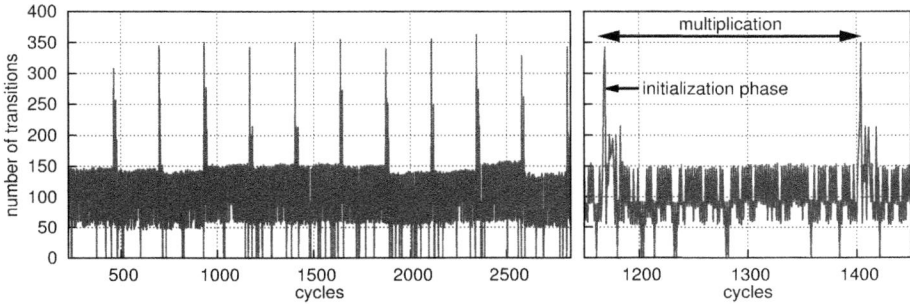

Fig. 5. Useful activity measurement results for random GF(2^m) multiplications with classical algorithm (left). Extract for a single representative multiplication (right).

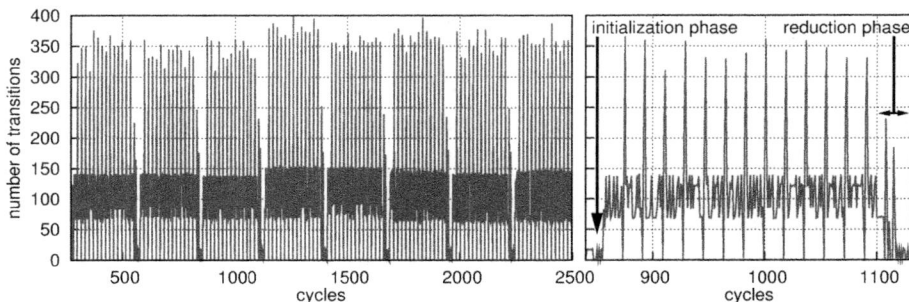

Fig. 6. Useful activity measurement results for random $GF(2^m)$ multiplications with Montgomery algorithm (left). Extract for a single representative multiplication (right).

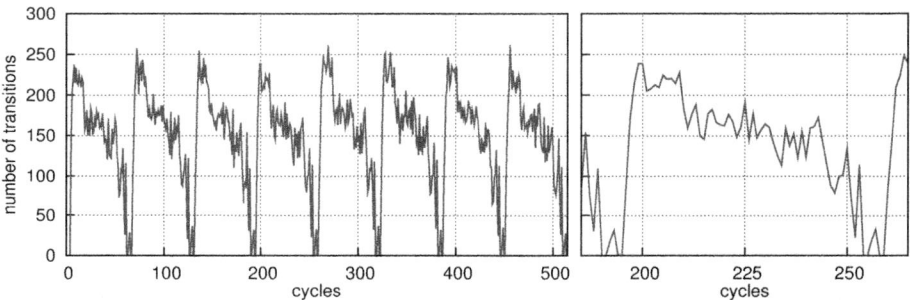

Fig. 7. Useful activity measurement results for random $GF(2^m)$ multiplications with Mastrovito algorithm (left). Extract for a single representative multiplication (right).

Peaks due to the initialization phase at the beginning of operations in Figure 5 are not related to the selected algorithm but to the implemented architecture and especially its control. Resetting all internal registers generates a lot of activity. Then this specific different shape for the initialization phase may occur for other algorithms and architectures. We will see in next section how this problem can be fixed using architecture modifications.

Activity drops at the end of operations in Figure 6 are due to low-complexity reduction step for the considered irreducible polynomial compared to multiplication iterations complexity. We reported measurements for complete multiplication (with final reduction) for fair comparison with other algorithms. In practice, those drops would not appear since reduction is only used at the end of a sequence of operations (with operands in Montgomery domain).

The most problematic shape is the one for Mastrovito algorithm in Figure 7. The decreasing "step-wave" shape is due to variation of the computations quantity in the algorithm. In next section, we will see modifications of this multiplier at algorithmic and arithmetic levels to reduce information leakage.

5   Modifications on Multiplication Algorithms for Reducing Useful Activity Variations

Analyzing the obtained activity variations curves, we can define modification objectives. First, we have to suppress the peaks at the initialization phase. This is an architecture issue (i.e. modification of the operator control). All multiplication algorithms may benefit from this type of modification. Second, we have to take care of the activity drops during the reduction phase of Montgomery algorithm. But as stated in previous section, this phase is only used at the end of long sequence of operations in real ECC applications. Last, we have to make the "step-wave" shape of useful activity variations of Mastrovito algorithm less distinguishable. Below, we describe our modifications for each algorithm.

Classical two-step multiplication: The analysis shows that peaks at the beginning of each multiplication occur due to circuit initialization. To suppress them, we have modified initialization method. Now we do not reset all registers in the first cycle but we have spread the reset activity over several cycles. Figure 8 shows useful activity measurements for a sequence of random multiplications using the modified multiplier. To reduce activity variations, we have also optimized the reduction step by reducing number of registers involved in reduction and merging all the steps of algorithm presented on Figure 3 into a chain of XOR operations. In the modified multiplier the average activity varies between 100 and 120 transitions (see Figure 8) while it was about 150 transitions in the original one (see Figure 5). Our modifications reduce the number of active registers in the operator thus they reduce also a little the power consumption of the operator.

Comparing the original operator's useful activity variations (Figure 5) with variations of modified multiplier (Figure 8), we can notice the absence of high initialization peaks. For instance, between cycles 1500 and 2400 it is difficult to detect the executed operations boundaries.

Montgomery multiplication: If we do not consider the reduction step, we may say that the activity variations of Montgomery multiplier are more or less uniform

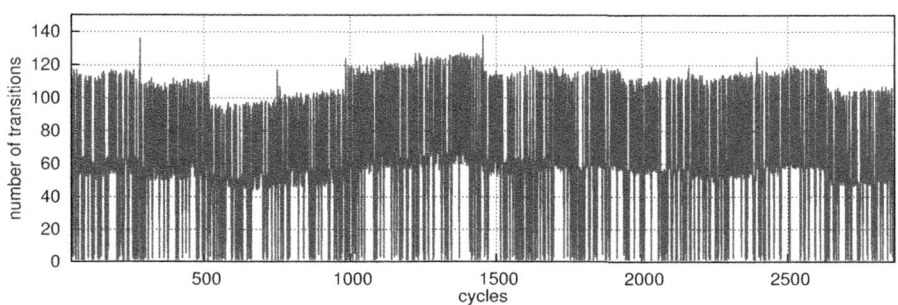

Fig. 8. Useful activity measurement results for random GF(2^m) multiplications with modified classical algorithm

(see Figure 6). The only thing, which may still give some information to the attacker is the initialization phase. Activity drops at this phase occur due to a specific way, in which the input data are fetched. Like for classical algorithm, a modification of the initialization control removes these drops.

Mastrovito multiplication: The "step-wave" shape of useful activity variations of Mastrovito multiplier in Figure 7 is distinguishable and can provide the attacker with a lot of information. Our objective is to modify the algorithm and the architecture in such a way that multiplications cannot be too easily distinguished.

We have investigated two types of modifications for Mastrovito multiplier: "uniformization" of the number of sub-multipliers' registers used in each clock cycle and "randomization" of the starting times of the operator sub-multipliers. We have derived many versions of those two types of modifications. The most worth showing according to us are variations of "randomization".

Figure 9 presents the way we have divided matrix M into sub-matrices (see Section 3.3). The boxes with same indices Mi denote blocks, which can be multiplied by parts of b, using the same sub-multiplier module.

M	0	1	2	3	4	5	6	7	8	9	10	11	12	13	14
0	M0	M0	M0	M0	M0	M0	M0	M0	M0	M0	M1	M1	M1	M1	Mc
1	M0	M0	M0	M0	M0	M0	M0	M0	M0	M0	M0	M1	M1	M1	Mc
2	M0	M0	M0	M0	M0	M0	M0	M0	M0	M0	M0	M0	M1	M1	Mc
3	M0	M0	M0	M0	M0	M0	M0	M0	M0	M0	M0	M0	M0	M1	Mc
4	M2	M2	M2	M2	M4	M4	M4	M4	M4	M4	M4	M4	M4	M4	Mc
5	M2	M2	M2	M2	M2	M4	M4	M4	M4	M4	M4	M4	M4	M4	Mc
6	M0	M2	M2	M2	M2	M2	M4	M4	M4	M4	M4	M4	M4	M4	Mc
7	M0	M0	M2	M2	M2	M2	M2	M4	M4	M4	M4	M4	M4	M4	Mc
8	M0	M0	M0	M2	M2	M2	M2	M2	M4	M4	M4	M4	M4	M4	Mc
9	M0	M0	M0	M0	M2	M2	M2	M2	M2	M3	M3	M3	M3	M3	Mc
10	M0	M0	M0	M0	M0	M2	M2	M2	M2	M2	M3	M3	M3	M3	Mc
11	M0	M0	M0	M0	M0	M0	M2	M2	M2	M2	M2	M3	M3	M3	Mc
12	M0	M0	M0	M0	M0	M0	M0	M2	M2	M2	M2	M2	M3	M3	Mc
13	M0	M0	M0	M0	M0	M0	M0	M0	M2	M2	M2	M2	M2	M3	Mc
14	Mr	Mr	Mr	Mr	Mr	Mr	Mr	Mr	Mr	Mr	Mr	Mr	Mr	Mr	Mr

Fig. 9. Mastrovito matrix for $m = 233$

It can be observed that some sub-multipliers are used more than the others. Thus if we start them all at the same time, the activity is higher at the beginning of the operation (where all sub-multipliers are used) and smaller at the end (almost all sub-multipliers are already switched off). Thus our first proposition is to make the utilization, in one clock cycle, of the number of internal registers

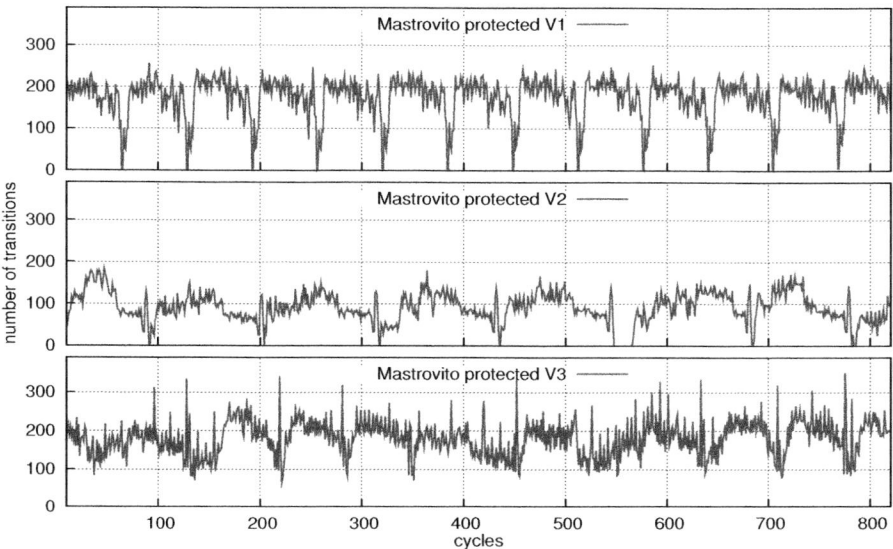

Fig. 10. Useful activity measurement results for random GF(2^m) multiplications with 3 versions of modified Mastrovito algorithm.

more uniform. For this we have modified the schedule of the sub-multipliers work but without changing the complete computation time. We have tried several schemes and the best yet obtained leads to the useful activity variations curve presented in Figure 10 (version v1 on top curve).

Activity shape for this modified Mastrovito multiplier shows small improvement. However we still try to improve our way of uniformization of utilization of internal registers used, taking into account also their sizes (each sub-multiplier uses different number of registers of various sizes). The dummy operations (shifting, incrementation) performed on some of those registers does not give visible improvements of activity variations shape. The dependencies between the subcomputations, limit the efficiency of this modification.

Our next objective was to randomize the starting moment of each submultiplier. This should "spread more" the activity over the whole computation. In order to randomize the beginning of sub-multiplications, we have used 8-bit LFSR register (Mastrovito v2) and pseudo random generator based on 4-bit LFSR (Mastrovito v3), which initialization values depend on some bits of a and b operands. In order to avoid blocking the multiplier we exchange the initialization values (seed) many times throughout multiplication operation. We have also tried other methods but the best results so far were achieved with use of LFSRs. Due to the randomization, the time needed to perform the complete multiplication, depending on which sub-multiplier is started first, will either decrease or increase randomly. The average number of clock cycles for Mastrovito v2 is 116 (minimal value: 98, maximal value: 126), whereas for Mastrovito v3

Table 2. FPGA implementation results of modified GF(2^m) multipliers with reduced activity variations

Algorithms	balanced area LUT	balanced freq. MHz	area area LUT	area freq. MHz	speed area LUT	speed freq. MHz	# clock cycles
Classical	2868	270	2778	228	3444	420	260
×α factor	×0.79	×0.89	×0.76	×0.75	×0.95	×1.39	×0.98
Montgomery	2099	323	2093	338	2099	423	264
×α factor	×0.96	×1.00	×0.96	×1.05	×0.96	×1.31	×0.98
Mastrovito v1	3463	414	3439	343	3489	384	75
×α factor	×0.99	×1.50	×0.98	×1.24	×0.94	×1.39	×1.00
Mastrovito v2	3700	306	3667	253	3717	388	avg. 116
×α factor	×1.06	×1.11	×1.05	×0.92	×1.06	×1.41	×1.55
							min.98, max.126
Mastrovito v3	3903	319	3837	250	4335	375	avg. 80
×α factor	×1.12	×1.16	×1.10	×0.91	×1.24	×1.36	×1.07
							min.64, max.108

average number of clock cycles needed is 80 (minimal value: 64, maximal value: 108). Useful activity measurements for v2 (middle curve) and v3 (bottom curve) modifications are presented in Figure 10. As one can observe on Figure 10, the shapes of useful activity variations are more irregular and not easily predictable compared to the curve for the initial version in Figure 7.

Implementation results for the modified multipliers: All modified multiplication algorithms have been implemented in FPGA. The corresponding results are reported in Table 2. Three optimization targets were used for the synthesis tool: balanced area/speed, area and speed optimizations. In order to compare the modified multipliers to the original ones (data from Table 1), we report a comparison factor α such as `modified` $= \alpha \times$ `original` both for area and frequency.

Evaluation of activity variation reduction: We used signal processing tools to evaluate our modifications. The measured activity traces, in time domain, are transformed into frequency domain using FFT (Fast Fourier Transform), see [20]. Figure 11 presents those results for unprotected and protected versions of some multipliers. It represents the mathematical power for each frequency bin and Y-axis uses the same logarithmic scale for all versions. Figure 11 shows an important reduction in the potential information leakage for all frequencies.

In order to numerically compare solutions, we have computed the spectral flatness measure (SFM) [20]:

$$\text{SFM} = \frac{\sqrt[n]{\prod_{i=1}^{n} p(i)}}{\frac{1}{n}\sum_{i=1}^{n} p(i)} \qquad \in [0,1]$$

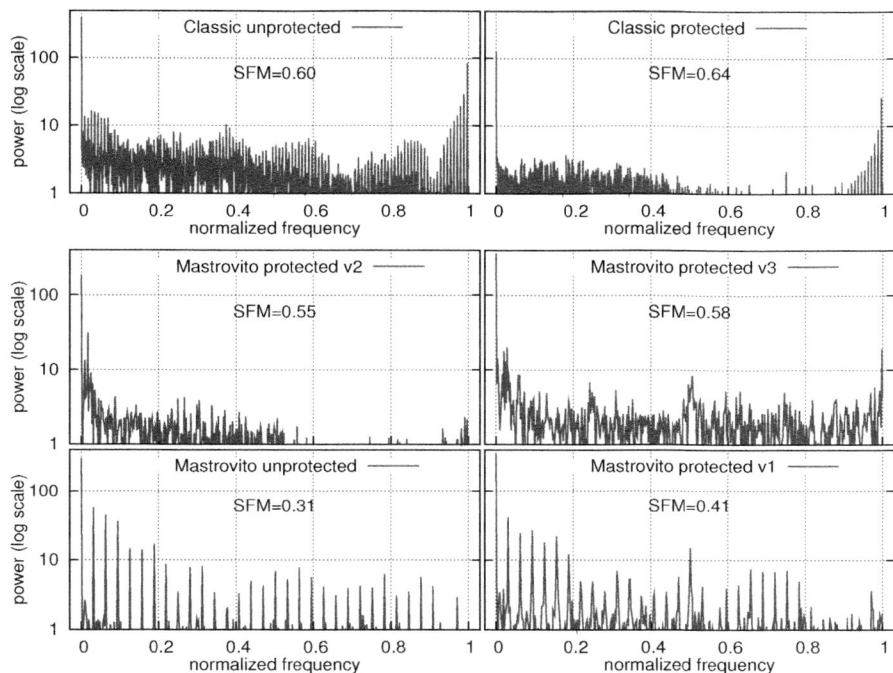

Fig. 11. FFT analysis results for unprotected and protected versions of multipliers (top: classic algorithm, middle and bottom: Mastrovito algorithm for various versions)

SFM is the ratio of the geometric mean to the arithmetic mean for a collection of n frequency bins $p(i)$ (power for frequency bin i). A SFM close to 1 indicates a spectrum with power well distributed in all frequency bins (flat curve) while a SFM close to 0 indicates that power is concentrated into a few bins (curve with peaks). SFM values are reported on Figure 11. Improvement is limited for classic algorithm, but for Mastrovito, our modifications lead to significant improvement (from 0.31 for unprotected version to 0.58 for the best protected version).

The obtained results are rather satisfying. We can see that there is a way to reduce information leakage. In the future, we plan to study chain of operations in ECC primitives such as point addition, point doubling and scalar multiplication.

6   Conclusion

GF(2^m) multipliers with reduced useful activity variations have been proposed. Useful activity has been evaluated using accurate FPGA emulation and activity counters at every operation cycle. Measurement analysis shows that the implemented multiplication algorithms (classical, Montgomery and Mastrovito) lead to specific shapes for the curve of activity variations which may be used as a small source of information leakage for some side channel attacks.

We proposed modifications of selected $GF(2^m)$ multipliers to reduce this information leakage source at two levels: architecture level by removing activity peaks due to control (e.g. reset at initialization) and algorithmic level by modifying the shape of the activity variations curve. Due to very low-level optimizations there is no significant area and delay overhead.

We have to complete our theoretical analysis with physical measurements and parasitic activity effects to get accurate results. We will study similar issues for very advanced $GF(2^m)$ multiplication algorithms such as [23,6,17,1] and for other operations (e.g. addition, subtraction, inversion, multiplication by constant and scalar multiplication) used in ECC protocols.

Acknowledgment. This work has been supported in part by a French government scholarship (in French: *bourse du gouvernement français*).

References

1. Bajard, J.-C., Negre, C., Plantard, T.: Subquadratic space complexity binary field multiplier using double polynomial representation. IEEE Transactions on Computers 59(12), 1585–1597 (2010)
2. Byrne, A., Meloni, N., Tisserand, A., Popovici, E.M., Marnane, W.P.: Comparison of simple power analysis attack resistant algorithms for an elliptic curve cryptosystem. Journal of Computers 2(10), 52–62 (2007)
3. Chabrier, T., Pamula, D., Tisserand, A.: Hardware implementation of DBNS recoding for ECC processor. In: Proc. 44th Asilomar Conference on Signals, Systems and Computers, pp. 1129–1133. IEEE (November 2010)
4. Deschamps, J.-P., Imana, J.L., Sutter, G.D.: Hardware Implementation of Finite-Field Arithmetic. McGraw-Hill (2009)
5. Erdem, S.S., Yanik, T., Koc, C.K.: Polynomial basis multiplication over $GF(2^m)$. Acta Applicandae Mathematicae 93(1-3), 33–55 (2006)
6. Fan, H., Hasan, M.A.: A new approach to subquadratic space complexity parallel multipliers for extended binary fields. IEEE Transactions on Computers 56(2), 224–233 (2007)
7. Gandolfi, K., Mourtel, C., Olivier, F.: Electromagnetic Analysis: Concrete Results. In: Koç, Ç.K., Naccache, D., Paar, C. (eds.) CHES 2001. LNCS, vol. 2162, pp. 251–261. Springer, Heidelberg (2001)
8. Guo, X., Schaumont, P.: Optimized system-on-chip integration of a programmable ECC coprocessor. ACM Transactions on Reconfigurable Technology and Systems 4(1), 6:1–6:21 (2010)
9. Hankerson, D., Menezes, A., Vanstone, S.: Guide to Elliptic Curve Cryptography. Springer (2004)
10. Joye, M.: Defenses Against Side-Channel Analysis. In: Advances in Elliptic Curve Cryptography. London Mathematical Society Lecture Note Series, vol. 317, pp. 87–100. Cambridge University Press (April 2005)
11. Karatsuba, A., Ofman, Y.: Multiplication of multi-digit numbers on automata. Doklady Akad. Nauk SSSR 145(2), 293–294 (1962) (in Russian); Translation in Soviet Physics-Doklady 44(7), 595–596 (1963)
12. Koc, C.K., Acar, T.: Montgomery multiplication in $GF(2^k)$. Designs, Codes and Cryptography 14(1), 57–69 (1998)

13. Lidl, R., Niederreiter, H.: Introduction to Finite Fields and Their Applications, 2nd edn. Cambridge University Press (1994)
14. Mangard, S., Oswald, E., Popp, T.: Power Analysis Attacks: Revealing the Secrets of Smart Cards. Springer (2007)
15. Mastrovito, E.: VLSI Architectures for Computation in Galois Fields. PhD thesis, Department of Electrical Engineering, Linkoping University, Sweden (1991)
16. Montgomery, P.L.: Modular multiplication without trial division. Mathematics of Computation 44(170), 519–521 (1985)
17. Namin, A.H., Huapeng, W., Ahmadi, M.: A high-speed word level finite field multiplier in F_{2^m} using redundant representation. IEEE Transactions on Very Large Scale Integration (VLSI) Systems 17(10), 1546–1550 (2009)
18. Orlando, G., Paar, C.: A High-Performance Reconfigurable Elliptic Curve Processor for GF(2^m). In: Paar, C., Koç, Ç.K. (eds.) CHES 2000. LNCS, vol. 1965, pp. 41–56. Springer, Heidelberg (2000)
19. Oswald, E.: Side Channel Analysis. In: Advances in Elliptic Curve Cryptography. London Mathematical Society Lecture Note Series, vol. 317, pp. 69–86. Cambridge University Press (April 2005)
20. Proakis, J.G., Manolakis, D.G.: Digital Signal Processing. Prentice Hall (1996)
21. Rodriguez-Henriquez, F., Saqib, N.A., Diaz-Perez, A., Koc, C.K.: Cryptographic Algorithms on Reconfigurable Hardware. Springer (2007)
22. Savas, E., Koc, C.K.: Finite field arithmetic for cryptography. IEEE Circuits and Systems Magazine 10(2), 40–56 (2010)
23. Sunar, B.: A generalized method for constructing subquadratic complexity GF(2^k) multipliers. IEEE Transactions on Computers 53(9), 1097–1105 (2004)
24. Tisserand, A.: Low-power arithmetic operators. In: Piguet, C. (ed.) Low Power Electronics Design, ch. 9. CRC Press (November 2004)
25. Tisserand, A.: Fast and accurate activity evaluation in multipliers. In: Proc. 42nd Asilomar Conference on Signals, Systems and Computers, pp. 757–761. IEEE (October 2008)
26. Weste, N.H.E., Harris, D.: CMOS VLSI Design: A Circuits and Systems Perspective, 3rd edn. Addison Wesley (2004)

Finding Optimal Formulae for Bilinear Maps

Razvan Barbulescu, Jérémie Detrey, Nicolas Estibals, and Paul Zimmermann

CARAMEL project-team, LORIA, Université de Lorraine / INRIA / CNRS,
Campus Scientifique, BP 239, 54506 Vandœuvre-lès-Nancy Cedex, France

Abstract. We describe a unified framework to search for optimal formulae evaluating bilinear or quadratic maps. This framework applies to polynomial multiplication and squaring, finite field arithmetic, matrix multiplication, etc. We then propose a new algorithm to solve problems in this unified framework. With an implementation of this algorithm, we prove the optimality of various published upper bounds, and find improved upper bounds.

Keywords: optimal algorithms, polynomial multiplication and squaring, finite field arithmetic, tensor rank, bilinear map, bilinear rank.

1 Introduction

This article studies the *bilinear rank problem*. Given a field K, this problem naturally arises when considering the computational complexity of formulae (or algorithms) for evaluating bilinear maps over K [4, Ch. 14]. For instance, typical bilinear maps include — but are not limited to — the multiplication of two n-term polynomials of $K[X]$, or of two $r \times r$ square matrices of $K^{r \times r}$.

Note that, given an algorithm for computing a bilinear map over K, this algorithm can be naturally extended to compute the same bilinear map over any larger K-algebra $\mathbb{K}$. For instance, given a formula for computing the product of two n-term polynomials over $K = \mathbb{F}_2$, the exact same formula can also be used to compute the product of two n-term polynomials over any field extension $\mathbb{K} = \mathbb{F}_{2^m}$, or even the product of two n-term polynomials of $\mathbb{F}_2^{r \times r}[X]$, *i.e.*, polynomials having binary $r \times r$ matrices as coefficients. A nice illustration of the latter is given by Albrecht in [1], where the multiplication of large $r \times r$ matrices over $\mathbb{F}_{2^n}$ is implemented by representing $\mathbb{F}_{2^n}^{r \times r}$ as $\mathbb{F}_2^{r \times r}[X]/\langle f(X) \cdot \mathbb{F}_2^{r \times r}[X]\rangle$, with f an irreducible polynomial over $\mathbb{F}_2$ of degree n, and then using Montgomery's n-term polynomial multiplication formulae [16].

Therefore, when dealing with such formulae, a crucial distinction is made between

- a *full multiplication* $a \cdot b$ of two elements a and b derived from the inputs of the bilinear map, and thus possibly living in the larger algebra $\mathbb{K}$; and
- an *addition* $a + b$ or a *scalar multiplication* λa, where λ is given by the bilinear map only, and thus belongs to the smaller coefficient field K.

Intuitively, the first one is expected to have a much higher computational cost than the last two. Table 3 in [1] shows that the time needed to multiply two

F. Özbudak and F. Rodríguez-Henríquez (Eds.): WAIFI 2012, LNCS 7369, pp. 168–186, 2012.
© Springer-Verlag Berlin Heidelberg 2012

matrices over fields of characteristic 2 is proportional to the number of full multiplications in the formulae used for this purpose. A sensible way to optimize the overall complexity of the formula at hand is therefore to minimize the number of these full multiplications, which corresponds to solving a particular instance of the bilinear rank problem.

Most of the fast algorithms for polynomial or integer multiplication can be expressed in terms of the bilinear rank problem, for example Karatsuba's algorithm [15] and Toom's algorithm [20]. But this applies to other problems as well, like the middle product [13], or matrix multiplication [3,11,19].

Definition of the Problem. Let K be a field. Given an $n \times m$ *bilinear map* $\phi : K^n \times K^m \to K^\ell$, the bilinear rank problem consists in finding formulae for evaluating ϕ involving a minimal number k of full multiplications. This optimal k is called the *bilinear rank* of the map ϕ.

More formally, consider $\mathcal{B}(K^n, K^m; K)$ the set of $n \times m$ *bilinear forms* from $K^n \times K^m$ to K. Each such bilinear form γ can be written as $\gamma : \mathbf{a}, \mathbf{b} \mapsto \sum_{i,j} \gamma_{i,j} a_i b_j$, with the two input vectors $\mathbf{a} = (a_0, a_1, \ldots, a_{n-1}) \in K^n$, $\mathbf{b} = (b_0, b_1, \ldots, b_{m-1}) \in K^m$, and the coefficients $\gamma_{i,j} \in K$. As γ is uniquely determined by the $\gamma_{i,j}$'s, we can represent $\mathcal{B}(K^n, K^m; K)$ as the vector space of dimension nm over K, which we will denote by V in the following. The representation of the bilinear form γ in V is therefore the nm-dimensional vector $\gamma = (\gamma_{0,0}, \gamma_{0,1}, \ldots, \gamma_{n-1,m-1})$, and a bilinear map ϕ from $K^n \times K^m$ to K^ℓ can then be represented as a tuple of ℓ vectors of V: $\phi = (\phi_0, \phi_1, \ldots, \phi_{\ell-1}) \in V^\ell$.

Consider now an $n \times m$ bilinear form γ. We say that γ is *of rank* 1 if there exist two linear forms $\alpha : K^n \to K, \mathbf{a} \mapsto \sum_i \alpha_i a_i$ and $\beta : K^m \to K, \mathbf{b} \mapsto \sum_j \beta_j b_j$ such that $\gamma(\mathbf{a}, \mathbf{b}) = \alpha(\mathbf{a}) \cdot \beta(\mathbf{b})$ for all $\mathbf{a} \in K^n$ and $\mathbf{b} \in K^m$.

Rank-1 bilinear forms are particularly relevant in the context of the bilinear rank problem: indeed, each such form can be evaluated using only one full multiplication, as evaluating the two linear forms α and β at $\mathbf{a}$ and $\mathbf{b}$, respectively, only requires scalar multiplications and additions. Furthermore, since the maps $e_{i,j} : \mathbf{a}, \mathbf{b} \mapsto a_i b_j$ are all rank-1 bilinear forms, any bilinear form can then be written as a linear combination of rank-1 bilinear forms: $\gamma = \sum_{i,j} \gamma_{i,j} e_{i,j}$. Therefore, one can ask the question: What is the minimal number of rank-1 bilinear forms necessary to evaluate a given bilinear form? A given bilinear map?

In the case of a (single) bilinear form γ, the answer is easy: the bilinear rank is given by the actual rank of the $n \times m$ matrix $(\gamma_{i,j})_{0 \leq i < n, 0 \leq j < m}$ formed by the coefficients of γ. However, the minimal number of full multiplications is much more difficult to compute when evaluating $\ell \geq 2$ bilinear forms simultaneously, such as is the case with bilinear maps: this is the aforementioned *bilinear rank problem*, formalized below.

Definition 1 (Bilinear rank problem [4, Ch. 14]). *Using the above notations, and given a finite generator set $\mathcal{G} \subset V$ composed of rank-1 $n \times m$ bilinear forms, along with a finite target set of ℓ bilinear forms $\mathcal{T} = \{t_0, t_1, \ldots, t_{\ell-1}\} \subseteq \mathrm{Span}(\mathcal{G})$, the bilinear rank problem consists in generating all the elements of $\mathcal{T}$ by K-linear combinations of a minimal number k of elements of $\mathcal{G}$ or, alternatively, to find all solutions for that optimal k.*

Without loss of generality, the target vectors t_i can further be assumed to be linearly independent, as reconstructing an extra target vector by linear combination would only require scalar multiplications and additions.

Note that this problem can be adapted to also encompass the case of sets of *quadratic* forms. Indeed, any quadratic form σ from K^n to K, $\sigma : \mathbf{a} \mapsto \sum_{0 \leq i \leq j < n} \sigma_{i,j} a_i a_j$, is uniquely represented by its coefficients $\sigma_{i,j} \in K$. The set of n-ary quadratic forms $\mathcal{Q}(K^n; K)$ can thus be seen as a vector space of dimension $n(n+1)/2$ over K. It suffices then to take V to be this vector space, and to define the rank-1 quadratic forms to be the quadratic forms σ for which there exist two linear forms over K^n, $\alpha : \mathbf{a} \mapsto \sum_i \alpha_i a_i$ and $\alpha' : \mathbf{a} \mapsto \sum_i \alpha'_i a_i$, such that $\sigma(\mathbf{a}) = \alpha(\mathbf{a}) \cdot \alpha'(\mathbf{a})$ for all $\mathbf{a} \in K^n$.

In fact, any $n \times m$ bilinear form can be seen as an $(n+m)$-ary quadratic form σ such that no product $a_i a_j$ occurs between the first n or between the last m input variables: $\sigma_{i,j} = 0$ for all $0 \leq i \leq j < n$ and for all $n \leq i \leq j < n + m$. Hence, computing the rank of a set of quadratic forms is more general than computing the bilinear rank.

Also, the bilinear rank problem is NP-hard. Indeed, Bürgisser *et al.* show in [4, Sec. 14.2] its equivalence to the decomposition of an order-3 tensor, known to be NP-hard [14].

Notation. In the rest of this document, by an abuse of notation, we will omit writing the input vectors of the considered bilinear or quadratic forms, as they will always be $\mathbf{a}, \mathbf{b}$ or $\mathbf{a}$, respectively. Hence, we will simply write $\gamma = \sum_{i,j} \gamma_{i,j} a_i b_j$ or $\sigma = \sum_{i \leq j} \sigma_{i,j} a_i a_j$ to refer to the corresponding forms in the following. This amounts to implicitly considering the a_i's and b_j's as formal variables over K.

Related Results. Several authors have considered special instances of the bilinear rank problem. We do not claim an exhaustive state-of-the-art here, but we mention the main results related to our work.

For polynomial multiplication, evaluation–interpolation algorithms like Karatsuba's and Toom's algorithms [15,20] are special cases of the problem, where the only full multiplications involved correspond to evaluations of the product of the two polynomials at different points.

Even though the bilinear rank problem was already well known in the algebraic complexity community [4], until only recently, all the formulae used in the computer arithmetic community were based on this evaluation–interpolation scheme. In [16], instead of using only the evaluation products, *i.e.*, products of the form $(\sum_i \kappa^i a_i)(\sum_j \kappa^j b_j)$ for some $\kappa \in K$, Montgomery considered other rank-1 bilinear forms. This allowed him to find formulae with less products for the 5-, 6-, and 7-term polynomial multiplications, which do not fit in any evaluation–interpolation scheme. Montgomery however restricted the exploration to the set of "symmetric" generators of the form $(\sum_i \alpha_i a_i)(\sum_j \alpha_j b_j)$. Our work shows how to improve Montgomery's exploration and proves that the formulae in [16] are optimal for the 5-term polynomial multiplication.

In [9], Chung and Hasan propose asymmetric squaring formulae over $\mathbb{Q}$, found using an exhaustive search method similar to that of Montgomery, but starting from an ad-hoc set of rank-1 generators. In [12], Fan and Hasan use Montgomery's formulae and the Chinese Remainder Theorem together with a short product for the high degree terms to improve the bounds over $\mathbb{F}_2$; those results were further improved by Cenk and Özbudak [5]. In [18], using both exhaustive and heuristic search algorithms, Oseledets considers the multiplication of n-term polynomials, and also their product modulo X^n — which corresponds to Mulders' short product [17] — and modulo $X^n + 1$. Some of the linear algebra routines he proposed are a building block in our method.

Multiplication in finite field extensions $\mathbb{F}_{q^n}$ for $n \geq 2$ reduces to polynomial multiplication modulo a degree-n irreducible polynomial f over $\mathbb{F}_q$. Thus, one could first compute a full product of two n-term polynomials, then reduce it modulo f using only scalar multiplications; however it is sometimes faster to directly compute the n terms of the product modulo f [7].

Fast matrix multiplication is another application of the bilinear rank problem, with the smallest unsolved problem being the multiplication of two 3×3 matrices (for two square matrices): we know that at least 19 products are needed, and that 23 are enough, but the minimal rank is still unknown [3,11].

Contributions. The contributions of the article are the following. First, we present the bilinear rank problem, as it is known to the algebraic complexity community [4], showing why it is pertinent to the computer algebra and computer arithmetic communities. Following Montgomery [16], we see this problem in the light of linear algebra, which enables one to search for optimal formulae for a large set of applications such as polynomial multiplication or squaring, matrix multiplication, etc. Second, we propose a new algorithm to solve the bilinear rank problem; this algorithm is faster than Montgomery's exploration algorithm [16] and is an improvement on Oseledets' heuristic MINBAS [18], which might not find the minimal rank. Last, using this new algorithm we prove the optimality of several known formulae, and exhibit new bounds for some bilinear maps.

Roadmap. The article is organized as follows. After detailing a few instances of the bilinear rank problem in Section 2, we show in Section 3 how this problem can be translated into a linear algebra problem. Section 4 gives an efficient algorithm solving this linear algebra problem. Finally Section 5 presents experimental results proving that some bounds from the literature are optimal, along with an improved bound for the multiplication in $\mathbb{F}_{3^5}$.

2   Some Instances of the Bilinear Rank Problem

Throughout the rest of this paper, we will use the following running example:

Example 1 (2 × 3-term polynomial product). We want to multiply $A = a_1 X + a_0$ by $B = b_2 X^2 + b_1 X + b_0$ in $K[X]$. The naive algorithm requires 6 products, while only 5 products are necessary when using Karatsuba's trick:

$$A \cdot B = g_3 X^3 + (g_1 + g_2) X^2 + (g_4 - g_2 - g_0) X + g_0,$$

where $g_0 = a_0 b_0$, $g_1 = a_0 b_2$, $g_2 = a_1 b_1$, $g_3 = a_1 b_2$, and $g_4 = (a_0 + a_1)(b_0 + b_1)$. Since $A \cdot B$ has 4 terms, if the base field K contains at least 3 elements, evaluating $A \cdot B$ at those elements and at infinity — i.e., the $a_1 b_2$ product — enables to recover $A \cdot B$ by Lagrange interpolation. However, if $K = \mathbb{F}_2$, are 5 products optimal?

To illustrate the generality of the proposed framework, we give a few more examples:

Example 2 (3-term polynomial squaring [9]). We want to square the polynomial $A = a_2 X^2 + a_1 X + a_0 \in K[X]$. For this quadratic map, we have $n = 3$, the generator set $\mathcal{G}$ consists of the 28 products

$a_0^2,\ a_0 a_1,\ a_0 a_2,\ a_0(a_0 + a_1),\ a_0(a_0 + a_2),\ a_0(a_1 + a_2),\ a_0(a_0 + a_1 + a_2),$
$a_1^2,\ a_1 a_2,\ a_1(a_0 + a_1),\ a_1(a_0 + a_2),\ a_1(a_1 + a_2),\ a_1(a_0 + a_1 + a_2),$
$a_2^2,\ a_2(a_0 + a_1),\ a_2(a_0 + a_2),\ a_2(a_1 + a_2),\ a_2(a_0 + a_1 + a_2),$
$(a_0 + a_1)^2,\ (a_0 + a_1)(a_0 + a_2),\ (a_0 + a_1)(a_1 + a_2),\ (a_0 + a_1)(a_0 + a_1 + a_2),$
$(a_0 + a_2)^2,\ (a_0 + a_2)(a_1 + a_2),\ (a_0 + a_2)(a_0 + a_1 + a_2),$
$(a_1 + a_2)^2,\ (a_1 + a_2)(a_0 + a_1 + a_2),$
$(a_0 + a_1 + a_2)^2,$

and the target set is

$$\mathcal{T} = \{t_0 := a_0^2,\ t_1 := 2a_0 a_1,\ t_2 := a_1^2 + 2a_0 a_2,\ t_3 := 2a_1 a_2,\ t_4 := a_2^2\},$$

which are the 5 quadratic forms corresponding to the coefficients of A^2.

Example 3 (Middle product [13]). Assume we only want to compute the degree-1 and -2 coefficients of the product $A \cdot B$ from Example 1. We have $n = 2$, $m = 3$, and $\mathcal{G}$ is the set of all rank-1 bilinear forms of the form $(\alpha_0 a_0 + \alpha_1 a_1)(\beta_0 b_0 + \beta_1 b_1 + \beta_2 b_2)$ for $\alpha_i, \beta_j \in K$. The target set is

$$\mathcal{T} = \{t_0 := a_0 b_1 + a_1 b_0,\ t_1 := a_0 b_2 + a_1 b_1\}.$$

A solution with $k = 3$ considers the subset

$$\{g_0 := a_0(b_1 + b_2),\ g_1 := a_1(b_0 + b_1),\ g_2 := (a_1 - a_0)b_1\} \subseteq \mathcal{G},$$

which gives the formulae $t_0 = g_1 - g_2$ and $t_1 = g_0 + g_2$.

Example 4 (3 × 3 matrix multiplication). Here, $n = m = 9$, and we consider the product $A \cdot B$ over $K^{3 \times 3}$ of the two matrices

$$A = \begin{pmatrix} a_0 & a_1 & a_2 \\ a_3 & a_4 & a_5 \\ a_6 & a_7 & a_8 \end{pmatrix} \quad \text{and} \quad B = \begin{pmatrix} b_0 & b_1 & b_2 \\ b_3 & b_4 & b_5 \\ b_6 & b_7 & b_8 \end{pmatrix}.$$

The target set $\mathcal{T}$ consists of the 9 bilinear forms $\{a_0 b_0 + a_1 b_3 + a_2 b_6, ..., a_6 b_2 + a_7 b_5 + a_8 b_8\}$, and the generator set $\mathcal{G}$ consists of the non-zero rank-1 bilinear forms $(\alpha_0 a_0 + \cdots + \alpha_8 a_8)(\beta_0 b_0 + \cdots + \beta_8 b_8)$ for $\alpha_i, \beta_j \in K$, i.e., $(2^9 - 1)^2 = 261\,121$ forms for $K = \mathbb{F}_2$.

3   From Bilinear Applications to Linear Algebra

We focus in the rest of the paper on bilinear maps, the case of quadratic maps being similar. Recall that we denote by V the vector space of dimension nm over K isomorphic to the space of $n \times m$ bilinear forms $\mathcal{B}(K^n, K^m; K)$. Thus, each element of $\mathcal{T}$ and $\mathcal{G}$ (see Definition 1) being such a bilinear form, it can be written as a row vector of dimension nm, where the $(im + j)$-th column corresponds to the coefficient of the $a_i b_j$ term.

Example 1 (Cont'd). For our running example of the 2×3-term polynomial multiplication over $K = \mathbb{F}_2$, the canonical basis of the vector space V is $(a_0 b_0, a_0 b_1, a_0 b_2, a_1 b_0, a_1 b_1, a_1 b_2)$. The target set is $\mathcal{T} = \{a_0 b_0, a_0 b_1 + a_1 b_0, a_0 b_2 + a_1 b_1, a_1 b_2\}$, and the set of generators $\mathcal{G}$ consists of the 21 products:

$$
\mathcal{G} = \{\
\begin{aligned}
& a_0 b_0, && a_1 b_0, && (a_0 + a_1) b_0, \\
& a_0 b_1, && a_1 b_1, && (a_0 + a_1) b_1, \\
& a_0 (b_0 + b_1), && a_1 (b_0 + b_1), && (a_0 + a_1)(b_0 + b_1), \\
& a_0 b_2, && a_1 b_2, && (a_0 + a_1) b_2, \\
& a_0 (b_0 + b_2), && a_1 (b_0 + b_2), && (a_0 + a_1)(b_0 + b_2), \\
& a_0 (b_1 + b_2), && a_1 (b_1 + b_2), && (a_0 + a_1)(b_1 + b_2), \\
& a_0 (b_0 + b_1 + b_2), && a_1 (b_0 + b_1 + b_2), && (a_0 + a_1)(b_0 + b_1 + b_2)\ \}.
\end{aligned}
$$

We can then rewrite $\mathcal{T}$ as the following matrix of 4 row vectors:

$$
\mathcal{T} =
\begin{matrix}
& a_0 b_0\ \ a_0 b_1\ \ a_0 b_2\ \ a_1 b_0\ \ a_1 b_1\ \ a_1 b_2 & \\
\begin{pmatrix}
1 & 0 & 0 & 0 & 0 & 0 \\
0 & 1 & 0 & 1 & 0 & 0 \\
0 & 0 & 1 & 0 & 1 & 0 \\
0 & 0 & 0 & 0 & 0 & 1
\end{pmatrix}
&
\begin{matrix}
a_0 b_0 \\
a_0 b_1 + a_1 b_0 \\
a_0 b_2 + a_1 b_1 \\
a_1 b_2
\end{matrix}
\end{matrix}
$$

The same applies to the set of generators $\mathcal{G}$, which gives the 21×6 matrix:

$$
\mathcal{G} =
\begin{matrix}
& a_0 b_0\ \ a_0 b_1\ \ a_0 b_2\ \ a_1 b_0\ \ a_1 b_1\ \ a_1 b_2 & \\
\begin{pmatrix}
1 & 0 & 0 & 0 & 0 & 0 \\
0 & 0 & 0 & 1 & 0 & 0 \\
1 & 0 & 0 & 1 & 0 & 0 \\
\vdots & \vdots & \vdots & \vdots & \vdots & \vdots \\
1 & 1 & 1 & 1 & 1 & 1
\end{pmatrix}
&
\begin{matrix}
a_0 b_0 \\
a_1 b_0 \\
(a_0 + a_1) b_0 \\
\vdots \\
(a_0 + a_1)(b_0 + b_1 + b_2)
\end{matrix}
\end{matrix}
$$

The bilinear rank problem is then stated as follows: For a given integer k, we want to find the subsets $\mathcal{W}$ of k elements of $\mathcal{G}$ such that the subspace W spanned by $\mathcal{W}$ contains $\mathcal{T}$. In linear algebra terms, we want to find the subsets of k rows of $\mathcal{G}$ whose K-linear span contains all the row vectors of $\mathcal{T}$, i.e., contains the subspace $T = \text{Span}(\mathcal{T})$.

4   Solving the Linear Algebra Problem

We recall the notations from Definition 1: V is the vector space of dimension nm spanned by the $a_i b_j$ products, $\mathcal{G} \subset V$ is a finite set of generators, $\mathcal{T} \subseteq \text{Span}(\mathcal{G})$

is the set of target vectors, and $T = \text{Span}(\mathcal{T})$ is the corresponding target space spanned by the target vectors, which has dimension ℓ.

In the previous section, we reduced our problem of finding all formulae with k full multiplications for evaluating a given bilinear map to a linear algebra problem: given a finite subset $\mathcal{G}$ of a K-vector space V and a target subspace $T = \text{Span}(\mathcal{T})$ of $\text{Span}(\mathcal{G})$, we want to find all the rank-k subspaces W of V containing T and which can be generated by elements of $\mathcal{G}$ only (*i.e.*, $\text{Span}(W \cap \mathcal{G}) = W$). One should note that this linear algebra problem is more general than the original problem of optimizing the computation of bilinear maps.

4.1   Naive Algorithm

A trivial approach to the linear algebra problem is to enumerate all the $\binom{\#\mathcal{G}}{k}$ subsets $\mathcal{W} = \{g_1, \ldots, g_k\}$ of $\mathcal{G}$ and, for each of them, test if its span covers T. The most efficient way to test this inclusion of spaces is to construct $W = \text{Span}(\mathcal{W})$, represented by its basis as a matrix in row echelon form, and then test if each vector $t \in \mathcal{T}$ reduces to 0 against this matrix.

For simplicity's sake, in this paper we focus on the *combinatorial complexity* of the algorithms, considering that all matrix operations (*e.g.*, computing the dimension of a vector space, checking if a vector is in a vector space, computing the row-echelon form of a matrix) have constant complexity. The complexity of the naive algorithm is thus

$$\binom{\#\mathcal{G}}{k}.$$

Example 1 (Cont'd). For the 2×3-term polynomial multiplication, $\#\mathcal{G} = 21$, $k = 5$. We therefore have to test $20\,349$ subsets $\mathcal{W}$.

4.2   Improved Algorithm

The main drawback of the naive algorithm is that different subsets of generators $\mathcal{W}_i$ may be linearly dependent and span the same vector space $W = \text{Span}(\mathcal{W}_i)$. For instance, for the multiplication of two 3-term polynomials over $\mathbb{Q}$, Montgomery [16] gives the following family of solutions:

$$
\begin{aligned}
(a_0 + a_1 X + a_2 X^2)(b_0 + b_1 X + b_2 X^2) &= a_0 b_0 (C + 1 - X - X^2) \quad\quad (1)\\
&+ a_1 b_1 (C - X + X^2 - X^3) + a_2 b_2 (C - X^2 - X^3 + X^4)\\
&+ (a_0 + a_1)(b_0 + b_1)(-C + X) + (a_0 + a_2)(b_0 + b_2)(-C + X^2)\\
&+ (a_1 + a_2)(b_1 + b_2)(-C + X^3) + (a_0 + a_1 + a_2)(b_0 + b_1 + b_2)C.
\end{aligned}
$$

Taking different values for the arbitrary polynomial $C \in \mathbb{Q}[X]$, we see that any subset of 6 out of the 7 generators $\mathcal{G} = \{a_0 b_0, a_1 b_1, a_2 b_2, (a_0 + a_1)(b_0 + b_1), (a_0 + a_2)(b_0 + b_2), (a_1 + a_2)(b_1 + b_2), (a_0 + a_1 + a_2)(b_0 + b_1 + b_2)\}$ yields a solution.

However, these 7 solutions correspond to the same vector space W, since the 7 generators are linearly dependent:

$$(a_0 + a_1 + a_2)(b_0 + b_1 + b_2) = (a_0 + a_1)(b_0 + b_1) + (a_0 + a_2)(b_0 + b_2)$$
$$+ (a_1 + a_2)(b_1 + b_2) - a_0 b_0 - a_1 b_1 - a_2 b_2.$$

We propose an algorithm that takes advantage of this redundancy by looking directly for the subspaces W — instead of the subsets of generators — that cover our target space T. More formally, we search all the subspaces W of V such that:

(i) $T \subset W$, i.e., the target space T is covered by W;
(ii) $\mathrm{Span}(W \cap \mathcal{G}) = W$, i.e., W is spanned by elements of $\mathcal{G}$; and
(iii) $\dim W = k$, i.e., only k generators are needed.

A first remark is that from (i), the target space T is contained in each solution space W. Thus we should look for all the W's by extending this vector space. Unfortunately, there are a lot of spaces above T, possibly more than there are subsets $\mathcal{W}$ of $\mathcal{G}$. Instead, we might consider only those spaces W which are obtained by adding generators to T. This technique has already been used by Oseledets [18] in some heuristic algorithms. In order to use this idea without losing the exhaustiveness of our algorithm, we introduce a new condition:

(ii') $\exists \mathcal{W}' \subset \mathcal{G}$ such that $W = T \oplus \mathrm{Span}(\mathcal{W}')$.

Lemma 1. *All subspaces W that verify (i) and (ii) also verify (ii').*

Proof. Let W a subspace verifying (i) and (ii), the result follows directly from Proposition 1 with $\mathcal{H}$ being $W \cap \mathcal{G}$, and $\mathcal{F}$ being the free family $\mathcal{T} \subset W$.   □

Proposition 1 ([2, Prop. 3.15]). *Let W be a vector space over a field K spanned by a finite set of generators $\mathcal{H}$. Let $\mathcal{F}$ be a free family of W. Then there exists a subset $\mathcal{J}$ of $\mathcal{H}$ such that $\mathcal{F} \cup \mathcal{J}$ is a basis of W.*

Proof. If $\mathcal{F}$ generates W then we take $\mathcal{J} = \emptyset$.
   Let now assume that $\mathcal{F}$ does not span W. If we had $\mathcal{H} \subset \mathrm{Span}(\mathcal{F})$, then it would follow that $W = \mathrm{Span}(\mathcal{H}) \subset \mathrm{Span}(\mathcal{F})$, which contradicts the assumption. Therefore, there exists $h \in \mathcal{H}$ such that $h \notin \mathrm{Span}(\mathcal{F})$, and thus the family $\mathcal{F}' := \mathcal{F} \cup \{h\}$ is also free. We then iterate the process with $\mathcal{F}'$, until it eventually spans the whole vector space W. Since one cannot choose the same element $h \in \mathcal{H}$ twice and since $\mathcal{H}$ is finite, the process terminates. The set $\mathcal{J}$ is then composed of all the generators h added to $\mathcal{F}$ when constructing $\mathcal{F}'$.   □

Thanks to Lemma 1, finding all subspaces W of V verifying conditions (i), (ii), and (iii) can be achieved by first enumerating all subsets $\mathcal{W}'$ of $\mathcal{G}$ such that $W = T \oplus \mathrm{Span}(\mathcal{W}')$ verifies conditions (i), (ii'), and (iii), and then keeping only those for which W also verifies condition (ii). This method is implemented in Algorithm 1, which is proven correct in Theorem 1.

Algorithm 1. Find all the subspaces of dimension k generated by elements of $\mathcal{G}$ only and that contain the target space T.

Input: A vector space V, a finite set $\mathcal{G}$ of elements of V, a subspace T of V included in $\mathrm{Span}(\mathcal{G})$, and an integer k such that $\dim T \leq k \leq \mathrm{rk}(\mathcal{G})$.

Output: The set $\mathcal{S}$ of all subspaces W of V such that $T \subset W$, $\mathrm{Span}(W \cap \mathcal{G}) = W$, and $\dim W = k$.

1. $\mathcal{S} \leftarrow \emptyset$
2. **procedure** expand_subspace(W)
3.   **if** $\dim W = k$ **and** $\mathrm{rk}(W \cap \mathcal{G}) = k$ **then**
4. $\mathcal{S} \leftarrow \mathcal{S} \cup \{W\}$
5.   **else if** $\dim W < k$ **then**
6.   **for each** $g \in \mathcal{G} \setminus W$ **do**
7.   expand_subspace($W \oplus \mathrm{Span}(g)$)
8. **end procedure**
9. expand_subspace(T)
10. **return** $\mathcal{S}$

Theorem 1 (Correctness of Algorithm 1). *For any given k, Algorithm 1 returns all the subspaces W of V verifying conditions (i), (ii) and (iii). In particular, when Algorithm 1 fails to find any solutions for all k under a bound k_0, then no solutions exist below this bound, and the bilinear rank is greater or equal to k_0.*

Proof. Let W be a subspace of V verifying (i), (ii) and (iii). We first prove that W is included in the output of Algorithm 1. By Lemma 1, W also satisfies (ii'). Thus there is a subset $\mathcal{W}' = \{g_1, ..., g_{k'}\}$ of $\mathcal{G}$ such that $W = T \oplus \mathrm{Span}(g_1, ..., g_{k'})$, and we can choose $g_1, ..., g_{k'}$ to be linearly independent. Therefore, Algorithm 1 will, at some point in the enumeration, consider the candidate subspace W. Since $\dim W = k$ and, by condition (ii) we have $\mathrm{rk}(W \cap \mathcal{G}) = \dim W = k$, Algorithm 1 will include W in $\mathcal{S}$.

Conversely, let W in the output of Algorithm 1; let us show that conditions (i), (ii), and (iii) are fulfilled. Condition (i) is trivially verified, since Algorithm 1 constructs W by expanding the initial subspace T. Furthermore, since W is in $\mathcal{S}$, we have $\dim W = k$, thus (iii) is also verified. Finally, we have $\mathrm{rk}(W \cap \mathcal{G}) = k$, which implies $\mathrm{Span}(W \cap \mathcal{G}) = W$, and thus W also fulfills condition (ii). $\square$

Example 1 (Cont'd). For our running example, Algorithm 1 immediately shows that no solutions exist with $k = 4$ over $\mathbb{F}_2$. Indeed, we have $\dim T = 4$, thus $\dim W = 4$ at the very first call of expand_subspace, but the rank of $W \cap \mathcal{G}$ is only 3.

Example 5 (Step-by-step illustration of Algorithm 1). Take $K = \mathbb{Q}$, $V = \mathbb{Q}^3$, $\mathcal{T} = \{t_0, t_1\}$ with $t_0 = (1, 1, 1)$ and $t_1 = (1, 1, 0)$; and $\mathcal{G} = \{g_0, g_1, g_2, g_3\}$ with $g_0 = (1, 0, 0)$, $g_1 = (0, 1, 0)$, $g_2 = (0, 0, 1)$, and $g_3 = (1, 0, 1)$, as depicted in Figure 1. In this example, for clarity's sake, we denote the recursive depth of the current call by a parenthesized superscript: *e.g.*, $W^{(i)}$ represents the subspace W at recursive depth i, starting at $i = 0$.

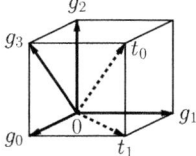

Fig. 1. The sets $\mathcal{G}$ and $\mathcal{T}$ from Example 5

Let us first consider the case $k = 2$. The algorithm initializes $W^{(0)}$ to $T = \text{Span}(t_0, t_1)$, which is already of dimension 2. It then computes $W \cap \mathcal{G} = \{g_2\}$, which is of rank 1. Therefore, there are no solutions for $k = 2$.

Consider now the case $k = 3$. Starting again with $W^{(0)} = T$, since $\dim T = 2$, the algorithm will then have to expand $W^{(0)}$ by adding a linear independent vector g from $\mathcal{G} \setminus W^{(0)} = \{g_0, g_1, g_3\}$. Taking $g = g_0$ and adding it to $W^{(0)}$, the algorithm recurses down with $W^{(1)} = T \oplus \text{Span}(g_0)$. We have $W^{(1)} \cap \mathcal{G} = \{g_0, g_1, g_2, g_3\}$, which has rank 3, meaning that this $W^{(1)}$ is a solution subspace, which is then added to the set $\mathcal{S}$. The algorithm can now reiterate the process for the subspaces $W^{(0)} \oplus \text{Span}(g_1)$ and $W^{(0)} \oplus \text{Span}(g_3)$, finding both of them to also be solutions.

Note that there is no point to try the algorithm for $k > 3$: since $\dim V = 3$, the algorithm will never find subspaces W of V of dimension k.

Remark also that this simple example illustrates *rank leaps* for $W \cap \mathcal{G}$: if $\text{rk}(W^{(0)} \cap \mathcal{G}) = 1$, when taking $W^{(1)} = W^{(0)} \oplus \text{Span}(g_0)$, the rank of $W^{(1)} \cap \mathcal{G}$ jumps directly to 3. In fact, since T cannot usually be generated by elements of $\mathcal{G}$ only, we have $\dim W^{(0)} > \text{rk}(W^{(0)} \cap \mathcal{G})$ at the beginning of the algorithm. Therefore, since $\dim W$ is only incremented by 1 at each recursive call, rank leaps between $W^{(i)} \cap \mathcal{G}$ and $W^{(i+1)} \cap \mathcal{G}$ gradually close the gap between $\dim W$ and $\text{rk}(W \cap \mathcal{G})$.

Example 6 (Evaluation–interpolation schemes). In the case of the n-term polynomial multiplication over a field K, evaluating the resulting polynomial at a point $\kappa \in K$ only involves a linear combination of the target coefficients t_i. Therefore, the rank-1 bilinear forms $(\sum_i \kappa^i a_i)(\sum_j \kappa^j b_j)$ corresponding to the evaluation of the two input polynomials at κ are also in T. This is also the case when evaluating "at infinity", *i.e.*, considering the leading coefficient $a_{n-1} b_{n-1}$. Therefore, since these rank-1 bilinear forms are linearly independent (by Vandermonde determinant), if $\#K \geq 2n - 2$, then we have $\text{rk}(T \cap \mathcal{G}) = 2n - 1 = \ell$, and T itself is a solution subspace.

4.3   Implementation Issues

Algorithm 1 relies extensively on operations on vector spaces, such as computing the rank of the intersection $W \cap \mathcal{G}$ (line 3), or excluding elements of W from $\mathcal{G}$ (line 6). In order to perform these operations efficiently, each subspace W is represented in the algorithm by the $r \times nm$ matrix of its basis in row echelon form, where $r = \dim W$. Note that we also denote by W the matrix in row echelon form corresponding to the subspace W.

Computing $W_{\mathcal{G}} = \mathrm{Span}(W \cap \mathcal{G})$ is done as follows: first reduce every vector $g \in \mathcal{G}$ by the matrix W; if g reduces to the null vector, then $g \in W$. For each such g, reduce it by the current matrix $W_{\mathcal{G}}$ (initially empty), and if the reduced vector is not null, add it to $W_{\mathcal{G}}$. Once $W_{\mathcal{G}}$ is constructed, the rank of $W \cap \mathcal{G}$ is then given by the number of rows of the matrix.

Enumerating all the generators $g \in \mathcal{G} \setminus W$ (line 6) can be achieved using a similar technique. However, it is more efficient to maintain the set $\mathcal{H}$ of all generators $g \in \mathcal{G}$ reduced by W, and pick only from these reduced generators for the recursive calls. Indeed, it might very well be the case that two generators g and $g' \in \mathcal{G}$ are in fact the same vector after reduction by W. In that case, the two subspaces $W \oplus \mathrm{Span}(g)$ and $W \oplus \mathrm{Span}(g')$ are identical, and recursing down the algorithm for both of them would just be a waste of time. Maintaining a set $\mathcal{H}$ of reduced generators prevents this to happen, as g and g' will then correspond to a single reduced generator in $\mathcal{H}$. Furthermore, at recursion depth i, since all elements g of $\mathcal{H}^{(i)}$ are reduced by $W^{(i)}$, when recursing down the algorithm:

- constructing $W^{(i+1)} = W^{(i)} \oplus \mathrm{Span}(g)$ only requires inserting the row vector g into the matrix $W^{(i)}$; and
- reducing each element $h \in \mathcal{H}^{(i)}$ by $W^{(i+1)}$ to construct $\mathcal{H}^{(i+1)}$ only requires reducing h by $\mathrm{Span}(g)$, as it is already reduced by $W^{(i)}$.

Example 1 (Cont'd). On our running example, $T = \mathrm{Span}(\mathcal{T})$ is the 4×6 matrix — which is already in row echelon form — and $\mathcal{G}$ consists of the 21 generators from Section 3. Let us follow the execution of the algorithm in the case $k = 5$.

We first construct $\mathcal{H}^{(0)}$ by reducing all generators of $\mathcal{G}$ by $W^{(0)} = T$. Only 3 unique reduced generators remain:

$$
\mathcal{H}^{(0)} = \begin{array}{c} a_0b_0 \ a_0b_1 \ a_0b_2 \ a_1b_0 \ a_1b_1 \ a_1b_2 \\ \left(\begin{array}{cccccc} 0 & 0 & 0 & 1 & 0 & 0 \\ 0 & 0 & 0 & 0 & 1 & 0 \\ 0 & 0 & 0 & 1 & 1 & 0 \end{array} \right) \end{array} \begin{array}{l} a_1b_0 \\ a_1b_1 \\ a_1(b_0+b_1) \end{array}
$$

Since $\dim W^{(0)} = 4 < k$, these generators in $\mathcal{H}^{(0)}$ are then enumerated, and the procedure expand_subspace is called recursively for each of them.

For instance, when taking $g = a_1b_0$, the subspace $W^{(1)}$ is constructed by inserting g into the matrix $W^{(0)}$:

$$
W^{(1)} = \begin{array}{c} a_0b_0 \ a_0b_1 \ a_0b_2 \ a_1b_0 \ a_1b_1 \ a_1b_2 \\ \left(\begin{array}{cccccc} 1 & 0 & 0 & 0 & 0 & 0 \\ 0 & 1 & 0 & 1 & 0 & 0 \\ 0 & 0 & 1 & 0 & 1 & 0 \\ 0 & 0 & 0 & 1 & 0 & 0 \\ 0 & 0 & 0 & 0 & 0 & 1 \end{array} \right) \end{array} \begin{array}{l} a_0b_0 \\ a_0b_1 + a_1b_0 \\ a_0b_2 + a_1b_1 \\ g = a_1b_0 \\ a_1b_2 \end{array}
$$

Computing $W^{(1)} \cap \mathcal{G}$ yields a set of 9 generators and of rank 5; $W^{(1)}$ is thus a solution subspace.

Finally, continuing the enumeration for $g = a_1 b_1$ then $g = a_1(b_0 + b_1)$, the algorithm will find two other solution subspaces of rank 5.

4.4   From Solution Subspaces back to Formulae

Given a solution subspace W verifying conditions (i), (ii), and (iii), we still need to retrieve the corresponding formulae. Note that since $T \subset W$, each formula corresponds to a basis of W using only generators from $\mathcal{G}$. A simple solution is thus to first compute the set $W \cap \mathcal{G}$, then enumerate all subsets of k linearly independent vectors in $W \cap \mathcal{G}$.

For every basis of W, obtaining the corresponding formula is then just a matter of finding the coordinates of the vectors of $\mathcal{T}$ in this basis, which will give us the corresponding linear combinations of generators required to compute them.

Example 1 (Cont'd). In our running example, each of the 3 solution subspaces yields 54 different 5-subsets of $\mathcal{G}$ that span T. We thus find a total of 162 formulae for evaluating 2×3-term polynomial multiplications over binary fields.

4.5   Complexity Analysis

Algorithm 1 allows us to drastically reduce the number of operations required to find all the formulae for evaluating a given bilinear map thanks to the two following key ideas:

1. searching for subspaces W of $\mathrm{Span}(\mathcal{G})$ instead of subsets $\mathcal{W}$ of $\mathcal{G}$; and
2. constructing these subspaces starting from T instead of $\{0\}$.

The algorithm consists in expanding T to a vector space of dimension k by adding $k - \ell$ vectors from $\mathcal{G}$, where $\ell = \dim T$. The combinatorial complexity of the algorithm is thus bounded by

$$\binom{\#\mathcal{G}}{k - \ell}.$$

Note however that this complexity bound does not reflect the gain due to the first idea — searching for subspaces instead of subsets — since that gain depends on the particular problem to be solved, and cannot be expressed simply. In Algorithm 1, once say $g_1, g_2, \ldots$ have been added to the initial $W^{(0)} = T$, this idea will prune the remaining set of independent generators $\mathcal{G} \setminus W$ (without this idea, instead of considering all $g \in \mathcal{G} \setminus W$ at line 6 of Algorithm 1, we would consider all $g \in \mathcal{G}$, which would be very similar to the naive algorithm). Note however that Algorithm 1 does not guarantee that all subspaces W explored and found to be solutions are different. Duplicates have to be detected and removed upon adding solution subspaces to the set $\mathcal{S}$.

In practice, this combinatorial complexity is usually not attained, as can be seen in the experimental results reported in Section 5.

Example 1 (Cont'd). For the 2×3-term polynomial multiplication over $\mathbb{F}_2$, with $\#\mathcal{G} = 21$, $k = 5$, and $\ell = 4$, the complexity estimate predicts 21 subspaces W to consider, whereas in practice, only 3 of these subspaces need be explored by the algorithm. This has to be compared with the 20 349 subsets explored by the naive algorithm.

4.6   Special Case of Symmetric Bilinear Maps

An $n \times n$ bilinear form $\gamma : \mathbf{a}, \mathbf{b} \mapsto \sum_{i,j} \gamma_{i,j} a_i b_j$ is said to be *symmetric* if $\gamma_{i,j} = \gamma_{j,i}$ for all $0 \leq i, j < n$. Similarly, an $n \times n$ bilinear map ϕ is said to be *symmetric* if the bilinear forms corresponding to its coefficients are all symmetric. For instance, it is the case for the n-term polynomial multiplication.

Example 7 (3-term polynomial multiplication). Given $A = a_2 X^2 + a_1 X + a_0$ and $B = b_2 X^2 + b_1 X + b_0 \in K[X]$, the target set for the product $A \cdot B$ is

$$\mathcal{T} = \{a_0 b_0, a_0 b_1 + a_1 b_0, a_0 b_2 + a_1 b_1 + a_2 b_0, a_1 b_2 + a_2 b_1, a_2 b_2\},$$

all of whose bilinear forms are symmetric.

In the case where the target set $\mathcal{T}$ is composed only of symmetric forms, an idea to accelerate the search for formulae is to restrict $\mathcal{G}$ to be the set of rank-1 symmetric bilinear forms, which are of the form $(\sum_i \alpha_i a_i)(\sum_j \alpha_j b_j)$. Indeed, while there are $(\#K)^{n^2} - 1$ non-zero bilinear forms of rank 1 in $\mathcal{B}(K^n, K^n; K)$, only $(\#K)^n - 1$ of them are symmetric, which yields a huge speedup.

However, the symmetric rank-1 forms alone may fail to produce optimal formulae for symmetric maps: for instance, $a_0 b_1 + a_1 b_0$ cannot be computed with 2 symmetric products over $\mathbb{F}_2$. (For an introduction on the topic of decomposing symmetric maps as sums of symmetric tensors see for instance [10].) Therefore, restricting to symmetric forms has to be considered as a heuristic, and cannot be used to prove lower bounds on the bilinear rank of symmetric maps.

5   Experimental Results

We present here some results obtained for various instances of the bilinear rank problem using a multi-threaded C implementation of Algorithm 1. The reported timings correspond to the total execution time on a single core of a 2.2 GHz Xeon L5640. We reproduce or improve some known results [7,9,16] but also find the complete list of solutions for each problem or prove lower bounds thanks to Theorem 1.

In all the following tables, only one value for k is reported for each problem. It is the smallest possible value for which solutions were found or, if no solutions were found, it is the largest value which was tried with our algorithm. In both cases, it means that there are no solutions for smaller values of k.

Under the heading "# of tests" are reported the number of subspaces W for which the algorithm had to check whether $\mathrm{rk}(W \cap \mathcal{G}) = k$ or not. It corresponds to the number of leaves in the tree of recursive calls. The column "# of solutions" reports the number $\#\mathcal{S}$ of solution subspaces returned by Algorithm 1. Finally, we indicate in the column "# of formulae" the corresponding number of formulae, or ∞ when this number is really large.

Finally, where applicable and when an exhaustive search using all rank-1 bilinear forms for $\mathcal{G}$ was too expensive, we enumerated subspaces using only symmetric generators. This is indicated by a "(Sym.)" mention next to the cardinality of $\mathcal{G}$.

Thus, Table 1 gives experimental results for $n \times m$-term polynomial multiplication over $\mathbb{F}_2$. For instance, the 6×6 row indicates that there are no formulae with only $k = 14$ full multiplications over $\mathbb{F}_2$; however, symmetric formulae exist for $k = 17$. Similarly, the 5×5 row shows that Montgomery's bound of $k = 13$ is optimal over $\mathbb{F}_2$ [16]. Table 2 gives similar results over $\mathbb{F}_3$. For example, the 5×5 row with no solution for $k = 11$ proves that the bound $M_3(5) = 12$ from [6, Table 2] is optimal.

Table 3 corresponds to small extensions of $\mathbb{F}_2$ or $\mathbb{F}_3$, where we consider the multiplication of two elements in polynomial basis. This is relevant for example for multiplication of matrices over $\mathbb{F}_{2^e}$ [1]. In particular the rows $\mathbb{F}_{2^4}$ with $k = 9$ and $\mathbb{F}_{3^3}$ with $k = 6$ confirm the values from [7, Table 1].

Furthermore, according to the literature, the best known formula for computing the product of two elements over the finite field $\mathbb{F}_{3^5}$ uses 12 full multiplications [7]. As indicated in Table 3, our algorithm found 121 symmetric formulae using only 11 full multiplications. One such formula is given in Algorithm 2.

Finally, Table 4 considers the product of two n-term polynomials modulo X^n and $X^n - 1$ over $\mathbb{F}_2$ — as in [18] — and also over $\mathbb{F}_3$. This proves the tensor rank from [18] is the optimal one for $n = 2, 3, 4$ over $\mathbb{F}_2$ modulo both X^n and $X^n - 1$. One should note that over $\mathbb{F}_3$, computing a product modulo $X^4 - 1$ can be done with the same number of products (5) than modulo $X^3 - 1$.

Algorithm 2. Multiplication over $\mathbb{F}_{3^5} \cong \mathbb{F}_3[X]/(X^5 - X + 1)$.

Input: $A = a_4 X^4 + a_3 X^3 + a_2 X^2 + a_1 X + a_0 \in \mathbb{F}_{3^5}$ and $B = b_4 X^4 + b_3 X^3 + b_2 X^2 + b_1 X + b_0 \in \mathbb{F}_{3^5}$.

Output: $A \cdot B \bmod (X^5 - X + 1) = t_4 X^4 + t_3 X^3 + t_2 X^2 + t_1 X + t_0 \in \mathbb{F}_{3^5}$.

1. $g_0 \leftarrow a_4 b_4$ $g_5 \leftarrow (a_0 + a_1 - a_2)(b_0 + b_1 - b_2)$
2. $g_1 \leftarrow (a_0 + a_2)(b_0 + b_2)$ $g_6 \leftarrow (a_1 - a_3 + a_4)(b_1 - b_3 + b_4)$
3. $g_2 \leftarrow (a_0 - a_3)(b_0 - b_3)$ $g_7 \leftarrow (a_2 + a_3 - a_4)(b_2 + b_3 - b_4)$
4. $g_3 \leftarrow (a_1 - a_4)(b_1 - b_4)$ $g_8 \leftarrow (a_0 + a_1 - a_2 - a_3)(b_0 + b_1 - b_2 - b_3)$
5. $g_4 \leftarrow (a_3 - a_4)(b_3 - b_4)$ $g_9 \leftarrow (a_0 + a_1 - a_3 + a_4)(b_0 + b_1 - b_3 + b_4)$
6. $g_{10} \leftarrow (a_1 + a_2 - a_3 - a_4)(b_1 + b_2 - b_3 - b_4)$
7. $t_0 \leftarrow g_0 + g_2 - g_3 - g_4 + g_5 + g_6 - g_8$
8. $t_1 \leftarrow g_1 + g_2 + g_3 + g_4 - g_5 - g_8 + g_{10}$
9. $t_2 \leftarrow g_1 - g_2 - g_3 - g_5 - g_6 - g_7 + g_8$
10. $t_3 \leftarrow -g_0 - g_3 + g_4 + g_5 - g_7 - g_8 + g_{10}$
11. $t_4 \leftarrow -g_1 + g_2 - g_4 + g_5 + g_6 + g_7 + g_8 + g_9 - g_{10}$
12. **return** $t_4 X^4 + t_3 X^3 + t_2 X^2 + t_1 X + t_0$

Table 1. Experimental results for $n \times m$ polynomial multiplication over $\mathbb{F}_2$

$n \times m$	$\#\mathcal{G}$		k	# of tests	# of solutions	# of formulae	Calculation time [s]
2×2	9		3	1	1	1	0.00
3×2	21		5	2	3	162	0.00
3×3	49		6	9	3	9	0.00
4×2	45		6	5	4	108	0.00
4×3	105		8	700	33	423	0.00
4×4	225		9	$6.60 \cdot 10^3$	4	4	0.03
5×2	93		8	56	28	790 272	0.00
5×3	217		10	$1.46 \cdot 10^5$	366	48 195	0.51
5×4	465		12	$3.13 \cdot 10^8$	4 113	66 153	$2.86 \cdot 10^3$
5×5	961		13	$9.65 \cdot 10^9$	27	27	$2.28 \cdot 10^5$
6×2	189		9	250	64	1 404 928	0.00
6×3	441		11	$2.05 \cdot 10^6$	3	243	11.5
6×4	945		13	$7.69 \cdot 10^9$	9	15	$1.62 \cdot 10^5$
6×5	1 953		14	$2.01 \cdot 10^{11}$	—	—	$9.97 \cdot 10^6$
6×6	3 969		14	$4.37 \cdot 10^9$	—	—	$6.03 \cdot 10^5$
	63	(Sym.)	17	$8.08 \cdot 10^6$	6	54	17.7
7×2	381		11	$9.14 \cdot 10^3$	960	∞	0.07
7×3	889		13	$2.52 \cdot 10^9$	87	63 423	$3.66 \cdot 10^4$
7×4	1 905		14	$1.47 \cdot 10^{11}$	—	—	$6.34 \cdot 10^6$
7×7	127	(Sym.)	22	$3.38 \cdot 10^{12}$	2 618	19 550	$1.59 \cdot 10^7$
8×2	765		12	$7.80 \cdot 10^4$	4 096	∞	0.75
8×3	1 785		14	$5.27 \cdot 10^{10}$	—	—	—

Table 2. Experimental results for $n \times m$ polynomial multiplication over $\mathbb{F}_3$

$n \times m$	$\#\mathcal{G}$		k	# of tests	# of solutions	# of formulae	Calculation time [s]
2×2	16		3	1	1	4	0.00
3×2	52		4	1	1	1	0.00
3×3	169		6	24	22	1 493	0.00
4×2	160		6	9	13	38 880	0.00
4×3	520		7	164	12	48	0.00
4×4	1 600		9	$4.11 \cdot 10^5$	726	50 640	14.9
5×2	484		7	24	36	93 312	0.00
5×3	1 573		9	$2.81 \cdot 10^5$	1 116	94 629	9.33
5×4	4 840		10	$4.75 \cdot 10^6$	48	768	$1.01 \cdot 10^3$
5×5	14 641		11	$4.89 \cdot 10^7$	—	—	$4.02 \cdot 10^4$
	121	(Sym.)	12	$3.93 \cdot 10^4$	31	6 460	0.14
6×2	1 456		8	69	81	104 976	0.01
6×3	4 732		10	$3.24 \cdot 10^6$	240	4 272	566
6×4	14 560		11	$4.55 \cdot 10^7$	—	—	$3.31 \cdot 10^4$
6×5	44 044		12	$4.58 \cdot 10^8$	—	—	$1.31 \cdot 10^6$
6×6	364	(Sym.)	15	$2.37 \cdot 10^8$	4	1 024	$3.79 \cdot 10^3$
7×2	4 372		10	$2.27 \cdot 10^4$	10 530	∞	2.84
7×3	14 209		11	$3.15 \cdot 10^7$	—	—	$1.84 \cdot 10^4$
7×4	43 720		12	$4.16 \cdot 10^8$	—	—	$1.03 \cdot 10^6$
7×7	1 093	(Sym.)	17	$2.69 \cdot 10^{10}$	—	—	$1.50 \cdot 10^6$
8×2	13 120		11	$2.01 \cdot 10^5$	85 293	∞	53.6
8×3	42 640		12	$2.90 \cdot 10^8$	—	—	$5.46 \cdot 10^5$

Table 3. Experimental results for multiplication over small extensions of $\mathbb{F}_2$ and $\mathbb{F}_3$

Finite field	$\#\mathcal{G}$		k	# of tests	# of solutions	# of formulae	Calculation time [s]
$\mathbb{F}_{2^2}$		9	3	3	3	3	0.00
$\mathbb{F}_{2^3}$		49	6	$7.03 \cdot 10^3$	105	147	0.01
$\mathbb{F}_{2^4}$		225	9	$2.57 \cdot 10^9$	2 025	2 025	$1.13 \cdot 10^4$
$\mathbb{F}_{2^5}$		961	9	$3.10 \cdot 10^{10}$	—	—	$8.11 \cdot 10^5$
	(Sym.)	31	13	$3.49 \cdot 10^6$	2 015	2 015	6.24
$\mathbb{F}_{2^6}$	(Sym.)	63	15	$2.21 \cdot 10^{10}$	21	21	$6.63 \cdot 10^4$
$\mathbb{F}_{2^7}$	(Sym.)	127	15	$1.34 \cdot 10^{12}$	—	—	$6.17 \cdot 10^6$
$\mathbb{F}_{3^2}$		16	3	3	4	16	0.00
$\mathbb{F}_{3^3}$		169	6	$2.42 \cdot 10^5$	11 843	105 963	1.08
$\mathbb{F}_{3^4}$		1 600	8	$2.27 \cdot 10^{11}$	—	—	$1.08 \cdot 10^7$
	(Sym.)	40	9	$1.10 \cdot 10^5$	234	615 240	0.45
$\mathbb{F}_{3^5}$	(Sym.)	121	11	$2.66 \cdot 10^9$	121	121	$1.45 \cdot 10^4$
$\mathbb{F}_{3^6}$	(Sym.)	364	12	$3.01 \cdot 10^{12}$	—	—	$4.50 \cdot 10^7$

Table 4. Experimental results for the multiplication of two n-term polynomials in $\mathbb{F}_p[X]/(X^n)$ and $\mathbb{F}_p[X]/(X^n - 1)$, with $p = 2$ and 3

Ring	n	$\#\mathcal{G}$		k	# of tests	# of solutions	# of formulae	Calculation time [s]
$\mathbb{F}_2[X]/(X^n)$	2		9	3	3	3	10	0.00
	3		49	5	590	12	40	0.00
	4		225	8	$5.17 \cdot 10^7$	1 440	9 248	230
	5		961	9	$2.66 \cdot 10^{10}$	—	—	$6.70 \cdot 10^5$
		(Sym.)	31	11	$3.64 \cdot 10^5$	112	736	0.48
	6	(Sym.)	63	14	$2.63 \cdot 10^9$	384	2 816	$7.66 \cdot 10^3$
	7	(Sym.)	127	15	$1.16 \cdot 10^{12}$	—	—	$5.46 \cdot 10^6$
$\mathbb{F}_2[X]/(X^n - 1)$	2		9	3	3	3	10	0.00
	3		49	4	21	3	3	0.00
	4		225	8	$2.69 \cdot 10^7$	1 440	9 248	124
	5		961	9	$1.39 \cdot 10^{10}$	—	—	$3.65 \cdot 10^5$
		(Sym.)	31	10	$7.46 \cdot 10^4$	25	25	0.09
	6	(Sym.)	63	12	$2.33 \cdot 10^7$	31	148	50.0
	7	(Sym.)	127	13	$2.55 \cdot 10^9$	1	49	$1.24 \cdot 10^4$
$\mathbb{F}_3[X]/(X^n)$	2		16	3	4	4	39	0.00
	3		169	5	$7.94 \cdot 10^3$	90	1 539	0.07
	4		1 600	7	$5.54 \cdot 10^8$	—	—	$3.22 \cdot 10^4$
		(Sym.)	40	8	$3.17 \cdot 10^5$	252	40 095	0.14
	5	(Sym.)	121	10	$1.45 \cdot 10^8$	243	13 122	$2.28 \cdot 10^3$
	6	(Sym.)	364	11	$4.79 \cdot 10^{10}$	—	—	$8.22 \cdot 10^5$
$\mathbb{F}_3[X]/(X^n - 1)$	2		16	2	1	1	1	0.00
	3		169	5	$4.45 \cdot 10^3$	90	1 539	0.04
	4		1 600	5	767	4	16	0.07
	5	(Sym.)	121	10	$8.74 \cdot 10^7$	234	615 240	$1.39 \cdot 10^3$
	6	(Sym.)	364	10	$3.37 \cdot 10^8$	9	2 025	$1.68 \cdot 10^4$

6   Conclusion

Long considered to be "magical", Karatsuba's and Strassen's formulae can be understood in a unified framework: the bilinear rank problem. Small instances of this NP-hard problem can be tackled by exhaustive search methods, as proposed by Montgomery [16]. Our improved algorithm proved that many known formulae from the literature, some of which discovered by ad-hoc methods, are optimal. Not relying on heuristic methods, the algorithm we presented here finds *all* the formulae using the optimal number of full multiplications, and gives lower bounds. For example, for the product of 3-term polynomials over $\mathbb{F}_2$, in addition to the 7 possible formulae from Eq. (1) already found by Montgomery in [16], we found two new asymmetric formulae, the first one using the generators

$$g_1 := a_0 b_0, \ g_2 := a_2 b_2, \ g_3 := (a_0 + a_1)(b_0 + b_2), \ g_4 := (a_0 + a_2)(b_1 + b_2),$$
$$g_5 := (a_1 + a_2)(b_0 + b_1), \ g_6 := (a_0 + a_1 + a_2)(b_0 + b_1 + b_2),$$

with $(a_0 + a_1 X + a_2 X^2)(b_0 + b_1 X + b_2 X^2)$ being equal to:

$$g_1 + (g_2 + g_3 + g_5 + g_6)X + (g_3 + g_4 + g_6)X^2 + (g_1 + g_4 + g_5 + g_6)X^3 + g_2 X^4,$$

and the second one using the generators

$$a_0 b_0, a_2 b_2, (a_0 + a_1)(b_1 + b_2), (a_0 + a_2)(b_0 + b_1), (a_1 + a_2)(b_0 + b_2), \text{ and}$$
$$(a_0 + a_1 + a_2)(b_0 + b_1 + b_2).$$

It would be interesting to give a simple mathematical explanation for these new formulae.

We were also able to improve the bound from [8] for the multiplication in $\mathbb{F}_{3^5}$ from 12 to 11 multiplications, using again a completely generic method (cf. Algorithm 2).

If one wants to go further, one has to either speed up the computations or to use more heuristic or proven techniques borrowed to the algebraic complexity community: restricting the set $\mathcal{G}$ of products, *e.g.*, to symmetric products for polynomial multiplication, enlarging the target set $\mathcal{T}$ by imposing that some products have to occur in the formula or using the symmetries of the problem.

Acknowledgements. The authors would like to thank Marc Mezzarobba for the interesting and fruitful discussions we had on the subject, especially for his knowledge of the many results on this topic.

The authors would also like to thank the anonymous referees for their encouraging reviews.

References

1. Albrecht, M.R.: The M4RIE library for dense linear algebra over small fields with even characteristic (2011) (preprint), http://arxiv.org/abs/1111.6900
2. Artin, M.: Algebra. Prentice-Hall, Inc. (1991)

3. Bläser, M.: On the complexity of the multiplication of matrices of small formats. Journal of Complexity 19, 43–60 (2003)
4. Bürgisser, P., Clausen, M., Shokrollahi, M.: Algebraic complexity theory, vol. 315. Springer (1997)
5. Cenk, M., Özbudak, F.: Improved polynomial multiplication formulas over $\mathbb{F}_2$ using Chinese remainder theorem. IEEE Trans. Comput. 58(4), 572–576 (2009)
6. Cenk, M., Özbudak, F.: Efficient Multiplication in $\mathbb{F}_{3^{\ell m}}$, $m \geq 1$ and $5 \leq \ell \leq 18$. In: Vaudenay, S. (ed.) AFRICACRYPT 2008. LNCS, vol. 5023, pp. 406–414. Springer, Heidelberg (2008)
7. Cenk, M., Özbudak, F.: On multiplication in finite fields. J. Complexity 26, 172–186 (2010)
8. Cenk, M., Özbudak, F.: Multiplication of polynomials modulo x^n. Theoret. Comput. Sci. 412, 3451–3462 (2011)
9. Chung, J., Hasan, M.A.: Asymmetric squaring formulae. In: Kornerup, P., Muller, J.M. (eds.) Proc. ARITH 18, pp. 113–122 (2007)
10. Comon, P., Golub, G., Lim, L., Mourrain, B.: Symmetric tensors and symmetric tensor rank. SIAM J. Matrix Anal. & Appl. 30(3), 1254–1279 (2008)
11. Courtois, N.T., Bard, G.V., Hulme, D.: A new general-purpose method to multiply 3×3 matrices using only 23 multiplications (2011) (preprint), http://arxiv.org/abs/1108.2830
12. Fan, H., Hasan, A.: Comments on five, six, and seven-term Karatsuba-like formulae. IEEE Trans. Comput. 56(5), 716–717 (2007)
13. Hanrot, G., Quercia, M., Zimmermann, P.: The middle product algorithm, I. Speeding up the division and square root of power series. Appl. Algebra Engrg. Comm. Comput. 14(6), 415–438 (2004)
14. Hastad, J.: Tensor rank is NP-complete. J. Algorithms 11(4), 644–654 (1990)
15. Karatsuba, A.A., Ofman, Y.: Multiplication of multi-digit numbers on automata. Doklady Akad. Nauk SSSR 145(2), 293–294 (1962) (in Russian); translation in Soviet Physics-Doklady 7, 595–596 (1963)
16. Montgomery, P.: Five, six, and seven-term Karatsuba-like formulae. IEEE Trans. Comput. 54(3), 362–369 (2005)
17. Mulders, T.: On short multiplications and divisions. Appl. Algebra Engrg. Comm. Comput. 11(1), 69–88 (2000)
18. Oseledets, I.: Optimal Karatsuba-like formulae for certain bilinear forms in GF(2). Linear Algebra and its Applications 429, 2052–2066 (2008)
19. Strassen, V.: Gaussian elimination is not optimal. Numerische Mathematik 13(4), 354–356 (1969)
20. Toom, A.: The complexity of a scheme of functional elements realizing the multiplication of integers. Soviet Mathematics Doklady 3, 714–716 (1963)

Appendix

We compare here Algorithm 1 to Oseledets' work. The main similarity is that both methods use a linear algebra setting and search for solution spaces W rather than solution sets $\mathcal{W}$ of generators. In terms of differences, we note that Algorithm 1 is proven, making it a tool for proving lower bounds (cf. Section 5). Oseledets' work presents a main algorithm called MINBAS, a subroutine called RANK-1 and some more tricks.

MINBAS starts by computing the rank of all $\#K^{\dim T} - 1$ elements of T (denoted C in [18]), where the rank of a bilinear form is the minimal k such that $t = g_1 + g_2 + \cdots + g_k$ with $g_i \in \mathcal{G}$. MINBAS continues by constructing a basis of T by adding in order linearly independent elements of rank 1, 2 and so on. In our setting, the rank-1 elements from [18] are elements of the target space which are also in the set $\mathcal{G}$ of generators. Since Algorithm 1 starts by $W = T$, those forms are automatically included in our algorithm. From this point on, the two algorithms diverge since Algorithm 1 explores spaces W of higher dimension whereas MINBAS adds forms of rank 2 or more.

The second algorithm of Oseledets' work is called RANK-1 and can be used as a subroutine in algorithms solving the bilinear rank problem. Given a candidate space W of dimension k lying above T, the RANK-1 procedure tests if W contains a basis made out of elements in $\mathcal{G}$ by writing the lines of W and $\mathcal{G}$ in row echelon form. In our setting this is equivalent to the test $\operatorname{rk}(W \cap \mathcal{G}) = k$.

Without giving a general description of the method, nor analyzing its correctness, Oseledets implements the idea which we presented in Section 4.2. For example, in the problem of 5-term polynomial multiplication we have $\dim T = 9$ and we search solution spaces of dimension $k = 13$; Oseledets notes that "we have to add 3 [there is a typo here, read 4] matrices to the current basis. There are $31 - 3 = 28$ possible symmetric rank-1 matrices to add". The implicit idea is that we restrict our search to spaces which can be written as $W = T \oplus \langle g_1, ..., g_{k-\dim T} \rangle$. One recognizes the idea of replacing condition (ii) by (ii') in Section 4.2. Oseledets did not realize this change still performs an exhaustive search, as demonstrated by Theorem 1, since he says on page 2061 "Partial exhaustive search – search only for the complement space C of R".

Solving Binary Linear Equation Systems over the Rationals and Binaries

Benedikt Driessen and Christof Paar

Horst-Goertz Institute for IT Security,
Ruhr-University Bochum, Germany
{benedikt.driessen,christof.paar}@rub.de

Abstract. This paper presents intermediate results of our investigations into the potential of analog hardware for the purpose of solving linear equation (LES) systems which are of quadratic form and binary. Based on the assumption that we can efficiently solve binary LES over the rationals with sufficient precision, we present a generic method to map a rational solution to a solution which solves the equation system over $\mathbb{F}_2$. We show that, in order to perform this mapping, we only need to look at two bits of the binary expansion of each of the elements of the rational solution vector.

1 Introduction

Solving binary linear equation systems (LES) of the form $Ax = b$ with $A \in \mathbb{F}_2^{n \times n}$, $b, x \in \mathbb{F}_2^n$ and n unknowns is a common problem and appears in numerous research and technical disciplines. In the field of cryptography, a special form of this problem arises when attacking stream ciphers. Certain attacks, such as attacks on A5/1 and A5/2 in the extremely wide-spread GSM-standard [PFS00,BBK03,GNR08,Gol97,PS00] require solving of a very large number of LES over $\mathbb{F}_2$.

These LES can be solved with the help of Gaussian elimination, which can easily be implemented in soft- and hardware. However, for some implementations this approach is unsatisfying in practice due to its cubic complexity. In spirit of unconventional cryptanalytical computing devices such as TWIRL [Sha99,LS00] and TWINKLE [ST03] we have explored the potential of a hypothetical device which solves a particular LES over $\mathbb{Q}$.

Based on the assumption that we are given a solution to $Au = b$ with $A \in \mathbb{F}_2^{n \times n}$, $b \in \mathbb{F}_2^n$ and $u \in \mathbb{Q}$, we have developed a method to convert the rational solution which allows us to solve quadratic LES over $\mathbb{F}_2$. We explicitly stress that the presented method is generic and not dependent on the conceptualized device, which is not in the focus of this paper.

2 Background

In the following we will shortly elaborate on what inspired our overall work and how this resulted in this paper. The computing device TWINKLE (and TWIRL)

F. Özbudak and F. Rodríguez-Henríquez (Eds.): WAIFI 2012, LNCS 7369, pp. 187–195, 2012.
© Springer-Verlag Berlin Heidelberg 2012

is used for the sieving step of the Number Field Sieve (NFS) [LLMP93] and –
in combination with "classical hardware" – assumed to be able to factor 512-bit
RSA moduli. Although the device has never been built, it is supposedly possible
to do so for $5000.

The idea of the hypothetical devices TWIRL and TWINKLE is to shift the
most expensive part of the NFS method from digital hardware into the analog
domain. More specifically, finding appropriate smooth numbers for the sieving
step is done with the help of an array of LEDs and a light sensor. In TWINKLE,
the LEDs, which are switched on and off in a specific manner, are emitting light
proportional to the logarithms of successive primes numbers. The sensor operates
as instantaneous adder of these intensities and signals when a certain threshold
has been reached. The LEDs switched on in this moment, together with the
product of their associated primes, indicate a smooth number.

The advantage of TWINKLE/TWIRL is that they exploit a physical property
to offload computation, which inspired us to experiment with an electrical device we call the Analog Solver. The core idea of our device is to use a network of
switched Operational AMPlifiers (OPAMPs) to solve binary linear equation systems. The basic principle of our circuit is given by the observation that OPAMPs
can be used as inverting adders for input voltages u_i, i.e.,

$$u_{\text{out}} = -R_{\text{add}} \left(\frac{u_1}{R_1} + \frac{u_2}{R_2} + \frac{u_3}{R_3} + \cdots + \frac{u_n}{R_n} \right). \tag{1}$$

See Figure 1 for the corresponding circuit. By choosing all resistors to be equal

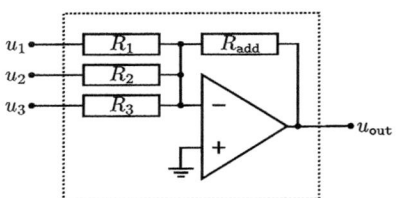

Fig. 1. Inverting adder with single OPAMP

and inverting the polarity of the input voltages, we can simplify Equation 1 to
the following form,

$$u_{\text{out}} = u_1 + u_2 + u_3 + \cdots + u_n$$

which is a simple addition of the input voltages. Based on this idea, we can construct a circuit of n OPAMPs with a switched feedback network, which computes
a solution to the equation system

$$A\boldsymbol{u} = -\boldsymbol{b}U_{\text{in}}, \tag{2}$$

where U_{in} is the input voltage of the circuit.

In this network, feedback loops between the OPAMPs are closed according
to coefficients of an equation system which is to be solved – the solution to the

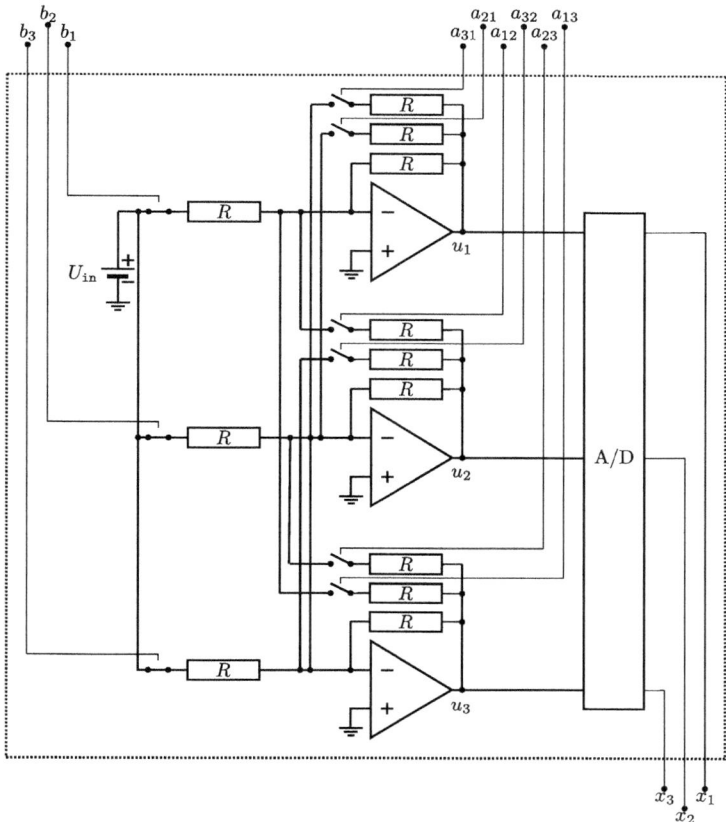

Fig. 2. Basic principle of constructing the Analog Solver (for three unknowns)

equation system can be measured (after some oscillation) as voltage output of
the OPAMPs. An example circuit for $n = 3$ is shown in Figure 2. The setting
of the switches is determined by the corresponding binary coefficients of A and
b (an open switch represents the binary value 0, and a closed one the value 1).
Looking at each of the OPAMPs separately, we can write down equations for
the expected output voltage of each OPAMP:

$$u_1 = -b_1 U_{\text{in}} - a_{1,2} u_2 - a_{1,3} u_3$$
$$u_2 = -a_{2,1} u_1 - b_2 U_{\text{in}} - a_{2,3} u_3$$
$$u_3 = -a_{3,1} u_1 - a_{3,2} u_2 - b_3 U_{\text{in}}$$

By re-arranging each equation accordingly, we easily see that the circuit represents a LES of quadratic form where A has a non-zero diagonal, i.e.,

$$u_1 + a_{1,2} u_2 + a_{1,3} u_3 = -b_1 U_{\text{in}}$$
$$a_{2,1} u_1 + u_2 + a_{2,3} u_3 = -b_2 U_{\text{in}}$$

$$a_{3,1}u_1 + a_{3,2}u_2 + u_3 = -b_3 U_{\text{in}}$$

The example circuit solves a particular[1] equation system with three unknowns and binary coefficients which is exactly what we stated in Equation 2. When the circuit has converged after all relevant switches have been set, the output voltages of the OPAMPs will approximate the solution of the LES, i.e.,

$$\boldsymbol{u} = -U_{\text{in}}(A^{-1}\boldsymbol{b}).$$

Without delving further into the specifics of this device (and our attempts to control "unwanted" oscillation and find easy to compute criteria for predicting these behaviors) we know that the device might eventually solve binary equation systems over the rationals. Based on this premise, we have developed a way to convert a rational solution to a solution over $\mathbb{F}_2$, which will be presented now.

3   Deriving a Solution over $\mathbb{F}_2$

In this section we present a method to "interpret" a rational solution $\boldsymbol{u} = A^{-1}\boldsymbol{b}$ with $A \in \mathbb{F}_2^{n \times n}, \boldsymbol{b} \in \mathbb{F}_2^n$ and $\boldsymbol{u} \in \mathbb{Q}^n$ in order to find a binary solution for the same equation system. The rational solution $\boldsymbol{u}$ can be obtained either as voltages measured in our hypothetical device, or by other means. The only thing that matters is that the solution is "sufficiently" precise, a requirement that will become clear in Section 4. Furthermore, we assume that the LES is uniquely solvable over $\mathbb{F}_2$, therefore the determinant of A must be $|A| = 1$ over $\mathbb{F}_2$, and thus $|A| \equiv 1 \bmod 2$ when computing the determinant over the rationals. The latter fact is important and will be used extensively in the remainder of this section.

Now we will describe how we can convert the rational solution to a binary vector $\boldsymbol{x} \in \mathbb{F}_2^n$ which satisfies $A\boldsymbol{x} = \boldsymbol{b}$ over $\mathbb{F}_2$. We will do this with the help of three lemmata which will lead to the conversion method.

Lemma 1. *If the LES $A\boldsymbol{u} = \boldsymbol{b}$, $A \in \mathbb{F}_2^{n \times n}, \boldsymbol{b} \in \mathbb{F}_2^n$ is uniquely solvable over $\mathbb{F}_2$ and $\mathbb{Q}$, then computing the solution over $\mathbb{Q}$ yields a vector $\boldsymbol{u} \in \mathbb{Q}_\nabla^n$ with*

$$\mathbb{Q}_\nabla = \left\{ \frac{p}{q} : p, q \in \mathbb{Z}, q \equiv 1 \bmod 2 \right\}.$$

Proof. Due to Cramer's Rule we know that a rational solution $\boldsymbol{u} = A^{-1}\boldsymbol{b}$ can be computed by the use of determinants, i.e.,

$$\boldsymbol{u} = \begin{pmatrix} u_0 \\ u_1 \\ \vdots \\ u_n \end{pmatrix} = \frac{1}{|A|} \begin{pmatrix} |A_1| \\ |A_2| \\ \vdots \\ |A_n| \end{pmatrix}$$

---

[1] The configuration of the switches indicates that the LES solved by the *Analog Solver* shown in the figure is $I\boldsymbol{u} = (1, 1, 1)^T$, where I is the 3×3 identity matrix.

where $|A| \in \mathbb{Z} \backslash \{0\}$ denotes the determinant of the binary matrix A and $|A_i| \in \mathbb{Z}$ denotes the determinant of A where the i-th column has been replaced by $\boldsymbol{b}$. Given the condition of unique solvability of the equation system over $\mathbb{F}_2$, computing $|A|$ over $\mathbb{F}_2$ cannot be zero. This must also hold when computing $|A|$ over $\mathbb{Z}$ and applying the modulo operator only as last step – since both procedures must yield the same result.

Therefore $|A| \equiv 1 \bmod 2$ holds and thus all potential rational solutions $\boldsymbol{u}$ must be in the set $\mathbb{Q}_{\nabla}^n$.

Considering Lemma 1 and with the help of the ring homomorphism $\varphi(\cdot)$ which is defined as

$$\varphi : \mathbb{Q}_{\nabla} \mapsto \mathbb{F}_2 \quad \text{with} \quad \varphi\left(\frac{p}{q}\right) = \frac{p \bmod 2}{q \bmod 2} = p \bmod 2$$

we could directly deduce a solution $\boldsymbol{x} \in \mathbb{F}_2^n$ for a particular solution $\boldsymbol{u} \in \mathbb{Q}_{\nabla}^n$ by simply applying the $\varphi(\cdot)$-operator to the vector of quotients component-wise, i.e.,

$$\boldsymbol{x} = \varphi(\boldsymbol{u}) = \begin{pmatrix} \varphi(|A_1|) \\ \varphi(|A_2|) \\ \vdots \\ \varphi(|A_n|) \end{pmatrix} \equiv \begin{pmatrix} |A_1| \bmod 2 \\ |A_2| \bmod 2 \\ \vdots \\ |A_n| \bmod 2 \end{pmatrix}.$$

However, when measuring the voltage output of our Analog Solver, we do only have fixed-point number representations of the solution vector $\boldsymbol{u}$ and no information about its quotients. Therefore no direct modulo-reduction is possible and we have to find another way to compute $\varphi(\boldsymbol{u})$.

Let us now consider how to convert $u_i \in \mathbb{Q}_{\nabla}$, which is the i-th element of $\boldsymbol{u}$ given as rational number in base-2, to a representation $x_i \in \mathbb{F}_2$. Let u_i be given in the following form which we will call the binary expansion of u_i:

$$u_i = c.d_0 d_1 d_2 d_3 \cdots, \quad c = \sum_{i=0}^{l-1} 2^i c_i \in \mathbb{N}, \quad c_i, d_i \in \{0,1\}. \tag{3}$$

If for some fixed $k \in \mathbb{N}$ and $\forall i \in \mathbb{N} \backslash \{0\}$ we have

$$d_{ik} = d_0, d_{ik+1} = d_1, \cdots, d_{(i+1)k-1} = d_{k-1},$$

we call the representation above the purely periodic binary expansion of u_i and can re-write Equation 3 to

$$u_i = c.\overline{d_0 d_1 d_2 d_3 \cdots d_{k-1}},$$

as there are no non-periodic digit-patterns after the decimal point.

Lemma 2. *If the LES $A\boldsymbol{u} = \boldsymbol{b}$, $A \in \mathbb{F}_2^{n \times n}, \boldsymbol{b} \in \mathbb{F}_2^n$ is uniquely solvable over $\mathbb{F}_2$ and $\mathbb{Q}$, then the binary expansion of any element u_i of the rational solution $\boldsymbol{u} \in \mathbb{Q}_{\nabla}^n$ is purely periodic.*

Proof. This proof follows the argumentation found in [YP04]. Suppose we have $u_i = p/q \in \mathbb{Q}_\nabla$ where $p = |A_i| = cq + r_0$ and $q = |A|$ is odd. Since q is odd, it holds that $q \mid (2^k - 1)$ for some k and hence

$$q = \frac{2^k - 1}{l}, \quad k, l \in \mathbb{N}\backslash\{0\}.$$

Since

$$0 \le \frac{r_0}{q} \le 1 \quad \text{and} \quad 0 \le (2^k - 1)\frac{r_0}{q} \le 2^k - 1 \quad \text{with} \quad (2^k - 1)\frac{r_0}{q} = lr_0$$

where $d = lr_0$ is an integer we can write

$$d = \sum_{i=1}^{k} 2^{k-i}d_{i-1} = d_0 d_1 d_2 \cdots d_{k-1}, \quad d_i \in \{0, 1\}$$

as a bit-string with k bits. Since

$$d = (2^k - 1)\frac{r_0}{q} \quad \Leftrightarrow \quad \frac{r_0}{q} = 2^{-k}d + 2^{-k}\frac{r_0}{q} \quad \text{and} \quad 2^{-k}d = 0.d_0 d_1 d_2 \cdots d_{k-1}$$

we easily see that the quotient of r_0 and q exhibits a recursive behavior

$$\frac{r_0}{q} = 0.d_0 d_1 d_2 \cdots d_{k-1} + 2^{-k}\frac{r_0}{q} = 0.d_0 d_1 d_2 \cdots d_{k-1}d_0 d_1 d_2 \cdots d_{k-1} + 2^{-2k}\frac{r_0}{q},$$

and therefore the binary expansion of

$$u_i = \frac{|A_i|}{|A|}, \quad u_i \in \mathbb{Q}_\nabla$$

is purely periodic.

If we actually measure a voltage representation of the solution for a given LES over the rationals and the results are given as purely periodic binary expansions of the form

$$c_{l-1}c_{l-2} \cdots c_0.d_0 d_1 d_2 d_3 \cdots d_{k-1}d_0 d_1 d_2 d_3 \cdots d_{k-1} \cdots,$$

we "only" need to recover the c_0 and d_k bit of all u_i in order to interpret $\boldsymbol{u} = A^{-1}\boldsymbol{b}$ as a solution over $\mathbb{F}_2^n$. Given c_0 and d_k of a particular u_i we have an alternative way to compute $\varphi(u_i) = \varphi(p/q) \equiv p \bmod 2$, which will be discussed now.

Lemma 3. *Given* $A\boldsymbol{u} = \boldsymbol{b}$, $A \in \mathbb{F}_2^{n \times n}, \boldsymbol{b} \in \mathbb{F}_2^n$, *a rational solution* $\boldsymbol{u} \in \mathbb{Q}_\nabla^n$ *and also a binary solution* $\boldsymbol{x} \in \mathbb{F}_2$, *the following holds for any pair of elements* u_i, x_i, *where*

$$c_{l-1}c_{l-2} \cdots c_0.\overline{d_0 d_1 d_2 d_3 \cdots d_{k-1}}, \quad c_i, d_i \in \{0, 1\}, \quad l, k \in \mathbb{N}$$

is the purely periodic binary expansion of u_i:

$$x_i = \varphi(u_i) = \varphi(p/q) \equiv p \bmod 2 = c_0 \oplus d_{k-1}.$$

Proof. Suppose we have $u_i \in \mathbb{Q}_\nabla^n$ in purely periodic form with

$$u_i = c_{l-1} \cdots c_0 . \overline{d_0 \cdots d_{k-1}}$$

which we convert to a quotient via

$$2^k u_i = c_{l-1} \cdots c_0 d_0 \cdots d_{k-1} . \overline{d_0 \cdots d_{k-1}}$$
$$\Leftrightarrow 2^k u_i - u_i = c_{l-1} \cdots c_0 d_0 \cdots d_{k-1} - c_{l-1} \cdots c_0$$
$$\Leftrightarrow u_i = \frac{c_{l-1} \cdots c_0 d_0 \cdots d_{k-1} - c_{l-1} \cdots c_0}{2^k - 1}.$$

Now we know that

$$u_i = \frac{|A_i|}{|A|} = \frac{c_{l-1} \cdots c_0 d_0 \cdots d_{k-1} - c_{l-1} \cdots c_0}{2^k - 1}$$

where both denominators are odd and therefore

$$\varphi(u_i) \equiv |A_i| \bmod 2 = c_{l-1} \cdots c_0 d_0 \cdots d_{k-1} - c_{l-1} \cdots c_0 \bmod 2 = c_0 \oplus d_{k-1}$$

holds.

There is also a special case, where deducing x is easy: In case $A \in \mathbb{F}_2^{n \times n}$ is in upper triangular form, we know that the determinant is the product of its diagonal elements. When the corresponding equation system is uniquely solvable over $\mathbb{F}_2$, it holds that

$$|A| = 1.$$

Therefore, for any $u_i \in \mathbb{Z}$ we can compute $\varphi(u_i)$ by looking at the least significant bit of u_i, which is an integer, i.e.,

$$x_i = \varphi(|A_i|) \equiv u_i \bmod 2.$$

4   Discussion

While our method is certainly interesting, there are some limitations. First of all, computing over the rationals with sufficient precision must be more desirable than direct and digital computation over $\mathbb{F}_2$. We were motivated to assume this by the prospect of a device, which promises to compute rational solutions very fast. However, the clear limitation, given a device which is indeed really fast, is the requirement of sufficient output precision – a notion which has been mentioned a few times already. "Sufficient" precision allows us to obtain the last bit of the binary expansion of each element of the rational vector u. Our argumentation in Section 3 implies that the smaller the determinant of A (over the rationals), the shorter the length of the binary expansion and hence the lower the required precision of a physical device.

Since we ultimately want to speed up cryptographic attacks and to examine the feasibility of our approach, we have experimentally generated some of the

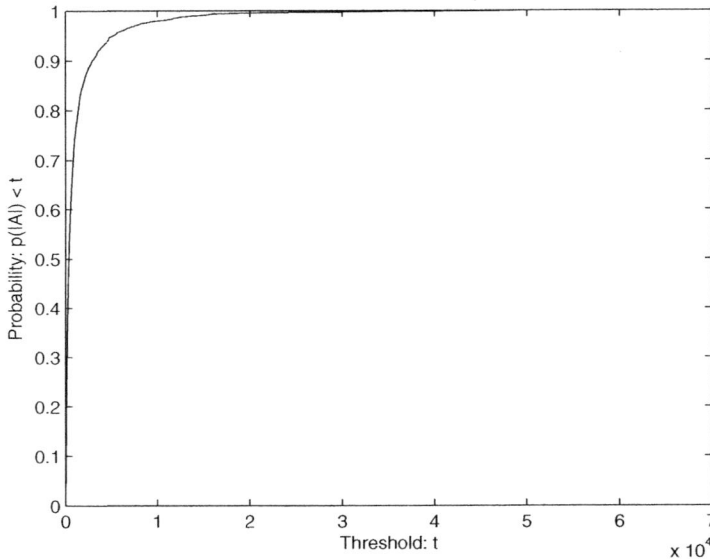

Fig. 3. Probability that the determinant in Golic's A5/1 attack is below a threshold t

binary linear equation systems which are used to execute Golic's attack [Gol97] against A5/1. For his attack, Golic needs to generate and solve more than 2^{40} equation systems of size $n = 64$ over $\mathbb{F}_2$. We have selected a random subset[2] of the generated equation systems and counted which determinants occur how often. The result of our examination is depicted in Figure 3 in a graph that shows the percentage of determinants (y-axis) which is lower than a specific value (x-axis). Observing the graph quite surprisingly[3] reveals that nearly 50% of all matrices have a determinant below 300, while 90% are still below 3000.

We deduce that there are indeed existing and practically relevant instances of binary linear equation systems in which the arising determinants are small.

5   Conclusion

We have presented a surprising method that allows us to convert a solution of a particular, binary and quadratic LES over $\mathbb{Q}$ to a solution over $\mathbb{F}_2$. Our investigations were motivated by the idea of a hardware, which is only sketched in this paper. We emphasize that our method is generic in that it can be used in any other scenario where deriving a solution over $\mathbb{Q}$ is more desirable than direct computation over $\mathbb{F}_2$, given that the rational solution is sufficiently precise. We have discussed latter requirement and shown that linear equation systems with a low determinant also occur in cryptanalytic attacks.

---

[2] Of the more than 2^{40} equation systems 10 000 were chosen.

[3] We assume that this result is due to the inherent structure of the generated matrices.

References

BBK03. Barkan, E., Biham, E., Keller, N.: Instant Ciphertext-Only Cryptanalysis of GSM Encrypted Communication. In: Boneh, D. (ed.) CRYPTO 2003. LNCS, vol. 2729, pp. 600–616. Springer, Heidelberg (2003)

GNR08. Gendrullis, T., Novotný, M., Rupp, A.: A Real-World Attack Breaking A5/1 within Hours. In: Oswald, E., Rohatgi, P. (eds.) CHES 2008. LNCS, vol. 5154, pp. 266–282. Springer, Heidelberg (2008)

Gol97. Golić, J.D.: Cryptanalysis of Alleged A5 Stream Cipher. In: Fumy, W. (ed.) EUROCRYPT 1997. LNCS, vol. 1233, pp. 239–255. Springer, Heidelberg (1997)

LLMP93. Lenstra, A., Lenstra, H., Manasse, M., Pollard, J.: The number field sieve. In: The Development of the Number Field Sieve, pp. 11–42 (1993)

LS00. Lenstra, A.K., Shamir, A.: Analysis and Optimization of the TWINKLE Factoring Device. In: Preneel, B. (ed.) EUROCRYPT 2000. LNCS, vol. 1807, pp. 35–52. Springer, Heidelberg (2000)

PFS00. Petrovic, S., Fuster-Sabater, A.: Cryptanalysis of the A5/2 Algorithm. Technical report (2000), http://eprint.iacr.org/

PS00. Pornin, T., Stern, J.: Software-Hardware Trade-Offs: Application to A5/1 Cryptanalysis. In: Paar, C., Koç, Ç.K. (eds.) CHES 2000. LNCS, vol. 1965, pp. 318–327. Springer, Heidelberg (2000)

Sha99. Shamir, A.: Factoring Large Numbers with the TWINKLE Device (Extended Abstract). In: Koç, Ç.K., Paar, C. (eds.) CHES 1999. LNCS, vol. 1717, pp. 2–12. Springer, Heidelberg (1999)

ST03. Shamir, A., Tromer, E.: Factoring Large Numbers with the TWIRL Device. In: Boneh, D. (ed.) CRYPTO 2003. LNCS, vol. 2729, pp. 1–26. Springer, Heidelberg (2003)

YP04. Yeung, T., Poon, E.: Binary decimal numbers and decimal numbers other than base ten. In: Proceedings of the International Conference on The Future of Mathematics Education (2004)

Hashing with Elliptic Curve L-Functions

Sami Omar[1,2], Raouf Ouni[1], and Saber Bouanani[1]

[1] Faculty of Science of Tunis, Department of Mathematics, 2092 Campus
Universitaire El Manar, Tunisia
[2] King Khalid University, Department of Mathematics, Abha 9004, Saudi Arabia
{Sami.Omar,Raouf.Ouni,Saber.Bouanani}@fst.rnu.tn

Abstract. In this paper, we show that the L-functions attached to Elliptic curves are good candidates to construct a hash function which can
be used as a MAC for cryptographic purpose. This is actually due to the
fact that they present a one way computation of the coefficients a_L. In
this work, we present some cryptographic preliminaries and we propose
a new protocol for hashing using L-functions of elliptic curves. We also
study the security of this protocol and its resistance to collision.

Keywords: L-functions, hash function, MAC, one way function, elliptic
curve.

1 Introduction

Information is an important commodity in the world of electronic communication. To achieve a secure communication between communicating parties, the
protection of authenticity and integrity of information are necessary. Cryptographic hash functions are a precious tool for this purpose and play a central
role in cryptology essentially to provide certain security goals such as authenticity, digital signatures, digital time stamping, and entity authentication. They
are also closely connected to other important cryptographic tools such as block
ciphers and pseudorandom functions. A hash function H is a transformation that
takes an input m and returns a fixed-size string, which is called the hash value
h (that is, $h = H(m)$). Hash functions with just this property have a variety of
general computational uses, but when they come to be employed in cryptography, the hash functions are usually chosen to have some additional properties.
The basic requirements for a cryptographic hash function are the following.

- The input can be of any length.
- The output has a fixed length.
- $H(x)$ is relatively easy to compute for any given x.
- $H(x)$ is one-way.
- $H(x)$ is collision-free.

A hash function H is considered to be one-way if it is hard to invert, where "hard
to invert" means that given a hash value h, it is computationally infeasible to find
some input x such that $H(x) = h$. If, given a message x, it is computationally

F. Özbudak and F. Rodríguez-Henríquez (Eds.): WAIFI 2012, LNCS 7369, pp. 196–207, 2012.
© Springer-Verlag Berlin Heidelberg 2012

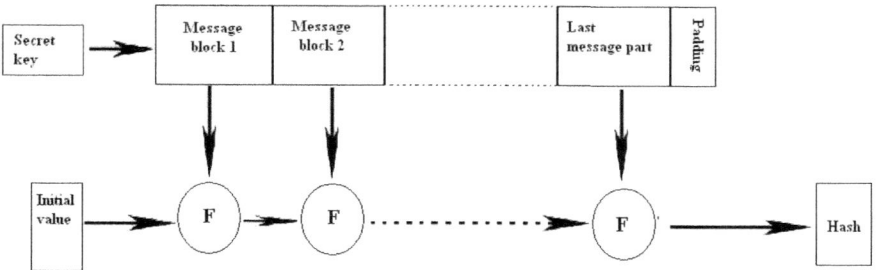

Fig. 1. Damgard and Merkle iterative structure for hash functions; F is a compression function

infeasible to find a message y not equal to x such that $H(x) = H(y)$, then H is said to be a weakly collision-free hash function. A strongly collision-free hash function H is one for which it is computationally infeasible to find any two messages x and y such that $H(x) = H(y)$. To have an extra information about that thorough study of hash functions, see Preneel [1]. The hash value represents concisely the longer message or document through which it was computed; this value is called the message digest. One can think of a message digest as a "digital fingerprint" of the larger document. Examples of well known hash functions are MD2 and MD5 and SHA . Perhaps the major prominent role of a cryptographic hash function is presented in the provision of message integrity checks and digital signatures. Since hash functions are generally faster than encryption or digital signature algorithms, it is typical to compute the digital signature or integrity check to some document by applying cryptographic processing to document's hash value, which is small compared to the document itself. Furthermore, a digest can be made public without revealing the contents of the document from which it is derived. This is basically important in digital timestamping in which, using hash functions, one can get a document timestamped without revealing its contents to the timestamping service.

Damgard and Merkle [2][3] greatly influenced cryptographic hash function design by defining a hash function in terms of what is called a compression function. A compression function takes a fixed-length input and returns a shorter fixed-length output with a slight and smooth transformation. Given a compression function, a hash function can be defined by repeated applications of the compression function until the entire message has been processed. In this process, a message of arbitrary length is broken into blocks whose length depends on the compression function and "padded" (for security reasons) so the message size is a multiple of the block size. The blocks are then processed sequentially by taking as input the result of the hash and the current message block with the final output which is the hash value of the message.

Related to hash functions are Message Authentication Codes (MACs). These are also functions that compress an input of arbitrary length into a fixed number of output bits. However, the computation depends on a secondary input of fixed length which is the key. Therefore, MACs are also referred to as keyed hash functions. In practical applications, the key on which the computation of a MAC depends is kept secret between two communicating parties. For a Message Authentication Code, the computation (and therefore the output or MAC result) depends on a secondary input, the secret key. The main idea is that an adversary without knowledge of this key should be unable to 'forge' the MAC result for any new message, even when many previous messages and their corresponding MAC results are known. A Message Authentication Code or MAC is a function h that satisfies the following conditions:

1. The input X can be of arbitrary length and the result $h(K; X)$ has a fixed length of n bits. The function has as secondary input the key K, with a fixed length of k bits.
2. Given h, K and an input X, the computation of $h(K; X)$ must be 'easy'.
3. Given a message X (but with unknown K), it must be 'hard' to determine $h(K; X)$. Even when a large set of pairs $X_i; h(K; X_i)$ is known, it is 'hard' to determine the key K or to compute $h(K; X')$ for any new message $X' \neq X_i, \forall i$.

The most commonly used MAC Algorithm based on a block cipher makes use of cipher-block-chaining is presented below and figure 2 presents the CBC-iteration process, for more details see [8]:

In this paper, we present a new protocol for hashing using a one way mathematical problem which is the computation of the coefficients a_L related to L-functions. More precisely, we have focussed our work on the elliptic curve L-functions in order to simplify the experimental computations. Our paper is organized as follows. In Section 2, we review some of the background on elliptic curves. In Section 3, we recall the specifications of the ECOH hash function, Elliptic Curve Only Hash. In Section 4, we introduce our protocol for hashing with L-functions and message authentication codes. Section 5 describes the security of our hash function. Finally in Section 6, we conclude by giving new prospects and continuity for the work we presented.

Algorithm 1. CBC-based MAC algorithm [8]

Require: Specification of the block cipher E, secret MAC key k for E and a message $x = x_1 x_2 ... x_t$

1: *Padding and blocking.* Pad x if necessary. Divide the padded text into n-bit blocks denoted $x = x_1 x_2 ... x_t$.
2: *CBC processing.* Letting E_k denote encryption using E with key k, compute the block H_t as follows: $H_1 \leftarrow E_k(x_1); H_i \leftarrow E_k(H_{i-1} \oplus x_i)$, $2 \leq i \leq t$. (This is standard cipher-block-chainig, $IV = 0$, discarding cipher text blocks $C_i = H_i$.)
3: *Optional process to increase strength of MAC.* Using a second secret key $k' \neq k$, optionally compute: $H'_t \leftarrow E_{k'}^{-1}(H_t)$, $H_t \leftarrow E_k(H'_t)$. (This amounts to using two key triple-encryption on the last block).
4: *Completion.* The MAC is the n-bit block H_t.
5: **return** n-bit MAC on x (n is the block length of E)

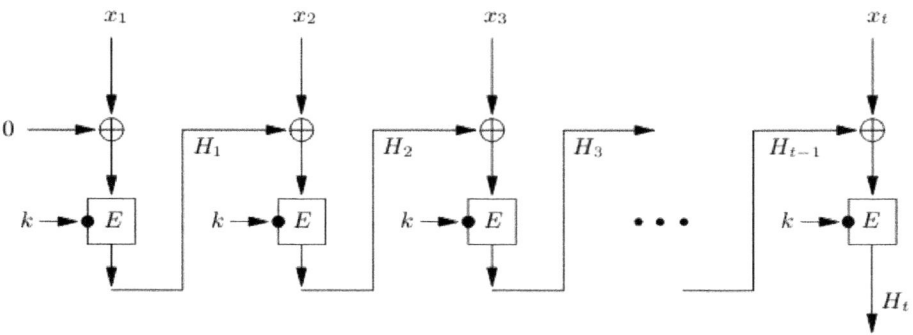

Fig. 2. CBC-based MAC iteration process [8]

2   Elliptic Curve L-Function

We begin by recalling some background related to elliptic curves and their attached L-functions.

An elliptic curve over $\mathbb{K}$ is defined by the equation

$$E \; : \; y^2 = x^3 + ax + b, \qquad a, b \in \mathbb{K},$$

and we define the discriminant of E by $\Delta(E) = -4a^3 - 27b^2$. The Mordell-Weil theorem says that the following set

$$E(\mathbb{K}) = \{(x, y) \in \mathbb{K} \times \mathbb{K} | y^2 = x^3 + ax + b\} \cup \mathcal{O}$$

is a finite abelian group with $\mathcal{O}$ is a point at ∞ as the identity element.

In elliptic curve cryptography, we are interested in case where $\mathbb{K}$ is a finite field, $\mathbb{K} = \mathbb{F}_q$, q is a prime number power. Suppose that E is defined over $\mathbb{Z}$, i.e. $a, b \in \mathbb{Z}$. For each prime p, we define

$$a_p = p + 1 - \sharp E(\mathbb{F}_p)$$

where $\mathbb{F}_p$ is a finite field with p elements and $\sharp E(\mathbb{F}_p)$ is the order of

$$E(\mathbb{F}_p) = \mathcal{O} \cup \{(x, y) \in \mathbb{F}_p \times \mathbb{F}_p | y^2 = x^3 + ax + b\}$$

The Hasse theorem says that for each prime p, a_p satisfies the following inequality

$$|a_p| \leq 2\sqrt{p}$$

When p divides $\Delta(E)$, one finds $a_p = 0, \pm 1$ where the exact value is determined according to the type of singularity of E modulo p. The L-function attached to E is defined as

$$L(s, E) = \prod_{p \mid \Delta(E)} \frac{1}{1 - a_p p^{-s}} \prod_{p \nmid \Delta(E)} \frac{1}{1 - a_p p^{-s} + p^{1-2s}}.$$

By the Hasse theorem, one can show that the Euler product $L(s, E)$ converges for all s with $Re(s) > \frac{3}{2}$ and it is given by the following absolutely convergent series

$$L(s, E) = \sum_{n=1}^{\infty} a_n n^{-s}.$$

3   ECOH

We first recall a description of the essential elements of ECOH. We restrict ourselves to messages that are an integral number of blocks. For full details, we refer to the ECOH specifications [9]. ECOH divides the message into blocks, maps each block to an elliptic curve point and adds these points together with two more points. One additional point depends on the message length and the exclusive-or of all message blocks. More formally, given n message blocks $M_0, M_1, ..., M_{n-1}$ we have

$$P_i := P(M_i, i) \quad for \; i = 0, ..., n-1$$

$$X_1 := P'(n)$$

$$X_2 := P''(\bigoplus_{i=0}^{n-1} M_i, n)$$

$$Q := \sum_{i=0}^{n-1} P_i + X_1 + X_2$$

$$R := f(Q)$$

Here, P is a function that maps a message block and an integer to an elliptic curve point. P' computes the padding point which depends only on the length of M. P'' computes the checksum point X_2 which depends on the exclusive-or of all message blocks and on the length of M. Finally, the $n + 2$ elliptic curve points are added, and the result Q is passed through an output transformation function f to get the hash result R.

4   Hashing with L-Functions

4.1   One Way Function

It is well known that the Schoof algorithm for point counting on elliptic curves over finite fields runs in polynomial time [4]. Therefore, for a given elliptic curve

E over $\mathbb{Q}$, the coefficients a_p of $L(s, E)$ can be also computed in polynomial time. Nevertheless, it is not easy to go in the other direction. Indeed, let T be a very large number and let α, β be fixed positive constants. Let we denote by $\mathcal{L}^T$ the set of all elliptic curves E over $\mathbb{Q}$ with $T \leq \Delta(E) \leq 2T$ and let k and m be integers satisfying

$$(\log T)^\alpha \leq k, \quad m \leq (\log T)^\beta.$$

If we assume the conditions presented in the work of Anshel and Goldfeld in their paper [5] then the following conjecture can be stated as follows

Conjecture : The following map

$$H : \quad \mathcal{L}^T \to \mathbb{Q}^{k+1}$$

$$E \mapsto (a_m, a_{m+1}, ..., a_{m+k})$$

where $L(s, E) = \sum_{n=1}^{\infty} a_n n^{-s}$, is a one way function.

Indeed, for a given elliptic curve that satisfies the above conditions, the above conjecture remains still unproved. So far, it is not known whether there exits an algorithm which computes H^{-1} in polynomial time. In connection with this conjecture, we mention the work of the authors [6, 7] in which they gave some numerical evidence for the partial Riemann hypothesis in conjunction with the coefficients of modular L-functions. For instance, it is expected that the conjecture would also hold for most L-functions used in number theory which share some basic analytic properties, in particular meromorphic continuation, an Euler product and a functional equation of a certain type.

Applications

1. Authentication
   (a) Both Alice and Bob are in possesion of a secret key E which is an elliptic curve.
   (b) Bob sends to Alice randomly chosen integers m and $b > 0$.
   (c) Alice computes and sends to Bob the vector $v = (a_m, a_{m+1}, ..., a_{m+b})$.
   (d) Bob checks if Alice's list is correct then she is an authenticated user.
2. Pseudorandom number generator
   A pseudorandom number generator (PRNG), also known as a deterministic random bit generator (DRBG), is an algorithm for generating a sequence of numbers that approximates the properties of random numbers. Pseudorandom numbers are important in practice to speed up the number generation and their reproducibility. The conjecture presented above is useful to create a such generator [5].
   (a) Choose $a, b \in \mathbb{Z}$ such that $4a^3 + 27b^2 \neq 0$ and where the degree of the splitting field of $X^3 + aX + b$ is of degree 6 over $\mathbb{Q}$. E the associated elliptic curve $E : y^2 = x^3 + ax + b$.
   (b) For each prime $3 = p_1 < p_2 < \cdots$, we compute

$$(a_{p_1}(mod2), a_{p_2}(mod2), \cdots .)$$

This is a pseudorandom sequence with probability distribution $(\frac{1}{3}, \frac{2}{3})$.

4.2   Hashing and MAC's Protocol

Most of the keyed hash functions also called MACs are designed as an iterative process which hashes arbitrary length inputs by processing successive fixed-size blocks of the input. This is illustrated in figure 3. For more details see [8]. First, we divide the input x of arbitray finite length into fixed-length r-bit blocks x_i. This preprocessing involves a necessary appending extra bits (padding: Algorithm 2) in order to attain a bitlength which is a multiple of the block length r. Every block x_i is an input for another fixed-size hash function f. The purpose of our work is to give a such simple and robust internal hash function, f computes a new result of bitlength n from a previous n-bit intermediate result and give the next input block x_i. If we denote by H_i the partial result after stage i, then the general process for an iterated hash function with input $x = x_1 x_2 ... x_t$ can be modeled as follows

$$H_0 = IV; \quad H_i = f(H_{i-1}, x_i), \quad 1 \leq i \leq t; \quad h(x) = H_t$$

H_{i-1} serves as the n-bit chaining variable between stage $(i-1)$ and stage i and H_0 is a pre-defined starting value or initializing value (IV). Therefore, the final result H_t of this process is the value of our hash function.

Algorithm 2. The padding algorithm [8]

Require: a non compression function f

1: Suppose f maps r-bit inputs to $n$bit outputs (for concreteness, consider $r=128$ or $r=512$). Construct a hash function h from f, yielding n-bit hash values, as follows.

2: Break an input x of bit length b into blocks $x_1 x_2 ... x_t$ each of bit length r, padding out the last block x_t with 0-bits if necessary.

3: Define an extra final block x_{t+1}, the length-block, to hold the right-justified binary representation of b (assume that $b < 2^r$).

4: Letting 0^j represent the bit string of j 0's, define the n-bit hash-value of x to be $h(x) = H_{t+1} = f(H_t, x_{t+1})$ computed from :

$$H_0 = 0^n; \quad H_i = f(H_{i-1}, x_i), \quad 1 \leq i \leq t+1$$

5: **return** a keyed hash function h which is collision resistant

---

The main idea of Algorithm 3 is that we use the one way function presented in paragraph 4.1 to compute $H_i = f(H_{i-1}, x_i) = x_i \oplus \pi(a_{k_i}, a_{k_i+1}, ..., a_{k_i+r})$ where $(a_{k_i}, a_{k_i+1}, ..., a_{k_i+r})$ are r-coefficients in stage i of an L-function of a secret elliptic curve, $\oplus$ is the XOR operator and π is a secret permutation used to avoid a differential attack on our hash function. These coefficients are defined in the first step of our protocol $H_0 = IV = (a_k, a_{k+1}, ..., a_{k+r})$. Actually, we can resume our hashing protocol in the following algorithm.

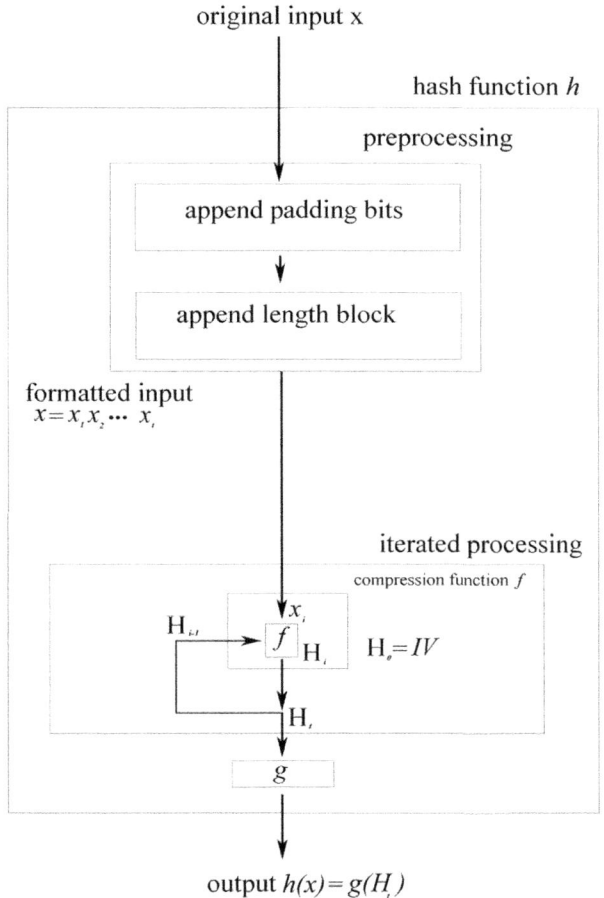

original input x

hash function h

preprocessing

append padding bits

append length block

formatted input
$x = x_1 x_2 \cdots x_t$

iterated processing

compression function f

H_{i-1} f x_i H_i $H_0 = IV$

H_i

g

output $h(x) = g(H_t)$

Fig. 3. General model for an iterated hash function [8]

Algorithm 3. MAC's elliptic curve L-function protocol

Require: a secret key (E, k, π) and a message $x = x_1 x_2 ... x_{t+1}$
1: use the padding algorithm with input x.
2: $H_0 = IV = (a_k, a_{k+1}, ..., a_{k+r})$.
3:   **For each** $i \in \ 1 \leq i \leq t+1$
       $H_i = f(H_{i-1}, x_i) = x_i \oplus \pi(a_{k_i}, a_{k_i+1}, ..., a_{k_i+r})$
  end
4: $h(x) = H_{t+1}$.
5: **return** a hash value of the message x

4.3   MAC's Elliptic Curve L-Function against ECOH

Elliptic Curve Only Hash function, ECOH, is based on the hardness of resolving the discrete logarithm problem. However, the MAC's we present in this paper is based on the fact it is hard to compute the coefficients a_k of the L-function of a secret elliptic curve E. In the literature, there are various algorithms and methods to attack DLP, such as index calculus or Pohlig-Hellman algorithm. However, there are no known attacks for the one way map constructed in paragraph 4.1. Both of these hash functions are robust against the exhaustive key search and the birthday attack because the output length of these functions can be chosen large enough so that these attacks become computationally infeasible. For instance, finding a second pre-image attack against ECOH and thus collisions can be reduced as an instance of the subset sum problem which is not doable in polynomial time. A second pre-image attack exists in the form of a generalized birthday attack proposed by Wagner in his paper [11]. Besides the latter attack, there exists another way to find second pre-images by solving the Semaev Summation Polynomial Problem which offers a construction of the index calculus algorithm for the DLP. For more details, see [10]. On the other side, for our MAC's these attacks are avoided due to the nature of our secret parameters $(a_k, a_{k+1}, , a_{k+r})$ and the use of the permutation π. This brings to our function an almost random feature joined to the mathematical difficulty of computations of the a_k coefficients.

Implementations: The implementation of the ECOH hash function needs computations in the elliptic curve group law and requires to find the inverses of elements in the field over which the curve is defined. This should be at least thousand times slower than SHA-1 on the same platform which is similar of our MAC's which is simple to be implemented either in the software or on hardware platforms.

5   Security

It is well known that a cryptographic hash function must be easy to compute in the directly and infeasible in the other direction. Furthermore, it must be able to withstand all known types of cryptanalytic attack. At least, it must have the following properties, for a keyed hash function h with inputs x, x' and outputs y, y':

1. **Preimage resistance :** for essentially all pre-specified outputs, it is computationally infeasible to find any input which hashes to that output, i.e., to find any preimage such that $h(x') = y$ when given any y for which a corresponding input is not known.

2. **2nd-preimage resistance :** it is computationally infeasible to find any second input which has the same output as any specified input, i.e., given x, to find a 2nd-preimage $x \neq x'$ such that $h(x) \neq h(x')$.

3. **Collision resistance** : it is computationally infeasible to find any two distinct inputs x, x' which hash to the same output, i.e. such that $h(x) = h(x')$.

The MAC function presented in the last section is simple to implement and easy to compute. On the other hand to compute H^{-1} is still infeasible. For instance, the naive algorithm which computes H^{-1} runs in exponential time. Let $L(s, E)$ be the L-function attached to an elliptic curve E over $\mathbb{Q}$ and let denote by c_n the coefficients of this function. By the Hasse theorem, we have

$$|c_n| \leq \sqrt{n}\, d(n), \tag{1}$$

where $d(n)$ is the number of positive divisors of n. Let B be a very large number and set $\mu = \nu = 1$ then we have

$$\log B \leq k \leq \sqrt{B}, \quad m \leq \log B.$$

Let C be a fixed positive constant with $C \geq \frac{\log B}{B}$. Let $\mathcal{L}_C^B$ be the subset of $\mathcal{L}^B$ consisting of all elliptic curves E: $y^2 = x^3 + ax + b$, $a, b \in \mathbb{Q}$ and $0 \leq |a|, |b| \leq CB$.

Theorem 1. *Let the projection map*

$$H_1: \quad \mathcal{L}_C^B \to \mathbb{Q}^{k+1}$$

$$E \mapsto (a_m, a_{m+1}, ..., a_{m+k})$$

where $L(s, E) = \sum_{n=1}^{\infty} \frac{a_n}{n^s}$. Then for any $(c_m, c_{m+1}, ..., c_{m+k}) \in \mathbb{Q}^{k+1}$ satisfying the inequality (1) and for any $\epsilon > 0$, there exits an algorithm which computes an elliptic curve E in $\mathcal{L}_C^B$ such that $a_p = c_p$ for all primes $m \leq p \leq m + k$ with a running time of order $(CB)^{\frac{3}{2}+\epsilon}$.

Proof. For each prime p, the number of isomorphism classes of an elliptic curve E over $\mathbb{F}_p$ satisfying $a_p = c_p$ is of order $O(\sqrt{p} \log p (\log \log p)^2)$. Since there are at most $\frac{p-1}{2}$ elliptic curves isomorphic to a given elliptic curve over $\mathbb{F}_p$, there are $O(p^{\frac{3}{2}+\epsilon})$ pairs of a, $b \in \mathbb{F}_p$ such that the elliptic curve $E: y^2 = x^3 + ax + b$ satisfies $a_p = c_p$. The following steps show how we can construct such a curve and give essentially the complexity in time.
Step 1. For each prime p with $m \leq p \leq m + k$, make a list $\mathcal{E}_p$ of all isomorphism classes of elliptic curves E over $\mathbb{F}_p$ satisfying $a_p = c_p$. Let p_1 be the smallest prime such that $p_1 \geq m$ and let p_2 be the smallest prime such that $p_2 > p_1$. In this way we denote the consecutive primes

$$m \leq p_1 < p_2 < p_3 < \cdots.$$

Let $\mathcal{E}_1 = \mathcal{E}_{p_1}$ and we inductively define $\mathcal{E}_M$ for each $M \geq 1$ as follows: for given $E: y^2 = x^3 + ax + b$ in $\mathcal{E}_M$ and $E': y^2 = x^3 + a'x + b'$ in $\mathcal{E}_{M+1}$, we use the Chinese remainder theorem to find α, β $(mod \prod_{i=1}^{M+1} p_i)$ such that

$$\alpha = a, a' (mod \prod_{i=1}^{M+1} p_i), \quad \beta = b, b' (mod \prod_{i=1}^{M+1} p_i).$$

Then the elliptic curve $E'' : y^2 = x^3 + ax + b$ satisfies $a_{p_i} = c_{p_i}$ for all i with $1 \leq i \leq M + 1$. Define $\mathcal{E}_{M+1}$ as the set of all such E''.

Step 2. Let M_B be an integer so that p_{M_B} would be the largest prime with $p_{M_B} \leq m + \log(CB)$. Then the coefficients of elliptic curves in $\mathcal{E}_{M_B}$ are determined modulo CB. Next, we make a list $\mathcal{E}'_{M_B}$ of all elliptic curves E in $\mathcal{E}_{M_B}$ satisfying $B \leq \Delta_E \leq 2B$.

Step 3. For each E in $\mathcal{E}'_{M_B}$, check whether $a_p = c_p$ for all primes p with $p_{M_B} < p \leq m + k$. If this is satisfied, then E is an elliptic curve in $\mathcal{L}_C^B$. Since there are at most $p^{\frac{3}{2}+\epsilon}$ elliptic curves E $(mod\,p)$ satisfying $a_p = c_p$ for each prime p. Therefore, the total number of elliptic curves constructed by this algorithm is at most $\theta = \prod_{m \leq p \leq m + \log CB} p^{\frac{3}{2}+\epsilon}$. As $B \to 0$, the prime number theorem yields

$$\log \theta = \sum_{m \leq p \leq m + \log CB} (\frac{3}{2} + \epsilon) \log p \approx (\frac{3}{2} + \epsilon) \log(CB).$$

Therefore, we have $\theta = O((CB)^{\frac{3}{2}+\epsilon})$.   □

Corollary 1. *For the hash function presented above, there is no known algorithm which can find a collision in polynomial time.*

Proof. According to the preceeding theorem, we deduce that the class of elliptic curves satisfying $a_p = c_p$ would require $\theta = O((CB)^{\frac{3}{2}+\epsilon})$. Moreover, we must find the secret permutation which requires at least $O(r!)$ time of computation.   □

6   Conclusion

In this work, we proposed a simple and efficient hash function based on the L-function attached to a given elliptic curve. It will be important to carry out the implementation aspect and complexity of this hashing protocol on hardware and software platforms. It may be also useful to do the same construction in the case of L-functions attached to hyperelliptic curves. More generally, we expect hopefully to apply the same procedure for an arbitrary L-function. For instance, it would be interesting to study what happens for a general family of L-functions called the Selberg class. This class contains most L-functions in number theory and share some basic analytic properties, in particular meromorphic continuation, an Euler product and a functional equation of a certain type.

References

1. Preneel, B.: Analysis and Design of Cryptographic Hash Functions. Ph.D. Thesis, Katholieke University Leuven (1993)
2. Damgård, I.B.: A Design Principle for Hash Functions. In: Brassard, G. (ed.) CRYPTO 1989. LNCS, vol. 435, pp. 416–427. Springer, Heidelberg (1990)
3. Merkle, R.C.: One Way Hash Functions and DES. In: Brassard, G. (ed.) CRYPTO 1989. LNCS, vol. 435, pp. 428–446. Springer, Heidelberg (1990)

4. Schoof, R.: Elliptic curves over finite fields and the computation of square root mod p. Math. Comp. 44, 483–494 (1985)
5. Anshel, M., Goldfeld, D.: Zeta functions, one-way functions, and pseudorandom number generators. Duke Math. J. 88, 371–390 (1997)
6. Omar, S., Ouni, R., Mazhouda, K.: On the zeros of Dirichlet L-functions. LMS Journal of Computation and Mathematics 14, 140–154 (2011)
7. Omar, S., Ouni, R., Mazhouda, K.: On the Li coefficients for the Hecke L-functions (preprint, 2012)
8. Menezes, A.J., van Oorschot, P.C., Vanstone, S.A.: Handbook of Applied Cryptography. CRC Press (2001)
9. Brown, D.R.L., Antipa, A., Campagna, M., Struik, R.: ECOH: the Elliptic Curve Only Hash, Tech. rep., Certicom Corp. First Round NIST SHA-3 Candidate (2008)
10. Semaev, I.: Summation polynomials and the discrete logarithm problem on elliptic curves. IACR Cryptology ePrint Archive 2004: 31 (2004)
11. Wagner, D.: A Generalized Birthday Problem. In: Yung, M. (ed.) CRYPTO 2002. LNCS, vol. 2442, pp. 288–303. Springer, Heidelberg (2002)

Square Root Algorithms
for the Number Field Sieve

Emmanuel Thomé

INRIA Nancy, Villers-lès-Nancy, France

Abstract. We review several methods for the square root step of the Number Field Sieve, and present an original one, based on the Chinese Remainder Theorem.

We consider in this article the final step of the Number Field Sieve (NFS) factoring algorithm [3], namely the algebraic square root computation. This problem is stated as follows. Let $K = \mathbb{Q}(\alpha)$ be a number field, where α is defined as a root of the irreducible polyomial $f(x) \in \mathbb{Z}[x]$. We further use the notation $d = \deg f = [K : \mathbb{Q}]$, and denote by $\mathcal{O}_K$ the ring of integers of K. For brevity, we assume throughout this article that f is a monic polynomial, although the more general case $f_d \neq 1$ can be treated similarly by considering $f_d\alpha$ in lieu of α. Let $\mathcal{S}$ be a set of pairs (a, b) such that $S(\alpha) = \prod_{(a,b) \in \mathcal{S}}(a - b\alpha)$ is known to be the square of an algebraic integer. Such a set $\mathcal{S}$ is the outcome of the linear algebra and character steps of the Number Field Sieve. The purpose of the algebraic square root step is the computation of a polynomial $T(x) \in \mathbb{Z}[x]$ such that $T(\alpha)^2 = f'(\alpha)^2 S(\alpha)$. Here, $f'(\alpha)^2$ is introduced merely in order to take into account the possibility that a square root for $S(\alpha)$, despite being an algebraic integer, needs not belong to the order $\mathbb{Z}[\alpha]$. For brevity again, this $f'(\alpha)^2$ term will be omitted throughout this article.

Two further characteristics of the problem, specifically related to the NFS context, are also important. The ideals $(a - b\alpha)\mathcal{O}_K$ always have a known factorization into prime ideals (the latter being for example readily available alongside with the pairs (a, b), e.g. in a file). Furthermore, the output defined as $T(x)$ above is not interesting *per se*. In the NFS, one intends to compute $T(m) \bmod N$, where m and N are known (N being the integer to factor). In some cases, it is possible to achieve this goal without explicitly computing $T(x)$.

This article reviews several approaches for the algebraic square root task. Most of them are classical, and date back to the early research on the NFS. Let us recall that back in 1993, this square root step was regarded as difficult. Attacking the problem "directly" by computing the algebraic number $S(\alpha)$ and later its square root, appeared a daunting task by then, and this justified the development of ad hoc algorithms, such as Montgomery's [14, 15] or Couveignes' [5] algorithms, which exploit the fact that the factorization of the ideals $(a - b\alpha)\mathcal{O}_K$ is known. The practicality of asymptotically fast methods appeared later and the direct approach then became realistic. Furthermore, while slower than e.g. Montgomery's algorithm, this method has turned out to be acceptably

F. Özbudak and F. Rodríguez-Henríquez (Eds.): WAIFI 2012, LNCS 7369, pp. 208–224, 2012.

fast, notably when compared with the overall cost of NFS. Because the direct approach embarks less algebraic number theory background than Montgomery's method, it is sometimes preferred in NFS implementations [17, 8]. However the direct approach, when stated in its simple form, is unable to tackle square root problems arising with number fields having no inert primes (e.g. with Galois group $\mathbb{Z}/2\mathbb{Z} \times \mathbb{Z}/2\mathbb{Z}$). Such fields are typically encountered with the Special Number Field Sieve, and provide one of the justifications for the Chinese Remainder Theorem (CRT)-based approach presented in this article.

The algorithms presented in this article are all relevant, when applicable, for modern integer factoring, and various working implementations can be found in publicly available software [8, 17, 13]. These different algorithms have actually been used or at least tested for the rsa768 factoring effort [11]. Furthermore, in the context of the oracle-assisted RSA problem addressed in [10], a variant of Montgomery's algorithm is used, and proves particularly well-adapted.

This article is organized as follows. The direct approach for computing the square root is described in Section 1. Couveignes' CRT-based method for the odd-degree case is described in Section 2. Section 3 describes Montgomery's algorithm. The new CRT-based approach we propose is presented in Section 4. Some viable approaches for the square root computation are not detailed here (e.g. the one proposed in [4, § 3.6.2]), since they offer no obvious advantage over the ones presented here.

Complexities of the algorithms presented are given as functions of the input size. In order to be able to provide comparable estimates, we afford some simplifying assumptions. We assume that the set of pairs $\mathcal{S}$ occupies n bits. This leads to coefficients of $S(\alpha)$ each having size roughly n bits as well. The bit size of $S(\alpha)$ is thus $O(dn)$ overall. Similarly, coefficients of $T(\alpha)$ have size roughly $n/2$ bits. We acknowledge that these estimates are slightly gross, but these do make sense for the NFS context, and hold in practice. Furthermore, the number field is considered constant, so that most dependencies on the number field parameters are deliberately ignored. The notation $\mathsf{M}(k)$ is used throughout the document to denote the time for multiplying k-bit integers.

1   The Direct (lifting) Approach

The direct method, which is also referred to as the p-adic, or lifting approach, applies when there exists an *inert* prime in the number field K. This is only possible when the Galois group of the polynomial f admits an element of order d. Generically, the Galois group of a polynomial used in the General Number Field Sieve (GNFS) algorithm is the full symmetric group $\mathfrak{S}_d$, which implies that this property is expected to be satisfied with overwhelming probability. This is not so, however, with polynomials considered in the context of the Special Number Field Sieve (SNFS). There, polynomials typically have some special shape, which makes all sorts of Galois groups plausible. For instance, a degree 4 polynomial with Galois group $\mathbb{Z}/2\mathbb{Z} \times \mathbb{Z}/2\mathbb{Z}$ can be encountered. This very case is too frequent to be completely neglected, which makes the direct approach only a partial solution to our problem.

1.1 Working p-Adically

Let thus p be an inert prime (such that $p\mathcal{O}_K$ is a prime ideal). To fix ideas, p is taken of manageable size (say at most 64 bits). The extension $K_p = \mathbb{Q}_p/f(x)$ is thus a degree d unramified extension of $\mathbb{Q}_p$. Let α_p be a root of f in K_p. We have a natural injective ring morphism from $\mathbb{Z}[\alpha]$ to $\mathbb{Z}_p[\alpha_p]$. The purpose of the lifting approach is to use this injective morphism in order to recognize the square root being sought.

The first step of the algorithm is the computation of a low-precision square root for $S(\alpha_p)$. Let $\mathbb{F}_p(\beta) = \mathbb{F}_{p^d}$ be the residue field of K_p, the projection $K_p \to \mathbb{F}_{p^d}$ being given by $\alpha_p \mapsto \beta$. By low-precision, we understand the computation of a square root of $S(\beta) \in \mathbb{F}_{p^d}$. Since $S(\beta) = \prod_{(a,b)\in\mathcal{S}}(a - b\beta)$ and all (a,b)'s are coprime, we know that $S(\beta) \neq 0$. Let $T(x) \in \mathbb{Z}[x]$ be, as above, a polynomial defining a square root $T(\alpha)$ of $S(\alpha)$. We know that $T(\beta)^2 = S(\beta)$, and our computation in $\mathbb{F}_{p^d}$ gives us the coefficients of $T(x)$ (or its opposite) modulo p. In the field K_p, we have

$$T(\alpha_p)^2 \in S(\alpha_p) + p\mathcal{O}_K.$$

We thus have low-precision knowledge on the coefficients of T, together with a defining equation. A Newton lifting approach then allows to recover all coefficients. For example, iterating the modification $T(\alpha_p) \leftarrow T(\alpha_p) + \frac{S(\alpha_p)-T^2(\alpha_p)}{2T(\alpha_p)}$ is sufficient. In practice, it is desirable to first compute the inverse square root instead, so that the iteration avoids inverse computations. Details are skipped, and can be found in [1, 2].

The key concern is the determination of the stopping point of the lifting process. Since coefficients of the desired solutions are known to be integers, we know that above a certain lifting step, the p-adic coefficients obtained for $T(x)$ no longer evolve. (The morphism $\mathbb{Z}[\alpha] \to \mathbb{Z}[\alpha_p]$ being injective, these coefficients are then the desired ones.) Therefore the number of lifting steps is controlled by a bound on the result coefficients, which can be provided with classical tools as we do now.

1.2 Bound on the Square Root Coefficients

Given a number field K, we denote by $\sigma_1, \ldots, \sigma_r$ and $\sigma_{r+1}, \ldots, \sigma_{r+s}$ the real and non-conjugate complex embeddings. We further denote by $\sigma_{r+s+k} = \overline{\sigma_{r+k}}$ for $k = 1, \ldots, s$. Let $U = \sum_{i=0}^{d-1} u_i \alpha^i \in K$. The coefficients u_i of U are related to the embedding values, since these are given by a polynomial expression with coefficients u_i. Namely, we have

$$(\sigma_1(U), \ldots, \sigma_d(U)) = (u_0, \ldots, u_{d-1}) \times V(\sigma_1(\alpha), \ldots, \sigma_d(\alpha))$$

where the matrix above is of Vandermonde type. Exploiting this relation in the opposite direction, it is possible to derive a bound on the $|u_i|$ from a bound on the $|\sigma_i(U)|$. A low precision computation of the inverse of the Vandermonde matrix above suffices to obtain a reasonable bound at a moderate cost.

Such a mechanism is typically used to bound the coefficients of an element U given the *logarithms* of the embeddings, where logarithms provide an additional guard against exponent overflow. This can precisely be done in the context of NFS square root computations. We know $S(\alpha)$, and look for $T(\alpha) \in \mathbb{Z}[\alpha]$ with $T(\alpha)^2 = S(\alpha)$. Hence we have $\log|\sigma_i(T(\alpha))| = \frac{1}{2}\log|\sigma_i(S(\alpha))|$. Furthermore, $S(\alpha)$ is also known in *product* form. The computation

$$\log|\sigma_i(S(\alpha))| = \sum_{(a,b)\in\mathcal{S}} \log|\sigma_i(a - b\alpha)|$$

therefore appears rather accessible, as its computational cost is essentially that of reading the input (set of (a, b) pairs).

The computation of the required number of lifting step therefore proceeds as follows.

- Compute the complex roots of f. A computation with limited precision suffices. The inverse of the matrix $V(\sigma_1(\alpha), \ldots, \sigma_d(\alpha))$ is then computed (this is a Lagrange interpolation matrix). It is important, in order for the bound to be valid, to use rounding towards $+\infty$ in the computations.
- Then, the $\log|\sigma_i(S(\alpha))|$ values may be computed.
- Given this data, deduce a bound on the coefficients $|t_i|$.

Given a bound M obtained by this method, lifting may stop at precision $k = \lceil\log_p M\rceil$. This implies in particular that in computations where $S(\alpha)$ appears, coefficients may be reduced modulo p^k, which provides noticeable savings in computation time.

One may notice that this bound computation requires no memory.

1.3   Complexity

The computation of a square root by Newton lifting is quasi-linear [1, 2]. This also holds in our case of interest here. We may write the complexity as $O(d^2\mathsf{M}(n))$, where n is the size in bits of the input coefficients, and where d^2 is taken for the cost of multiplying polynomials of degree $d - 1$. (One may of course use better algorithms than the naive one for this task. This is relevant only to a limited extent in our range of interest, since d denotes an NFS degree.) Note that this complexity is in fact dominated by the complexity of the preliminary computation of $S(\alpha)$ from the set $\mathcal{S}$, which claims $O(d^2\mathsf{M}(n)\log n)$ using a subproduct tree.

The space complexity of the direct approach is linear, namely $O(dn)$. In comparison to other approaches considered in this article, this approach does compute $T(x)$ as a prerequisite before computing $T(m)$ as required by NFS.

2   Couveignes' Algorithm

In [5], Couveignes proposes an algorithm which allows to avoid the space complexity of the direct method above, and allows some parallelism. This approach is based on the Chinese Remainder Theorem (CRT), and is only applicable under the simultaneous conditions that the number field degree d be odd, and that there exist inert primes[1].

The approach goes as follows. We intend to compute $T(\alpha)$ with a CRT approach. Let $\{p_i\}$ be a collection of inert primes. For each such p_i, we denote by β_i a root of $f(x)$ modulo p_i, and consider (often implicitly) the ring morphism from K to $\mathbb{F}_{p_i^d}$ sending α to β_i. We begin by computing a square root of $S(\beta_i)$ for all i. Such a square root is denoted by $T_i'(\beta_i)$, and writes as $T_i'(\beta_i) = \pm T(\beta_i)$. We then wish to reconstruct the result $T(x)$ as a polynomial with integer coefficients. For this purpose, we assume that a bound M on the coefficients of T has been obtained in a manner similar to the approach described in Subsection 1.2. We assume that the product of the chosen primes p_i exceeds the bound M. Suppose then that $T_i(x) = T(x) \bmod p_i$ is known. Let q_i be the smallest positive integer such that $q_i \bmod p_j = \delta_{i,j}$ (one may write $q_i = s \cdot (s^{-1} \bmod p_i)$, where $s = \prod_{j \neq i} p_j$). We have then:

$$T(x) = \sum_i q_i T_i(x).$$

The stumbling block for such an approach is related to the choice of roots. We have acquired knowledge of $\pm T(\beta_i)$, but this is only sufficient to determine $T_i(x)$ up to a sign. In presence of a large number of primes p_i, finding the correct sign combination in the expression $\sum_i \pm q_i T_i'(x)$ is intractable.

In order to overcome this problem, Couveignes' algorithm takes advantage of the degree of K being odd. Under this assumption, we have

$$\mathrm{Norm}_{K/\mathbb{Q}}(-\zeta) = -\mathrm{Norm}_{K/\mathbb{Q}}(\zeta)$$

for any $\zeta \in K$. In the particular case of the NFS square root, computing the absolute value $|\mathrm{Norm}_{K/\mathbb{Q}}(T(\alpha))|$ is rather easy. Indeed, we know the factored form of the principal ideal $S(\alpha)\mathcal{O}_K$. We may thus write:

$$S(\alpha)\mathcal{O}_K = \prod_{\mathfrak{p} \in \mathcal{F}} \mathfrak{p}^{2e_{\mathfrak{p}}} \text{ with } e_i \in \mathbb{Z},$$

$$|\mathrm{Norm}_{K/\mathbb{Q}}(S(\alpha))| = \prod_{\mathfrak{p} \in \mathcal{F}} \mathrm{Norm}(\mathfrak{p})^{2e_{\mathfrak{p}}},$$

$$\mathrm{Norm}_{K/\mathbb{Q}}(T(\alpha)) = \pm \prod_{\mathfrak{p} \in \mathcal{F}} \mathrm{Norm}(\mathfrak{p})^{e_{\mathfrak{p}}}.$$

---

[1] The existence of one or many inert primes are equivalent conditions, in virtue of Čebotarev's density theorem. This theorem appears in countless graduate level algebraic number theory textbooks. For an introduction to Čebotarev's density theorem, including also historical aspects and applications, the article [12] is an interesting read.

We acknowledge the fact that our chosen notation $T(x)$ for the solution being sought is ambiguous, given that there are two solutions. This accounts for the $\pm$ sign in the equation above. From now on, we intend to focus on the computation of one of these two solutions only, and we need to make one single consistent choice. To this end, we require that the computed $T(x)$ correspond to a positive norm above. Such an arbitrary choice is legitimate, provided it is done only once (if we were to combine mixed information related to either of the two different solutions for many primes, consistent reconstruction would be impossible). Let thus ν be this positive norm, which is accessible to calculation. We have $\mathrm{Norm}_{K/\mathbb{Q}}(T(\alpha)) = \nu$ and $\mathrm{Norm}_{K/\mathbb{Q}}(-T(\alpha)) = -\nu$. Modulo p_i, this property transfers conveniently. We have

$$\mathrm{Norm}_{\mathbb{F}_{p_i^d}/\mathbb{F}_{p_i}}(T(\beta_i)) = \nu \bmod p_i.$$

This implies that given $T_i'(\beta_i)$, it is possible to decide whether $T_i'(\beta_i) = T_i(\beta_i)$ or $T_i'(\beta_i) = -T_i(\beta_i)$, by comparing $\mathrm{Norm}_{\mathbb{F}_{p_i^d}/\mathbb{F}_{p_i}}(T_i'(\beta_i))$ with $\nu \bmod p_i$.

Once the sign problem has been solved, it is possible to use the expression $T(x) = \sum_i q_i T_i(x)$. Note though that for the NFS application, only the quantity $T(m) \bmod N$ is needed eventually. Therefore, computing $T(x)$ is unnecessary. It is sufficient (and considerably cheaper) to write

$$T(m) \bmod N \equiv \sum_i (q_i \bmod N)(T_i(m) \bmod N).$$

Complexity. Let us assume that primes of fixed size λ are considered[2] Since the result coefficients occupy $O(n)$ bits, it suffices to consider $O(n/\lambda)$ primes. We first consider the complexity of such an approach in the perspective of a constant space complexity. For each such prime, one has to read the input set $\mathcal{S}$, in order to compute the norm ν. This step dominates the complexity, since it takes $O(n^2)$. Furthermore, parallelizing this step is not necessarily obvious, since the complexity essentially consists of input-output operations (several threads of a single processor may benefit from a single read at the same time, but such a benefit does not scale well to a multi-machine setup).

This approach is amenable to a time-memory trade-off. It is possible to read the set $\mathcal{S}$ in $\sqrt{n}$ blocks of size $\sqrt{n}$. For each such block, an intermediary product of size $\sqrt{n}$ may be computed, and then reduced modulo each prime p_i. In such a way, the time complexity drops to $O(n^{3/2})$, for a space complexity in $O(n^{1/2})$. Such a modification could be considered in a distributed setting, as it offers more opportunities for parallelization.

3   Montgomery's Algorithm

Montgomery's algorithm [14, 15] for the square root step is the one which is most specially crafted for the NFS square root situation. This algorithm radically

---

[2] The obvious finiteness of the number of primes satisfying this criterion is an irrelevant concern here.

differs from the two algorithms presented above, as well as from the one which is described in Section 4. The original description of Montgomery's algorithm is given in [14], although the unpublished draft [15] is also worth reading in that it contains additional details. Nguyen also presented Montgomery's algorithm in [16].

Montgomery's algorithm is not dependent on strong assumptions as the algorithms presented above. In particular, the Galois group of the defining polynomial is not an obstacle for Montgomery's algorithm. However, the knowledge of the factorization of the ideal $S(\alpha)\mathcal{O}_K$ as a product involving only ideals of small norm is crucially important. Furthermore, as we will see, the "dependency on 2" is especially good with this algorithm. More precisely, this algorithm may be stated in a more general way as an algorithm for computing a λ-th root, where λ is any integer, provided that the factorization of $S(\alpha)$ remains known. Of course, in such a case, one expects the coefficients of $T(\alpha)$ to be λ times smaller than those of $S(\alpha)$. This generalization matters to the oracle-assisted computation of λ-th roots as detailed in [10], and accounts for our choice to present Montgomery's algorithm in the extended setting $\lambda \geq 2$.

We are interested in the factored form of the ideal $S(\alpha)\mathcal{O}_K$. In the NFS context, the computation of the maximal order $\mathcal{O}_K$ of the number field K is a hard problem, because the discriminant of polynomials used in the (general) NFS is possibly even harder to factor than the number which we intend to factor in the first place. Yet, given a monic polynomial $f(x)$, factoring its discriminant is necessary in order to decide which are the primes p for which the index $[\mathcal{O}_K : \mathbb{Z}[\alpha]]$ is divisible by p. For such primes, we say that *locally at p*, the order $\mathbb{Z}[\alpha]$ is not maximal, since $\mathcal{O}_K$ is larger.

A very important observation is that for the purpose of factoring ideals over a prescribed factor base, which consists of all primes ideals above a prescribed set of primes, it is sufficient to work with an order $\mathcal{O}$ which is maximal at those primes only, and needs not be maximal over all primes. For the ideals considered, factorization into $\mathcal{O}$-ideals or $\mathcal{O}_K$-ideals coincide. Extension of a starting order to a p-maximal one for a finite set of primes p can be done using Zassenhaus' Round-2 algorithm, as presented for example in [4]. The computation of such an order $\mathcal{O}$ is an inexpensive preliminary step for Montgomery's algorithm, and we assume it is done.

The following paragraphs use the notation $\mathcal{F}$ for the set of prime ideals occurring in the factorization of $S(\alpha)$ (considerations related to "large primes" aside, this can be thought of as the factor base).

3.1   Iterative Reduction

The main idea of the algorithm is the following. The computation has an iterative structure, and the successive steps are numbered from step number 0 onwards. Throughout the course of the computation, an expression such as the following one is maintained (we recall that λ denotes the order of the root which we intend to compute, e.g. classically $\lambda = 2$):

$$S(\alpha)(\gamma_0^{\epsilon_0} \cdots \gamma_{k-1}^{\epsilon_{k-1}})^{-\lambda}\mathcal{O} = \prod_{\mathfrak{p} \in \mathcal{F}} \mathfrak{p}^{\lambda \cdot e_{\mathfrak{p}}^{(k)}}$$

where the integer k denotes the step of the calculation, starting at $k = 0$. We use γ_i to denote some algebraic number which is computed at step i. We also denote by ϵ_i a sign, which is used as a notational convenience for a choice between numerator and denominator. The exponents $e_{\mathfrak{p}}^{(k)}$ may become negative in the course of the computation.

The aim of these reduction steps is to transform the left-hand side above into an algebraic number which has small coefficients. Not only do we wish to obtain an ideal factorization which is as small as possible (possibly trivial), but we also strive to minimize the *unit contribution* as well, which is important to minimize the size of the coefficients.

We denote by $S_k(\alpha) = S(\alpha)(\gamma_0^{\epsilon_0} \cdots \gamma_{k-1}^{\epsilon_{k-1}})^{-\lambda}$ (in particular, $S_0(\alpha) = S(\alpha)$). Step k of the algorithm chooses a subset of the ideals appearing in the factorization $\prod_{\mathfrak{p} \in \mathcal{F}} \mathfrak{p}^{\cdot e_{\mathfrak{p}}^{(k)}}$. This factorization naturally decomposes into a numerator (positive exponents) and a denominator (negative exponents). We choose a set $I_k = \mathfrak{p}_1 \ldots \mathfrak{p}_{n_k}$, which consists of ideals all appearing in the numerator, or all in the denominator (in this notation we do not forbid a given ideal to be selected several times, provided that its multiplicity in I_k does not exceed $|e_{\mathfrak{p}}^{(k)}|$). We intend to reduce the contribution of the the ideals $\mathfrak{p}_1, \ldots, \mathfrak{p}_{n_k}$ to the factorization of $S_k(\alpha)\mathcal{O}$.

A reduced basis (e.g., an LLL-reduced basis suffices) of the ideal I_k is computed. In this way, we obtain algebraic integers belonging to I_k. Such an algebraic integer v satisfies $\mathrm{Norm}_{K/\mathbb{Q}}(I_k) \mid \mathrm{Norm}_{K/\mathbb{Q}}(v)$, and also the following inequality:

$$m(v) \overset{\text{def}}{=} \left| \frac{\mathrm{Norm}_{K/\mathbb{Q}}(v)}{\mathrm{Norm}_{K/\mathbb{Q}}(I_k)} \right| \leq C_K,$$

where the constant C_K is effectively computable (it depends only on K). Suppose we are given such an element v. Without loss of generality, we consider the case that I_k is a factor of the numerator of $S_k(\alpha)\mathcal{O}$. We then have:

$$\mathrm{Norm}(S_k(\alpha)v^{-\lambda}) = \mathrm{Norm}(S_k(\alpha))\,\mathrm{Norm}(I_k)^{-\lambda}m(v)^{-\lambda}.$$

In the factorization of $S_k(\alpha)v^{-\lambda}\mathcal{O}_K$, the norm of the numerator is reduced by a factor $\mathrm{Norm}(I_k)^\lambda$ in comparison to $S_k(\alpha)\mathcal{O}_K$. At the same time, the norm of the denominator increases by $m(v)^\lambda \leq C_k^\lambda$. Therefore, provided that I_k is chosen to have a norm significantly larger than the constant C_K, we obtain in this way a reduction of the size of the expression. This reduction step is the workhorse of Montgomery's algorithm.

The iterative reduction step may possibly complete with a trivial ideal factorization, in other words with some $S_k(\alpha)$ being a unit. For this unit to be acceptably small, and accessible to direct λ-th root computation, it is important to "guide" the reduction step. The guiding principle is that the logarithms of the complex embeddings $\log|\sigma_i(S_k(\alpha)v^{-\lambda})|$ should become as balanced as possible:

among the different short vectors v formed by the reduced basis of I_k, some lead $S_k(\alpha)v^{-\lambda}\mathcal{O}_K$ to reduce the jitter in the logarithmic complex embeddings, while some others have the opposite effect. The former are favored. In practice this suffices to obtain a final $S_k(\alpha)$ which is a unit (thus the ideal factorization is indeed trivial), and furthermore has trivial logarithmic embeddings, so that it is actually a root of unity.

More details on Montgomery's algorithm, together with an example, can be found in [15]. An implementation of Montgomery's algorithm can be found in [13]. We also mention a MAGMA prototype by the author in [8].

3.2   Complexity

Complexity of Montgomery's algorithm, especially in the extended case $\lambda \geq 2$ which is useful for the context in [10], calls for a more precise consideration of the input and output sizes. We assume that the input set $\mathcal{S}$ occupies n bits. This set $\mathcal{S}$ is obtained in a manner similar to the NFS situation: the solution of some linear system is computed so as to force exponents in the corresponding ideal factorization to cancel modulo λ. Naturally, this linear system is defined modulo λ, and so are the coefficients of this combination. Therefore, our set $\mathcal{S}$ naturally consists of (a, b) pairs together with exponents for each pair, which contribute to the size of $S(\alpha)$. For an input size of n bits for $\mathcal{S}$, we thus expect $O(\lambda n)$ bits for the coefficients of $S(\alpha)$, and $O(n)$ bits for the coefficients of its λ-th root $T(\alpha)$.

Observe now that $S(\alpha)$ is never computed by Montgomery's algorithm. Each reduction step has a cost which is linear in the size of I_k, and more importantly in the size reduction obtained with respect to the input set. The calculations related to the unit contributions are also done incrementally, and have linear cost in the size of I_k. Therefore the complexity of the algorithm is linear, and the dependency on λ is of logarithmic type. The dependency on the parameters of the field K is not detailed here.

4   A New CRT-Based Lifting Approach

We describe here a new approach which is a mix between the direct approach (Section 1) and Couveignes' CRT algorithm (Section 2). This approach, just as Montgomery's algorithm, is free of limiting assumptions on the number field. In particular, we do not assume the existence of inert primes.

This work may be considered as connected to recent works by Enge and Sutherland [6], as well as by Sutherland [18], on the topic of the CRT-based computation of class polynomials in the context of the complex multiplication method for constructing elliptic curves over finite fields. In these works, as well as in the method described here, the "explicit" aspect of the CRT is particularly important.

Let $d = [K : \mathbb{Q}]$. We recall that for brevity we have restricted our presentation to the case where the polynomial f defining the number field K is monic.

As stated already before, we know a way to obtain a bound on the coefficients of the result $T(\alpha)$ which is being sought. We thus assume that such a bound M has been computed and is available.

Let $\mathcal{P} = \{p_i\}$ be a set of $\ell = t \times r$ totally split primes[3]. (The form $\ell = t \times r$ will allow us later to partition $\mathcal{P}$ into t distinct subsets of size r, for distribution purposes.) Let $P = \prod_{p \in \mathcal{P}} p$ and $\lambda = \left\lceil \frac{\log M/\epsilon}{\log P} \right\rceil$, where $\epsilon \leq 1$ is an arbitrarily chosen value discussed later. We thus have $M \leq \epsilon P^\lambda$. We denote by $B = \frac{\lambda}{\ell} \log_2 P$ the bit-size of p^λ for primes $p \in \mathcal{P}$.

Following our assumptions for estimating complexities, the expected bit size of the coefficients of $T(\alpha)$ is $n/2$, where as previously n denotes the size of the input set $\mathcal{S}$. We thus expect expect $\log_2 M \approx \ell B \approx n/2$.

4.1   CRT-Based Reconstruction

The square root is computed from several calculations done modulo p_i^λ, for each p_i. One of the difficulties is naturally linked to resolving the indetermination among the two possible choices of square roots in each of the possible residue fields $\mathbb{F}_{p_i}$. In total, the primes p_i being totally split, we have ℓd square roots to compute, and as many choices to make.

For each p_i, we denote by $(r_{i,j})_{j=1\dots d}$ the roots of f modulo p_i. Since p_i is totally split, the values $r_{i,j}$ are d distinct elements of $\mathbb{F}_{p_i}$. This assumption also allows to compute, corresponding to each $r_{i,j}$, a lift $\tilde{r}_{i,j}$ in $\mathbb{Z}/p_i^\lambda \mathbb{Z}$. We thus have $f(\tilde{r}_{i,j}) \equiv 0 \mod p_i^\lambda$.

For each p_i and each $r_{i,j}$, we compute a p_i-adic lift of $\sqrt{S(\tilde{r}_{i,j})}$, with precision λ. Let $T'_{i,j}$ be this lift. If $T(x)$ denotes as usual the expression of our desired square root in K, we have

$$T'_{i,j} = s_{i,j} T(\tilde{r}_{i,j}) \quad \text{where } s_{i,j} = \pm 1.$$

We wish to find $T(x)$ from the values $T(\tilde{r}_{i,j})$. The latter are, for now, known only up to the sign $s_{i,j}$, and we postpone this sign problem for later analysis. We consider the integers Q_i and polynomials $H_{i,j}(x) \in \mathbb{Z}[x]$ defined as follows:

$$Q_i \overset{\text{def}}{=} \left(\frac{P}{p_i} \right)^\lambda = \prod_{\substack{p \in \mathcal{P}, \\ p \neq p_i}} p^\lambda,$$

$$H_{i,j} \overset{\text{def}}{=} \frac{f(x)}{(x - \tilde{r}_{i,j})} = \prod_{j' \neq j} (x - \tilde{r}_{i,j'}).$$

These polynomials are chosen so as to verify:

Proposition 1. *Let $T_{i,j} = T(\tilde{r}_{i,j}) \mod p_i^\lambda$. Then:*

$$T(x) = \left(\sum_{i,j} Q_i H_{i,j}(x) T_{i,j} \frac{1}{Q_i f'(\tilde{r}_{i,j})} \right) \mod P^\lambda.$$

---

[3] The density of such primes is $1/\# \operatorname{Gal}(f)$, again by Čebotarev's density theorem.

In the statement above, the inverse is to be understood in the p_i-adic ring $\mathbb{Z}_{p_i}$, or more precisely modulo p_i^λ. The proof is straightforward. Remark that $Q_i \bmod p_{i'}^\lambda$ cancels for $i' \neq i$, and that in the case $i = i'$, $H_{i,j}(\tilde{r}_{i',j'}) \bmod p_{i'}^\lambda$ is exactly equal to $f'(\tilde{r}_{i,j})$ for $j' = j$, and cancels otherwise. Then the bound $M \leq P^\lambda$ on the coefficients of T yields the announced result.

4.2 Determining Signs

Proposition 1 allows to reconstruct $T(x)$ as soon as all $T_{i,j}$ are known. However, from the computation of roots modulo p_i, we only know $T'_{i,j} = s_{i,j}T_{i,j}$. Still, by examining the coefficient[4] of degree $d-1$ in $T(x)$, we obtain the following identity, which is derived from the expression in Proposition 1:

$$[x^{d-1}]T(x) = \sum_{i,j} Q_i T_{i,j} \frac{1}{Q_i f'(\tilde{r}_{i,j})} \bmod P^\lambda,$$

$$\frac{1}{P^\lambda}[x^{d-1}]T(x) = \sum_{i,j} \frac{1}{p_i^\lambda}\left(T_{i,j}\frac{1}{Q_i f'(\tilde{r}_{i,j})} \bmod p_i^\lambda\right) \bmod 1.$$

Let now $x_{i,j}$ and $y_{i,j}$ be the real numbers

$$x_{i,j} = \frac{1}{p_i^\lambda}\left(T_{i,j}\frac{1}{Q_i f'(\tilde{r}_{i,j})} \bmod p_i^\lambda\right) \in [0,1[,$$

$$y_{i,j} = \frac{1}{p_i^\lambda}\left(T'_{i,j}\frac{1}{Q_i f'(\tilde{r}_{i,j})} \bmod p_i^\lambda\right) \in [0,1[,$$

$$\equiv \pm x_{i,j} \quad \bmod 1.$$

We can show that the sum of the $x_{i,j}$ numbers is exceptionally close to an integer. Indeed, the coefficient $[x^{d-1}]T(x)$ is at most M. Thus we have:

$$\left|\frac{1}{P^\lambda}[x^{d-1}]T(x)\right| \leq MP^{-\lambda} \leq \epsilon,$$

where ϵ is the arbitrary parameter introduced above. As a consequence, we have:

$$\sum_{i,j} x_{i,j} \in [-\epsilon, \epsilon] + \mathbb{Z}, \qquad \sum_{i,j} s_{i,j}y_{i,j} \in [-\epsilon, \epsilon] + \mathbb{Z}.$$

This implies that by solving a knapsack-like problem, we may find which is the "right" combination of the $y_{i,j}$ coefficients, thereby solving the sign indetermination problem. Such problems are hard to solve, and are discussed further down. Let us only mention briefly that at least for modest numbers of primes, this problem remains practical.

---

[4] Any coefficient of $T(x)$ may be considered. The only special thing about degree $d-1$ is that the corresponding expression is shorter to write.

Once the $s_{i,j}$ have been computed, we deduce the integer κ_{d-1} close to $\sum_{i,j} x_{i,j}$. We also obtain $T_{i,j} = s_{i,j} T'_{i,j}$. This gives:

$$\frac{1}{P^\lambda}[x^{d-1}]T(x) = \sum_{i,j} \frac{1}{p_i^\lambda}\left(s_{i,j}T'_{i,j}\frac{1}{Q_i f'(\tilde{r}_{i,j})} \bmod p_i^\lambda\right) - \kappa_{d-1}.$$

We can generalize this approach to now derive information relative to all coefficients of $T(x)$. This requires taking into account the interpolating polynomials $H_{i,j}(x)$. We define

$$c_{i,j,k} = [x^k]\left(T'_{i,j}\frac{H_{i,j}(x)}{Q_i f'(\tilde{r}_{i,j})} \bmod p_i^\lambda\right),$$

$$c^*_{i,j,k} = s_{i,j}c_{i,j,k}.$$

The coefficients $c_{i,j,k}$ generalize the notations $x_{i,j}$ and $y_{i,j}$ above, since we have

$$x_{i,j} = \frac{c^*_{i,j,d-1}}{p_i^\lambda}, \quad \text{and} \quad y_{i,j} = \frac{c_{i,j,d-1}}{p_i^\lambda}.$$

We may use these notations to write a generalization of the expression above, and likewise for the expression of Proposition 1:

$$\frac{1}{P^\lambda}[x^k]T(x) = \sum_{i,j}\frac{1}{p_i^\lambda}[x^k]s_{i,j}c_{i,j,k} - \kappa_k,$$

$$T(x) = \sum_{i,j,k} x^k\left(Q_i s_{i,j}c_{i,j,k} - \kappa_k P^\lambda\right).$$

Finally, we recall that the aim of the square root computation step is not really to compute $T(x)$, but instead its evaluation $T(m) \bmod N$, as already stated previously. As a consequence of the formula above, we are thus interested in $c_{i,j,k}m^k \bmod N$, which is a priori significantly smaller than p_i^λ.

4.3   Strategies for Fast Computation

The algorithm sketched so far needs some tuning, because the split into several computations, if done incorrectly, may in fact do more harm than good with respect to the overall complexity. The first important concern is the computation of the values $S(\tilde{r}_{i,j})$. Recall that these are $d\ell$ B-bit integers. In order to compute these values efficiently, one may proceed as follows. First compute $S(\alpha)$ with a subproduct tree (see e.g. [7, § 10.1]). Then proceed by computing reductions of all d coefficients modulo the ℓ prime powers p_i^λ. This multimodular reduction step may again be achieved with a subproduct tree. Finally, the evaluations modulo all $\tilde{r}_{i,j}$ is again a multi-evaluation, albeit relatively shallow, since we have only d evaluation points.

For an n-bit input, recall that we have set $\ell B \approx n/2$. Hence the first two steps above respectively have complexity $O(d^2\mathsf{M}(n)\log n)$ and $O(d\mathsf{M}(n)\log n)$. The multi-evaluation at all $\tilde{r}_{i,j}$ has complexity $O(\ell d^2\mathsf{M}(B))$.

Algorithm crtalgsqrt($N, m, f, \mathcal{S}$)

INPUT. f monic irreducible, defining $K = \mathbb{Q}(\alpha)$,
                N integer,
                m a root of f modulo N.
                $\mathcal{S}$ set of pairs (a,b) such that $S(\alpha) = \prod(a - b\alpha) \in \mathbb{Z}[\alpha]^2$.
PARAMETERS. $\ell = r \times t$, number of primes to consider.
OUTPUT. $T(m) \bmod N$, where $T(\alpha)^2 = S(\alpha)$.

1. Choose t sets $\mathcal{P}_1, \ldots, \mathcal{P}_t$ of r primes totally split in K.
   Partition $\mathcal{S}$ into t disjoints subsets $\mathcal{S}_1, \ldots, \mathcal{S}_t$.
2. For $k = 1, \ldots, t$, read $\mathcal{S}_k$. Deduce M and λ. Compute $S_k(x) = \prod_{(a,b)\in\mathcal{S}_k}(a - bx) \bmod f$.
3. For $i = 1, \ldots, \ell$ compute: p_i^λ, Q_i, $\frac{1}{Q_i} \bmod p_i^\lambda$, as well as for $j = 1, \ldots, d$, the root $r_{i,j} \bmod p_i$ of f, the lift $\tilde{r}_{i,j}$ with precision λ, as well as $\frac{H_{i,j}(x)}{f'(\tilde{r}_{i,j})} \bmod p_i^\lambda$.
4. For $\tau = 1, \ldots, t$:
   4.1. Compute $(S_\sigma(x) \bmod \mathcal{P}_\tau \overset{\text{def}}{=} \{S_\sigma(x) \bmod p, \ p \in \mathcal{P}_\tau\})_{\sigma=1\ldots t}$.
   4.2. For each $p_i \in \mathcal{P}_\tau$, and for $j = 1, \ldots, d$, compute:
        4.2.1 the evaluations $S_\sigma(\tilde{r}_{i,j}) \bmod p_i^\lambda$ for $\sigma = 1, \ldots, t$,
        4.2.2 the products $S(\tilde{r}_{i,j})$,
        4.2.3 the square roots $T'_{i,j} = \sqrt{S(\tilde{r}_{i,j})} \bmod p_i^\lambda$.
        4.2.4 the coefficients $c_{i,j,k} = [x^k]\left(T'_{i,j}\frac{H_{i,j}(x)}{Q_i f'(\tilde{r}_{i,j})} \bmod p_i^\lambda\right)$, as well as $c_{i,j,k}/p_i^\lambda \in \mathbb{R}$, and $c_{i,j,k}m^k \bmod N$.
5. Find the signs $s_{i,j}$ and the integer κ_{d-1} such that $\left|\sum_{i,j} s_{i,j}(c_{i,j,d-1}/p_i^\lambda) - \kappa_{d-1}\right| \leq \epsilon$. Deduce the integers $\kappa_0, \ldots, \kappa_{d-2}$.
6. return $\sum_{i,j,k} Q_i s_{i,j} c_{i,j,k} m^k - \kappa_k P^\lambda \bmod N$.

Algorithm 1. NFS square root using lifting and CRT

This algorithm allows some trivial limited parallelism. We can achieve a t-fold reduction of the space complexity for the computation of the values $S(\tilde{r}_{i,j})$, and likewise for the time complexity, using t^2 nodes. This is done by splitting both $\mathcal{S}$ and $\mathcal{P}$ into t equally sized parts denoted by $\mathcal{S}_1, \ldots, \mathcal{S}_t$ and $\mathcal{P}_1, \ldots, \mathcal{P}_t$, respectively.

Corresponding to each set $\mathcal{S}_\sigma$, we define the polynomial $S_\sigma(x)$ as being the product

$$S_\sigma(x) = \prod_{(a,b)\in\mathcal{S}_\sigma}(a - bx) \bmod f.$$

For each pair of indices (σ, τ), we define the r-uple:

$$S_\sigma(x) \bmod \mathcal{P}_\tau \overset{\text{def}}{=} \{S_\sigma(x) \bmod p, \ p \in \mathcal{P}_\tau\}.$$

Each of t^2 nodes, indexed by (σ, τ), may compute the quantity above. The time and space complexity on each node are thus $(1/t)$-th of the total amount given above.

Table 1. Time and space complexity of the different steps of algorithm 1 (we have $rt = \ell$ and $rtB \approx n/2$)

Step	Time	Space
1	$O(\ell)$	$O(\ell)$
2	$O(d^2\mathsf{M}(n))$	$O(dn)$
3	$O(\ell d^2\mathsf{M}(B))$	$O(dn)$
4.1	$O(d\mathsf{M}(n)\log n)$	$O(dn\log n)$
4.2.1	$O(\ell d^2\mathsf{M}(B))$	$O(dn)$
4.2.2	$O(\ell d\mathsf{M}(B))$	$O(dn)$
4.2.3, 4.2.4	$O(\ell d\mathsf{M}(B))$	$O(dn)$
5	$O(2^{d\ell/2})$	$O(2^{d\ell/2})$
6	$O(\ell)$	$O(\ell)$

Once the values $S_\sigma(\tilde{r}_{i,j})$ have been computed, the values $S(\tilde{r}_{i,j})$ corresponding to the complete input set $\mathcal{S}$ can be obtained as the product over all σ.

The steps we have just detailed form the core of algorithm 1, namely steps 4.1, 4.2.1, and 4.2.2. Other steps of this algorithm are not detailed at length here. In algorithm 1, we have chosen to present the parallel version on up to t^2 nodes, but instantiating with $t = 1$ gives the version which is best suited for a sequential implementation.

4.4 Complexity

Table 1 gives complexity estimates for the different steps of the algorithm. Concerning parallelization, steps 3, 4.2.3, and 4.2.4 clearly scale to up to ℓ nodes easily. We have shown in the previous paragraphs how steps 4.1 to 4.2.2 may be improved t-fold in a quite simple way when run over t^2 nodes, neglecting communication costs.

The complexity of step 5 of algorithm 1 is related to a knapsack-like problem. It is straightforward to solve such a problem in time $2^{\frac{1}{2}d\ell}$, and the best known approach is $2^{0.313d\ell}$ [9].

In total, the algorithm proposed has a time and space complexity of the order of $O(d^2\mathsf{M}(n)\log n)$, which is similar to the lifting approach of Section 1 (when counting the computation of $S(\alpha)$ in the complexity).

4.5 Implementation and Experimental Data

Algorithm 1 has been implemented in `cado-nfs` [8]. For a practical application, the choice of parameters r and t is chiefly limited by the complexity of step 5 of the algorithm, since above some dimension, the knapsack problem becomes intractable. This algorithm has been run for the RSA-768 square root calculation, with the following parameters: $d = 6$, $t = 3$, $r = 2$, the total input size being 21 GB. On 18 nodes equipped with two 4-core Intel Xeon E5520 processors, and 32 GB of RAM, the computation claimed 6 hours. In comparison, using

Montgomery's algorithm, the same computation on different hardware, but with the same number of 144 cores, took 4 hours, as it was reported in [11].

The difference in timings illustrates the difference in complexity. Montgomery's algorithm is linear, while the CRT-based algorithm we propose is quasi-linear. Because Montgomery's algorithm needs to exploit more accurate input data (namely, the ideal factorization of each ideal $(a - b\alpha)\mathcal{O}_K$), the time for reading this extended input set is significant. If we were to ignore I/O costs, the difference would be even more visible. Overall, the timings here are thus unsurprising. The CRT algorithm proposed here, compared to Montgomery's, has the advantage of not requiring the knowledge of the complete ideal factorization. In the perspective of a complete NFS implementation, this is interesting in that it simplifies the data flow, and also removes the need for an implementation of accurate computation of ideal valuations at *all* prime ideals. In the context of some NFS implementations which so far have chosen to restrict to the direct approach and avoid this step [8, 17], the CRT algorithm presented here offers a viable alternative.

4.6   A Variant Using a Large Number of Primes

A variant of the algorithm above may be employed in order to handle a larger number of primes (for example up to 10 000). To this end, it is important to avoid the knapsack reconstruction step. An idea found in [6] allows to work around this issue in the case where inert primes can be found. Let $\mathcal{P}$ be the set of primes selected for the reconstruction. We assume each is chosen with exponent $\lambda = 1$, although varying this parameter is possible. As done previously, we denote by $T_p(x) = T(x) \bmod p$, for some $p \in \mathcal{P}$, and also $T_p'(x)$ the expression of the square root which is computed. Let s_p be the sign such that $T_p'(x) = s_p T(x) \bmod p$.

Let us focus on one particular coefficient of $T(x)$, for example that of degree $d-1$, which we denote by τ. Similarly to T_p and T_p', we define the notations $\tau_p = \tau \bmod p$ and $\tau_p' = s_p \tau \bmod p$. Lacking the knowledge of s_p, we cannot distinguish between $\tau \bmod p$ and $-\tau \bmod p$. However, the *set* $\{\tau \bmod p, -\tau \bmod p\}$ is well determined by the computation of τ_p'. We may thus compute unambiguously $\tau^2 \bmod p = (\tau_p')^2 \bmod p$. If this computation is done modulo many primes p, we can deduce the integer τ^2. To this end, we must have $\prod_{p \in \mathcal{P}} p \geq M^2$, where M is the bound on the coefficients of $T(x)$. The integer τ itself is then obtained by the square root of an n-bit integer, which is significantly cheaper than the algebraic square root we are computing here. A byproduct of this computation is the set of signs $\{s_p\}$, which allows to complete the calculation.

This variant may also apply in a case without inert primes. In such a case, let $H \subset \mathfrak{S}_d$ be a representative set for the quotient $\mathrm{Gal}(L/\mathbb{Q})/\mathrm{Gal}(L/K)$, where L is the normal closure of K. Let $\sigma \in H$ be an element decomposing into a minimal number of distinct cycles, and let γ be this minimum. We want to use primes whose splitting pattern in K matches the cycle decomposition of σ. From Čebotarev's density theorem, it follows that the density of such primes is the density of this cycle decomposition pattern in H, which in particular is positive. If, as before, we try to relate $\tau \bmod p$ with the γ distinct coefficients obtained modulo each of the prime ideals above p, we have $\tau = \pm\tau_p'^{(1)} \pm \ldots \pm \tau_p'^{(\gamma)}$. There are 2^γ possible combinations. By evaluating the elementary symmetric functions on the 2^γ possible solutions modulo each p, we obtain τ as the root of an integer polynomial of degree 2^γ. The worst case for this extension is when γ is large: γ may reach $d/2$ in the case of Swinnerton-Dyer polynomials. We recover unsurprisingly the hard case of factoring polynomials over number fields.

References

[1] Brent, R.P.: Multiple-precision zero-finding methods and the complexity of elementary function evaluation. In: Traub, J.F. (ed.) Analytic Computational Complexity, pp. 151–176. Academic Press, New York (1975), http://web.comlab.ox.ac.uk/oucl/work/richard.brent/ftp/rpb028.ps.gz

[2] Brent, R., Zimmermann, P.: Modern Computer Arithmetic. Cambridge Monographs on Applied and Computational Mathematics, vol. 18. Cambridge University Press (2010)

[3] Buhler, J.P., Lenstra, A.K., Pollard, J.M.: Factoring integers with the number field sieve. In: Lenstra, A.K., Lenstra Jr., H.W. (eds.) The Development of the Number Field Sieve. Lecture Notes in Math., vol. 1554, pp. 50–94. Springer (1993)

[4] Cohen, H.: A course in algorithmic algebraic number theory. Grad. Texts in Math., vol. 138. Springer (1993)

[5] Couveignes, J.-M.: Computing a square root for the number field sieve. In: Lenstra, A.K., Lenstra Jr., H.W. (eds.) The Development of the Number Field Sieve. Lecture Notes in Math., vol. 1554, pp. 95–102. Springer (1993)

[6] Enge, A., Sutherland, A.V.: Class Invariants by the CRT Method. In: Hanrot, G., Morain, F., Thomé, E. (eds.) ANTS-IX. LNCS, vol. 6197, pp. 142–156. Springer, Heidelberg (2010)

[7] von zur Gathen, J., Gerhard, J.: Modern computer algebra. Cambridge University Press, Cambridge (1999)

[8] Gaudry, P., Kruppa, A., Morain, F., Muller, L., Thomé, E., Zimmermann, P.: cado-nfs, An Implementation of the Number Field Sieve Algorithm (2011), http://cado-nfs.gforge.inria.fr/, Release 1.1

[9] Howgrave-Graham, N., Joux, A.: New Generic Algorithms for Hard Knapsacks. In: Gilbert, H. (ed.) EUROCRYPT 2010. LNCS, vol. 6110, pp. 235–256. Springer, Heidelberg (2010)

[10] Joux, A., Naccache, D., Thomé, E.: When e-th Roots Become Easier Than Factoring. In: Kurosawa, K. (ed.) ASIACRYPT 2007. LNCS, vol. 4833, pp. 13–28. Springer, Heidelberg (2007)

[11] Kleinjung, T., Aoki, K., Franke, J., Lenstra, A.K., Thomé, E., Bos, J.W., Gaudry, P., Kruppa, A., Montgomery, P.L., Osvik, D.A., te Riele, H., Timofeev, A., Zimmermann, P.: Factorization of a 768-Bit RSA Modulus. In: Rabin, T. (ed.) CRYPTO 2010. LNCS, vol. 6223, pp. 333–350. Springer, Heidelberg (2010)

[12] Lenstra Jr., H.W., Stevenhagen, P.: Chebotarëv and his density theorem. Math. Intelligencer 18(2), 26–37 (1996)

[13] Monico, C.: ggnfs, A Number Field Sieve Implementation (2004-2005), http://www.math.ttu.edu/~cmonico/software/ggnfs/, Release 0.77

[14] Montgomery, P.L.: Square roots of products of algebraic numbers. In: Gautschi, W. (ed.) Mathematics of Computation 1943-1993: a Half-Century of Computational Mathematics. Proc. Sympos. Appl. Math., vol. 48, pp. 567–571. Amer. Math. Soc. (1994)

[15] Montgomery, P.L.: Square roots of products of algebraic numbers (1997), unpublished draft, significantly different from published version [14] (May 16, 1997)

[16] Nguyên, P.Q.: A Montgomery-like Square Root for the Number Field Sieve. In: Buhler, J.P. (ed.) ANTS 1998. LNCS, vol. 1423, pp. 151–168. Springer, Heidelberg (1998)

[17] Papadopoulos, J.: msieve, A Library for Factoring Large Integers – release 1.50 (2004), http://www.boo.net/~jasonp, Release 1.50

[18] Sutherland, A.V.: Accelerating the CM method (2012) (preprint), http://arxiv.org/abs/1009.1082

Improving the Berlekamp Algorithm
for Binomials $x^n - a$

Ryuichi Harasawa, Yutaka Sueyoshi, and Aichi Kudo

Graduate School of Engineering, Nagasaki University,
1-14 Bunkyomachi, Nagasaki-shi, Nagasaki, 852-8521, Japan
{harasawa,sueyoshi,kudo}@cis.nagasaki-u.ac.jp

Abstract. In this paper, we describe an improvement of the Berlekamp algorithm, a method for factoring univariate polynomials over finite fields, for binomials $x^n - a$ over finite fields $\mathbb{F}_q$. More precisely, we give a deterministic algorithm for solving the equation $h(x)^q \equiv h(x) \pmod{x^n - a}$ directly without applying the sweeping-out method to the corresponding coefficient matrix. We show that the factorization of binomials using the proposed method is performed in $O^\sim(n \log q)$ operations in $\mathbb{F}_q$ if we apply a probabilistic version of the Berlekamp algorithm after the first step in which we propose an improvement. Our method is asymptotically faster than known methods in certain areas of q, n and as fast as them in other areas.

Keywords: finite field, polynomial factorization, Berlekamp algorithm, binomial.

1 Introduction

The factorization of univariate polynomials over finite fields is one of the interesting topics in computer algebra, for example, it is used to determine the decomposition of prime numbers in number fields and to construct (non-prime) finite fields and so on.

Applying the formal derivation, we can reduce the factorization of polynomials over finite fields to that of square-free polynomials (i.e., polynomials having no multiple factors) [10,11]. For the factorization of square-free polynomials over finite fields, the Berlekamp algorithm is well known [4,10].

In this paper, we propose an improvement of the Berlekamp algorithm for binomials $x^n - a$ over finite fields $\mathbb{F}_q$. More precisely, we give a deterministic algorithm for solving the equation $h(x)^q \equiv h(x) \pmod{x^n - a}$ directly without applying the sweeping-out method to the corresponding coefficient matrix, which is a generarization of the Prange method for $x^n - 1$ [16]. We show that the factorization of binomials using the proposed method is performed in $O^\sim(n \log q)$ operations in $\mathbb{F}_q$ if we apply a probabilistic version of the Berlekamp algorithm after the first step in which we propose an improvement. Our method is asymptotically faster than known methods (for example the Berlekamp method [4], the Gathen and Shoup method [9], and the Kaltofen and Shoup method [13],

F. Özbudak and F. Rodríguez-Henríquez (Eds.): WAIFI 2012, LNCS 7369, pp. 225–235, 2012.
© Springer-Verlag Berlin Heidelberg 2012

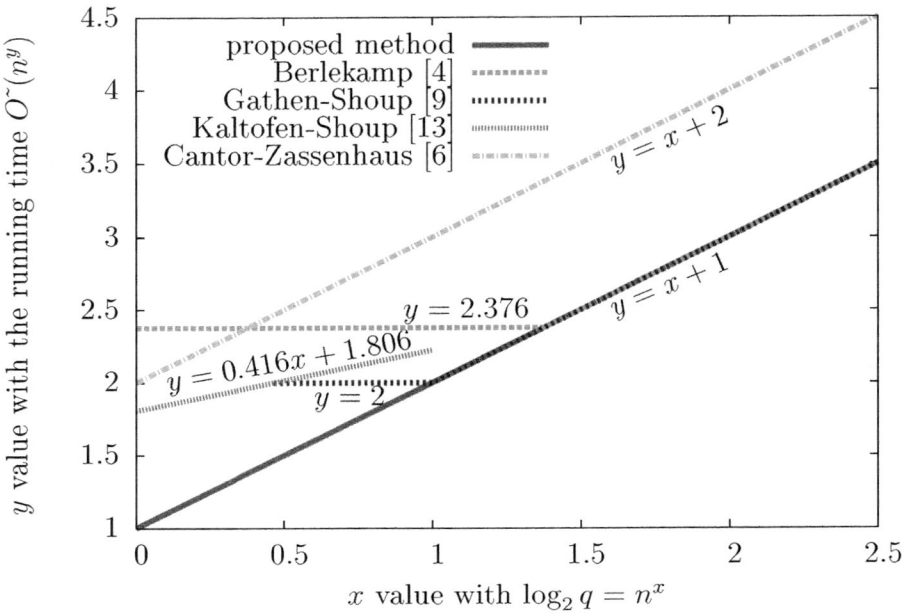

Fig. 1. Running times of some factoring algorithms

the Cantor and Zassenhaus method [6], [10, Figure 14.9]) in certain areas of q, n and as fast as them in other areas (Fig. 1). We mention that both the Gathen and Shoup method [9] and the Kaltofen and Shoup method [13] improve the distinct-degree factorization in the Cantor and Zassenhaus method, by using the "iterated Frobenius" method for the former and using fast matrix multiplication for the latter.

Note that there exist some efficient methods for computing the solution of $x^n = a$ over finite fields (e.g., [1,21]).

The remainder of this paper is organized as follows: In section 2, we introduce some theoretical results on binomials over finite fields. In Section 3, we describe the Berlekamp algorithm. In Section 4, we propose an improvement of the Berlekamp algorithm for binomials $x^n - a$. In section 5, we estimate the complexity of the proposed method. In Section 6, we give the conclusion and future works.

2   Binomials

In this section, we introduce some (theoretical) facts on binomials over finite fields $\mathbb{F}_q$, while we do not apply directly these facts to our proposed method. However, we use them to make a lot of examples $x^n - a$ for various cases q and n (that is, various patterns of factorizations).

At first, for the irreducibility of binomials, the following result is known:

Theorem 1 [14, Theorem 3.75]. *Let $n \geq 2$, $a \in \mathbb{F}_q^*$, and let e be the multiplicative order of a. Then we have*

$$x^n - a \text{ is irreducible in } \mathbb{F}_q[x]$$
$$\iff \begin{cases} (i) & \text{for each prime divisor } p \text{ of } n, \ p|e \text{ and } p \nmid \frac{q-1}{e}, \\ (ii) & q \equiv 1 \pmod 4 \qquad \text{if } n \equiv 0 \pmod 4. \end{cases}$$

We next describe a result on the number of irreducible factors of binomials. Let $\mu(\cdot)$ denote the Möbius function, that is,

$$\mu(x) = \begin{cases} 1 & (x = 1), \\ 0 & (x \text{ is not square-free}), \\ (-1)^s & (x \text{ is the product of } s \text{ distinct primes}). \end{cases}$$

For a binomial $x^n - a$ over $\mathbb{F}_q$, we denote by δ_i the number of roots of it in $\mathbb{F}_{q^i}$. If we put $d_i = \gcd(q^i - 1, n)$, then we easily see that

$$\delta_i = \begin{cases} d_i & (a^{(q^i - 1)/d_i} = 1), \\ 0 & (\text{otherwise}). \end{cases}$$

We get a result on the number of irreducible factors of $x^n - a$ as follows:

Theorem 2 [19]. *With the notation as above, we assume $\operatorname{char} \mathbb{F}_q \nmid n$ and $a \neq 0$. Let σ_t be the number of irreducible factors of degree t of $x^n - a$. Then we have*

$$\sigma_t = \frac{1}{t} \sum_{i|t} \mu\left(\frac{t}{i}\right) \delta_i.$$

Proof. We easily see that

$$\delta_t = \sum_{i|t} i\sigma_i.$$

Applying Möbius's inversion formula, we have

$$t\sigma_t = \sum_{i|t} \mu\left(\frac{t}{i}\right) \delta_i,$$

which implies the assertion of the theorem. □

Remark 1. *This theorem implies that, for given q and n with $\operatorname{char} \mathbb{F}_q \nmid n$, the values σ_t's depend only on the multiplicative order of a.*

In the rest of this section, we describe a result on the minimal degree (resp. the maximal degree) of irreducible factors of binomials.

Proposition 1. *With the notation as above, we assume* $\operatorname{char} \mathbb{F}_q \nmid n$ *and* $a \neq 0$. *Let* $\kappa = \min\{i \mid \delta_i \neq 0\}$, *that is, the extension field* $\mathbb{F}_{q^\kappa}$ *is the minimal field that contains some root of* $x^n - a$. *Let* ζ_n *be a primitive n-th root of unity, and* $\lambda = [\mathbb{F}_q(\zeta_n) : \mathbb{F}_q]$ *(in other words, the value* λ *is equal to the order of* q *in* $(\mathbb{Z}/n\mathbb{Z})^*$). *Then we have*

1. *the minimal degree of irreducible factors of* $x^n - a$ *is equal to* κ,
2. *the maximal degree of irreducible factors of* $x^n - a$ *is equal to* $\operatorname{lcm}(\kappa, \lambda)$.

Proof. It is obvious for the minimal degree.

For the maximal degree, let e be the multiplicative order of a and ζ_{en} a primitive en-th root of unity. Then there exists an element b in $\mathbb{F}_q$ with multiplicative order e such that ζ_{en} is a root of $x^n - b$. Indeed, since ζ_{en} is a root of $x^{en} - 1$ and $x^{en} - 1 = \prod_{0 \leq i < e}(x^n - a^i)$, there exists j such that ζ_{en} is a root of $x^n - a^j$. Setting $b = a^j$, we see that the multiplicative order of b is equal to e (equivalently $\gcd(j, e) = 1$), because ζ_{en} is a primitive en-th root of unity. We further see, from Theorem 2 (or Remark 1), that the pattern of the factorization of $x^n - a$ coincides with that of $x^n - b$. Namely, for each $i \geq 1$, the number of irreducible factors of degree i of $x^n - a$ is equal to that of $x^n - b$. Therefore it is sufficient to show our assertion for $x^n - b$.

Let u be a root in $\mathbb{F}_{q^\kappa}$ of $x^n - b$. We see that $u\zeta_n^i$ $(0 \leq i \leq n - 1)$ are the roots of $x^n - b$, which implies that the splitting field of $x^n - b$ over $\mathbb{F}_q$, say $\mathbb{K}$, becomes $\mathbb{K} = \mathbb{F}_q(u, \zeta_n)$. So the maximal degree of irreducible factors of $x^n - b$ is less than or equal to $[\mathbb{K} : \mathbb{F}_q] = \operatorname{lcm}(\kappa, \lambda)$. On the other hand, we see that $\{\zeta_{en}^{1+ie} \mid 0 \leq i \leq n - 1\}$ gives an alternative representation of the roots of $x^n - b$. Hence we have $\mathbb{K} = \mathbb{F}_q(\zeta_{en})$ by considering the spliting field for this representation of the root of $x^n - b$, which implies the degree of minimal polynomial, say $\psi(x)$, of ζ_{en} over $\mathbb{F}_q$ is equal to $[\mathbb{K} : \mathbb{F}_q] = \operatorname{lcm}(\kappa, \lambda)$. Therefore the maximal degree of irreducible factors of $x^n - b$ is greater than or equal to $\operatorname{lcm}(\kappa, \lambda)$ because $\psi(x)$ is an irreducible factor of $x^n - b$. So we get the desired result. $\qquad\square$

3   Berlekamp Algorithm

In this section, we assume that $\operatorname{char} \mathbb{F}_q > 2$ for simplicity. The Berlekamp algorithm [4,10] is a well-known algorithm for factoring square-free polynomials over finite fields. In Table 1, we describe its procedure [1]. From Step 3, we see this algorithm is probabilistic, which runs in polynomial time for the input size $O(n \log q)$ with n the degree of polynomial to be factored (Fig. 1). We note that there exists a deterministic procedure for Step 3, but the complexity is not the polynomial time for the input size [3]. So we use a probabilistic version of the Berlekamp method in this paper.

---

[1] If $q = 2^m$, we perform the same procedure as in $\operatorname{char} \mathbb{F}_q > 2$ except for Step 3. For Step 3, we compute $\gcd(\operatorname{Tr}(g(x)), v(x))$ with $g(x)$ a random element in V instead of $\gcd(g(x)^{(q-1)/2} - 1, v(x))$, where $\operatorname{Tr}(g(x)) = g(x) + g(x)^2 + \cdots + g(x)^{2^{m-1}} \bmod v(x)$ [9, Algorithm 3.6 (Step 4)].

Table 1. Berlekamp algorithm

Input: A square-free polynomial $f(x)$ over $\mathbb{F}_q$.
Output: The factorization of $f(x)$.
Step 1: Compute the polynomials $h(x)$ over $\mathbb{F}_q$ of degree less than $\deg f(x)$ such that $h(x)^q \equiv h(x) \pmod{f(x)}$. The set V of $h(x)$'s forms an $\mathbb{F}_q$-vector space. Let $\{h_1(x), \ldots, h_k(x)\}$ be a basis of V.
Step 2: $F \leftarrow \{f(x)\}$.   **if** $k = 1$, go to Step 4.

Step 3: while $\#F < k$
    Choose a random element $g(x)$ in V
    ($g(x)$ is of the form $\sum_{1 \le i \le k} \lambda_i h_i(x)$ with $\lambda_i \in \mathbb{F}_q$).
    For each $v(x) \in F$, do the following procedure:
    Compute $d(x) = \gcd(g(x), v(x))$.
    if $0 < \deg d(x) < \deg v(x)$
        $F \leftarrow (F \setminus \{v(x)\}) \cup \{d(x),\, v(x)/d(x)\}$.
        if $\#F = k$, go to Step 4.
        $v(x) \leftarrow v(x)/d(x)$.
    end if
    Compute $d(x) = \gcd(g(x)^{(q-1)/2} - 1,\, v(x))$.
    if $0 < \deg d(x) < \deg v(x)$
        $F \leftarrow (F \setminus \{v(x)\}) \cup \{d(x),\, v(x)/d(x)\}$.
        if $\#F = k$, go to Step 4.
    end if
    end while

Step 4: Return F (the product of the elements in F equals $f(x)$).

In the next section, we focus on Step 1 in Table 1. More precisely, we consider the equation

$$h(x)^q \equiv h(x) \pmod{f(x)} \tag{1}$$

for a square-free polynomial $f(x)$ over $\mathbb{F}_q$, i.e., the eigenspace V of the eigenvalue 1 for the linear transformation $h(x) \mapsto h(x)^q \pmod{f(x)}$ on the n-dimensional vector space $\mathbb{F}_q[x]/(f(x))$ over $\mathbb{F}_q$.

Let k denote the number of irreducible factors of $f(x)$ and $f(x) = \prod_{1 \le i \le k} f_i(x)$ be the factorization of $f(x)$. Then we see that the vector space $\mathbb{F}_q[x]/(\overline{f}(x))$ is isomorphic to $\bigoplus_{1 \le i \le k} \mathbb{F}_q[x]/(f_i(x))$ and that the solution space V of the equation (1) is isomorphic to the subspace of $\bigoplus_{1 \le i \le k} \mathbb{F}_q[x]/(f_i(x))$ consisting of $(a_1, \cdots, a_k)$ with each a_i in $\mathbb{F}_q$. Hence, the number of irreducible factors of $f(x)$ is equal to the dimension of V over $\mathbb{F}_q$.

Remark 2. *For Step 3 in Table 1, if the polynomial $v(x)$ is not irreducible, then the probability that $\gcd(g(x), v(x))$ or $\gcd(g(x)^{\frac{q-1}{2}} - 1, v(x))$ (or both) is a proper factor of $v(x)$ is at least $\frac{1}{2}$ from the description above.*

In the remainder of this section, for the solution $h(x)$ to the equation $h(x)^q \equiv h(x) \pmod{f(x)}$, we mention the method [14, Theorems 4.3 and 4.5] using the Prange method [16] for $f(x) = x^n - 1$.

Let $f(x)$ be a polynomial over $\mathbb{F}_q$ with $f(0) \neq 0$ and s the least positive integer satisfying $f(x) \mid x^s - 1$, which is called the order of $f(x)$. Applying the Prange method to $x^s - 1$, we then get the solution above [14, Theorems 4.3 and 4.5]. Namely, for each $\alpha \geq 0$, let l be the least positive integer such that

$$x^{\alpha q^l} \equiv x^\alpha \pmod{f(x)},$$

which is equivalent to

$$\alpha q^l \equiv \alpha \pmod{s}.$$

Defining

$$h_\alpha(x) = x^\alpha + x^{\alpha q} + \cdots + x^{\alpha q^{l-1}},$$

we see that $h_\alpha(x)^q \equiv h_\alpha(x) \pmod{f(x)}$. Moreover, the set $\{h_\alpha(x) \bmod x^s - 1 \mid \alpha \in \mathbb{Z}/s\mathbb{Z}\}$ forms a basis of the solution space of $h_\alpha(x)^q \equiv h_\alpha(x) \pmod{x^s - 1}$, which we see in the next section (see Remark 3. Also [10, p. 419 (Exercise 14.47)] or [16]). If $f(x) \neq x^s - 1$, then the set $\{h_\alpha(x) \bmod f(x) \mid \alpha \in \mathbb{Z}/s\mathbb{Z}\}$ forms a generator system of the solution space of $h_\alpha(x)^q \equiv h_\alpha(x) \pmod{f(x)}$.

Our main assertion in this paper is that we analyze the method above more strictly and simplify it in the case of binomials $f(x) = x^n - a$. We emphasize that our method does not need the computation of the value s, the order of $f(x)$. In the case $f(x) = x^n - a$, the order of $f(x)$ is $s = en$ with e the multiplicative order of a [14, Lemma 3.17]. So, if we need the value s, then we need to find the multiplicative order of a, for which we might compute the factorization of $q - 1$. However the task becomes extremely heavy as q becomes large.

4   Our Algorithm

In this section, we propose an improved method for obtaining a basis of V in Step 1 of Table 1 for $x^n - a$. Namely, we solve

$$h(x)^q \equiv h(x) \pmod{x^n - a} \tag{2}$$

by a new method. Since $x^n - a$ is assumed to be square-free, char $\mathbb{F}_q$ does not divide n.

For $f(x) = x^n - a$, instead of dealing with the coefficient matrix corresponding to the equation above, we consider the orbits in $\mathbb{Z}/n\mathbb{Z}$ according to the action of $\langle q \rangle$ by multiplication.

For α in $\mathbb{Z}/n\mathbb{Z} = \{0, 1, \ldots, n-1\}$, let l be the least positive integer such that $q^l \alpha \equiv \alpha \pmod{n}$ and let $\alpha_i = q^i \alpha \bmod n$ for $0 \leq i < l$. We denote the orbit of α by $\bar{\alpha} = \{\alpha_0, \alpha_1, \ldots, \alpha_{l-1}\}$. In particular, we have $\bar{0} = \{0\}$.

For each orbit $\bar{\alpha}$, we consider the subspace

$$T_{\bar{\alpha}} = \{\beta_0 x^{\alpha_0} + \beta_1 x^{\alpha_1} + \cdots + \beta_{l-1} x^{\alpha_{l-1}} \bmod x^n - a \mid \beta_i \in \mathbb{F}_q\}$$

of $\mathbb{F}_q[x]/(x^n - a)$.

Since $q\alpha_i \equiv \alpha_{i+1} \pmod{n}$ $(0 \le i < l - 1)$ and $q\alpha_{l-1} \equiv \alpha_0 \pmod{n}$, we put

$$q\alpha_i = c_i n + \alpha_{i+1} \ (0 \le i < l - 1) \text{ and } q\alpha_{l-1} = c_{l-1} n + \alpha_0$$

with integers c_i $(0 \le i < l)$. Then we see

$$(x^{\alpha_i})^q \equiv a^{c_i} x^{\alpha_{i+1}} \pmod{x^n - a} \ (0 \le i < l - 1) \text{ and}$$
$$(x^{\alpha_{l-1}})^q \equiv a^{c_{l-1}} x^{\alpha_0} \pmod{x^n - a}.$$

Hence, for $h_{\bar{\alpha}}(x) = \beta_0 x^{\alpha_0} + \beta_1 x^{\alpha_1} + \cdots + \beta_{l-1} x^{\alpha_{l-1}}$ in $T_{\bar{\alpha}}$, we have

$$h_{\bar{\alpha}}(x)^q \equiv a^{c_{l-1}}\beta_{l-1} x^{\alpha_0} + a^{c_0}\beta_0 x^{\alpha_1} + \cdots + a^{c_{l-2}}\beta_{l-2} x^{\alpha_{l-1}} \pmod{x^n - a}.$$

Therefore, for the linear transformation π_q of $\mathbb{F}_q[x]/(x^n - a)$ defined by $\pi_q(h(x)) = h(x)^q \bmod x^n - a$, the subspace $T_{\bar{\alpha}}$ is π_q-invariant and $\mathbb{F}_q[x]/(x^n - a) = \bigoplus_{\bar{\alpha}} T_{\bar{\alpha}}$. We put $V_{\bar{\alpha}} = V \cap T_{\bar{\alpha}}$, then $V = \bigoplus_{\bar{\alpha}} V_{\bar{\alpha}}$ and $V_{\bar{0}} = T_{\bar{0}} \simeq \mathbb{F}_q$.

If there exist w orbits $\bar{\alpha}$'s in $\mathbb{Z}/n\mathbb{Z}$, the equation (2) is divided into w equations

$$h_{\bar{\alpha}}(x)^q \equiv h_{\bar{\alpha}}(x) \pmod{x^n - a} \tag{3}$$

where $h(x) = \sum_{\bar{\alpha}} h_{\bar{\alpha}}(x)$ with $h_{\bar{\alpha}}(x)$ as above. For the orbit $\bar{0}$, the constant polynomial 1 forms a basis of one-dimensional vector space $V_{\bar{0}}$.

We consider the orbit $\bar{\alpha} \ne \bar{0}$. Then the equation (3) is written as

$$\begin{cases} \beta_0 = a^{c_{l-1}} \beta_{l-1} \\ \beta_1 = a^{c_0} \beta_0 \\ \quad \vdots \\ \beta_{l-1} = a^{c_{l-2}} \beta_{l-2}, \end{cases}$$

which leads to the relation

$$\beta_0 = a^{c_0 + c_1 + \cdots + c_{l-1}} \beta_0.$$

Therefore, we obtain the solution(s) of (3) as follows:

$$\begin{cases} \beta(x^{\alpha_0} + a^{c_0} x^{\alpha_1} + a^{c_0+c_1} x^{\alpha_2} + \cdots + a^{c_0+c_1+\cdots+c_{l-2}} x^{\alpha_{l-1}}) \\ \qquad\qquad\qquad\qquad (\text{if } a^{c_0+c_1+\cdots+c_{l-1}} = 1), \\ 0 \qquad\qquad\qquad\qquad (\text{otherwise}), \end{cases}$$

where β runs over all elements of $\mathbb{F}_q$. The solution space $V_{\bar{\alpha}}$ of the equation (3) is $\{0\}$ if $a^{c_0+c_1+\cdots+c_{l-1}} \ne 1$ and, otherwise, forms one-dimensional subspace of $T_{\bar{\alpha}}$ generated by $x^{\alpha_0} + a^{c_0} x^{\alpha_1} + a^{c_0+c_1} x^{\alpha_2} + \cdots + a^{c_0+c_1+\cdots+c_{l-2}} x^{\alpha_{l-1}}$.

We describe the proposed algorithm in Table 2.

Remark 3. *For $i \ge 1$, we see $x^{\alpha q^i} \equiv a^{c_0+c_1+\cdots+c_{i-1}} x^{\alpha_i} \pmod{x^n - a}$ by the definition of α_i and c_i. This implies that $a^{c_0+c_1+\cdots+c_{l-1}} = 1$ holds if and only if*

Table 2. Solutions of $h(x)^q \equiv h(x) \pmod{x^n - a}$

Input: A binomial $x^n - a$ over $\mathbb{F}_q$ with char $\mathbb{F}_q \nmid n$.   **Output:** A basis B of the solution space V of $\quad h(x)^q \equiv h(x) \pmod{x^n - a}$.
Step 1: $B \leftarrow \{1\}$, $\quad G \leftarrow \{1, 2, \ldots, n-1\}$.
Step 2: if $G = \emptyset$, return B.
Step 3: $i_0 \leftarrow \min\{i \mid i \in G\}$, $\quad G \leftarrow G \setminus \{i_0\}$, $\quad j \leftarrow i_0$, $\quad f \leftarrow x^j$, $\quad b \leftarrow 1$
Step 4: Compute the integers t, r $\quad$ such that $jq = tn + r$ with $0 \leq r < n$. $\quad b \leftarrow b \cdot a^t$.
Step 5: while $r \neq i_0$ $\quad G \leftarrow G \setminus \{r\}$, $\quad f \leftarrow f + b \cdot x^r$, $\quad j \leftarrow r$. $\quad$ Compute the integers t, r $\quad$ such that $jq = tn + r$ with $0 \leq r < n$. $\quad b \leftarrow b \cdot a^t$.   **end while**
Step 6: if $b = 1$, $B \leftarrow B \cup \{f\}$. $\quad$ **goto** Step 2.

$x^{\alpha q^l} \equiv x^\alpha \pmod{x^n - a}$, and then $\sum_{0 \leq i \leq l-1} x^{\alpha q^i} \bmod x^n - a$ is a solution of the equation (3). Especially, in the case of $a = 1$, the equation $a^{c_0 + c_1 + \cdots + c_{l-1}} = 1$ always holds for all $\alpha \in \mathbb{Z}/n\mathbb{Z}$, which implies that the set $\{\sum_{0 \leq i \leq l-1} x^{\alpha q^i} \bmod x^n - 1 \mid \alpha \in \mathbb{Z}/n\mathbb{Z}\}$ forms an $\mathbb{F}_q$-basis of the solution space V and that the dimension of V (i.e., the number of the irreducible factors of $x^n - 1$) is equal to the number of the orbits in $\mathbb{Z}/n\mathbb{Z}$ with respect to $\langle q \rangle$.

As a theoretical consideration about the proposed method, we get the following result on the congruence equation $x^{\alpha q^l} \equiv x^\alpha \pmod{x^n - a}$ in Remark 3.

Proposition 2. *With the notation as above, we put $d = \gcd(\alpha, n)$. Then we have*

1. *$\#\bar{d} = \#\bar{\alpha}(= l)$.*
2. *$x^{d q^l} \equiv x^d \pmod{x^n - a} \iff x^{\alpha q^l} \equiv x^\alpha \pmod{x^n - a}$.*

Proof. From the definition of d, we denote $\alpha = sd$ with $\gcd(s, \frac{n}{d}) = 1$.

For the first assertion, it is sufficient to show that $dq^i \equiv d \pmod{n}$ if and only if $\alpha q^i \equiv \alpha \pmod{n}$. By multiplying s (resp. $s^{-1} \bmod \frac{n}{d}$) to the both sides, we obtain the only if part (resp. the if part).

In the same way, we get the second assertion. $\qquad \square$

Proposition 2 implies that, for two orbits $\bar{\alpha}$ and $\bar{\alpha}'$ with $\gcd(\alpha, n) = \gcd(\alpha', n)$, $\sum_{0 \leq i \leq l-1} x^{\alpha q^i} \bmod x^n - a$ is a solution of the equation (3) if and only if so is $\sum_{0 \leq i \leq l-1} x^{\alpha' q^i} \bmod x^n - a$.

5   Complexity

We estimate the complexity of the factorization of binomials using the proposed procedure described in the previous section (Table 2). Namely, we describe the complexity not only of the proposed method but also of the procedures of square-free factorization and of finding irreducible factors using the basis which is obtained by the proposed method. The notation $O^\sim(x)$ means $O(x(\log x)^t)$ with some positive constant t.

We assume that the multiplication of l-bit integer and m-bit integer (resp. the division/remainder of l-bit integer by m-bit integer) needs $O(lm)$ bit operations (resp. $O(m(l - m))$ bit operations).

We first see that the square-free factorization of binomials $x^n - a$ over $\mathbb{F}_q$ of characteristic p is reduced to the computation of i and r such that $n = p^i r$ with $\gcd(p, r) = 1$ and the computation of the p^i-th root of a, say a'. Namely, the square-free factorization of $x^n - a$ becomes $x^n - a = (x^r - a')^{p^i}$. The computation of i and r takes at most

$$O(\log p \log \frac{n}{p}) + O(\log p \log \frac{n}{p^2}) + \cdots + O(\log p \log \frac{n}{p^v})$$
$$= O(v \log p \log n)   \text{(by } \log \frac{n}{p^i} \leq \log n)$$
$$= O(\log^2 n)   \text{(by } v \log p = \log p^v \leq \log n)$$

bit operations, where $v := \lfloor \log_p n \rfloor$. For the computation of a', if we let $q = p^m$ then we get the p^i-th root of a by $a' = a^{p^\eta}$ with $\eta := -i \bmod m$, from which we see the computation of a' takes at most

$$O(\log p^\eta) = O(\log q)   \text{(by } 0 \leq \eta < m \text{ and } q = p^m)$$

operations in $\mathbb{F}_q$.

With the notations q, n and a as before, we assume that the binomial $x^n - a$ to be factored is square-free. In order to perform the proposed method (Table 2) for Step 1 in Table 1, for each j with $1 \leq j \leq n - 1$, we must execute the following computations:

- the computation of the values t, r such that $jq = tn + r$ with $0 \leq r < n$;
- for the value t above, the computaion of $b \cdot a^t$ with b being a prescribed element in $\mathbb{F}_q$.

The former computation takes $O(\log j \cdot \log q) + O(\log t \cdot \log n)$ bit operations and the latter one takes $O(\log t)$ operations in $\mathbb{F}_q$. By the fact $t \leq \frac{jq}{n}$ and Stirling's formula $\log(n!) = O(n \cdot \log n)$, we estimate an upper bound of the complexity of the former task as

$$\sum_{1 \le j \le n-1} \{O(\log j \cdot \log q) + O(\log \frac{jq}{n} \cdot \log n)\}$$

$$= O(n \log n \cdot \log q) + O(n \log n \cdot \log q) \qquad (\text{by } \log \frac{jq}{n} \le \log q)$$

$$= O(n \log n \cdot \log q)$$

$$= O^{\tilde{}}(n \log q) \text{ bit operations,}$$

and an upper bound of the complexity of the latter task as

$$\sum_{1 \le j \le n-1} O(\log \frac{jq}{n})$$

$$= O(n \log q) \text{ operations in } \mathbb{F}_q \qquad (\text{by } \log \frac{jq}{n} \le \log q).$$

Therefore, the proposed method runs in $O^{\tilde{}}(n \log q)$ operations in $\mathbb{F}_q$.

We note that, for alternative methods, it takes $O(n^3)$ (resp. $O(n^{2.376})$) operations in $\mathbb{F}_q$ using the original Gaussian elimination (resp. the Gaussian elimination using a fast method for matrix multiplication [8]) and $O^{\tilde{}}(n^2)$ operations in $\mathbb{F}_q$ using the Kaltofen-Lobo method [12] based on the Wiedemann method [22], which is known as the fastest method for solving linear equations.

We additionally perform the final step (Step 3 in Table 1) in $O^{\tilde{}}(n \log q)$ operations in $\mathbb{F}_q$ on average [10, Theorems 14.11 and 14.32], assuming that one applies a fast arithmetic in $\mathbb{F}_q[x]$. Namely, for two polynomials in $\mathbb{F}_q[x]$ of degree at most n, the multiplication, the division with remainder and the greatest common divisor are performed in $O^{\tilde{}}(n)$ operations in $\mathbb{F}_q$ using the fast arithmetic in $\mathbb{F}_q[x]$ [2,7,17,18]. We therefore see that the factorization of binomials is performed in $O^{\tilde{}}(n \log q)$ operations in $\mathbb{F}_q$, which is asymptotically faster than known methods (e.g., [4,6,9,13] or [10, Figure 14.9]) in certain areas of q, n and as fast as them in other areas (Fig. 1 in Section 1).

6   Conclusion and Future Works

In this paper, we described an improvement of the Berlekamp algorithm for binomials $x^n - a$ over finite fields $\mathbb{F}_q$. More precisely, we proposed a method for solving the equation $h(x)^q \equiv h(x) \pmod{x^n - a}$ directly. We evaluate the complexity as $O^{\tilde{}}(n \log q)$ operations in $\mathbb{F}_q$. Our method is asymptotically faster than known methods in certain areas of q, n and as fast as them in other areas. Our future works include the experimental consideration, the detailed analysis of other methods (e.g., the Cantor and Zassenhaus method [6] and its improvements [5,9,13,15,20]) in the case of binomials, and the combination of our method with other ones as above.

Acknowledgments. We are grateful to the referees for giving a lot of useful comments and suggestions which make this paper more valuable. This work was partially supported by the Japan Society for the Promotion of Science (JSPS) under the Grant-in-Aid for challenging Exploratory Research No. 24650009.

References

1. Adleman, L., Menders, K., Miller, G.: On taking roots in finite fields. In: Proc. 18th IEEE Symposium on Foundations of Computer Science (FOCS), pp. 175–178 (1977)
2. Aho, A.V., Hopcroft, J.E., Ullman, J.D.: The Design and Analysis of Computer Algorithms. Addison-Wesley, Reading (1974)
3. Berlekamp, E.R.: Factoring polynomials over finite fields. Bell System Technical Journal 46, 1853–1859 (1967)
4. Berlekamp, E.R.: Factoring polynomials over large finite fields. Math. Comp. 24, 713–735 (1970)
5. Camion, P.: Improving an algorithm for factoring polynomials over a finite field and constructing large irreducible polynomials. IEEE Transactions on Information Theory 29(3), 378–385 (1983)
6. Cantor, D.G., Zassenhaus, H.: A new algorithm for factoring polynomials over finite fields. Math. Comp. 36, 587–592 (1981)
7. Cantor, D.G., Kaltofen, E.: On fast multiplication of polynomials over arbitrary algebras. Acta Inform. 28, 693–701 (1991)
8. Coppersmith, D., Winograd, S.: Matrix multiplication via arithmetic progressions. J. Symb. Comput. 9, 251–280 (1990)
9. von zur Gathen, J., Shoup, V.: Computing Frobenius maps and factoring polynomials. Comput. Complexity 2, 187–224 (1992)
10. von zur Gathen, J., Gerhard, J.: Modern Computer Algebra, 2nd edn., Cambridge (2003)
11. Geddes, K., Czapor, S., Labahn, G.: Algorithms for Computer Algebra. Kluwer Academic Publishers (1992)
12. Kaltofen, E., Lobo, A.: Factoring high-degree polynomials by black box Berlekamp algorithm. In: Proceedings of ISSAC 1994, pp. 90–98. ACM Press (1994)
13. Kaltofen, E., Shoup, V.: Subquadratic-time factoring of polynomials over finite fields. Math. Comp. 67, 1179–1197 (1998)
14. Lidl, R., Niederreiter, H.: Finite Fields. Addison-Wesley (1983)
15. McEliece, R.J.: Factorization of polynomials over finite fields. Math. Comp. 23, 861–867 (1969)
16. Prange, E.: An algorithm for factoring $X^n - 1$ over a finite field, Technical Report AFCRC-TN-59-775, Air Force Cambridge Research Center, Bedford, MA (1956)
17. Schönhage, A., Strassen, V.: Schnelle Multiplikation großer Zahlen. Computing 7, 281–292 (1971)
18. Schönhage, A.: Schnelle Multiplikation von Polynomen über Körpern der Charakteristik 2. Acta Inform. 7, 395–398 (1977)
19. Schwarz, Š.: On the reducibility of binomial congruences and on the bound of the least integer belonging to given exponent mod p. Časopis pro Pěstování Matematiky 74, 1–16 (1949), http://dml.cz/dmlcz/109143
20. Shoup, V.: On the deterministic complexity of factoring polynomials over finite fields. Information Processing Letters 33, 261–267 (1990)
21. Sze, T.W.: On solving univariate polynomial equations over finite fields and some related problem, preprint,
http://people.apache.org/~szetszwo/umd/papers/poly.pdf
22. Wiedemann, D.H.: Solving sparse linear equations over finite fields. IEEE Transactions on Information Theory 32, 54–62 (1986)

On Some Permutation Binomials of the Form $x^{\frac{2^n-1}{k}+1} + ax$ over $\mathbb{F}_{2^n}$: Existence and Count

Sumanta Sarkar[1], Srimanta Bhattacharya[2], and Ayça Çeşmelioğlu[3]

[1] Department of Computer Science, University of Calgary, Canada
sarkas@ucalgary.ca
[2] Applied Statistics Unit, Indian Statistical Institute, Kolkata, India
srimanta_r@isical.ac.in
[3] Faculty of Mathematics, Otto-von-Guericke University, Magdeburg, Germany
ayca.cesmelioglu@ovgu.de

Abstract. Based on a criterion of permutation polynomials of the form $x^r f(x^{\frac{q-1}{m}})$ by Wan and Lidl (1991) and some very elementary techniques we show existence of permutation binomials of the following forms

(i) $x(x^{\frac{2^n-1}{3}} + a) \in \mathbb{F}_{2^n}[x]$, for $n > 4$

(ii) $x^{\frac{2^{2n}-1}{2^n-1}+1} + ax = x^{2^n+2} + ax \in \mathbb{F}_{2^{2n}}[x]$, for $n \geq 3$.

In (i), we extend a result of Carlitz (1962) for even characteristic. Moreover we present the count of such permutation binomials when a is in a certain subfield of $\mathbb{F}_{2^n}$. In (ii), we reprove, using much simpler technique, a recent result of Charpin and Kyureghyan (2008) and give the number of permutation binomials of this form. Finally, we discuss some cryptographic relevance of these results.

Keywords: Finite field, permutation binomial, norm, trace, Boolean function.

1 Introduction

Let $\mathbb{F}_q$ denote the finite field of order q. A polynomial $F(x) \in \mathbb{F}_q[x]$ is called a *permutation polynomial* of $\mathbb{F}_q$ if the mapping $F : a \to F(a), a \in \mathbb{F}_q$, induced by $F(x)$ is a permutation of $\mathbb{F}_q$. Permutation polynomials have numerous applications in cryptography, coding theory and combinatorial designs. Finding new classes of permutation polynomials is therefore an interesting topic of research. The monomial $F(x) = x^d$ over $\mathbb{F}_q$ is a permutation polynomial of $\mathbb{F}_q$ if and only if $\gcd(d, q-1) = 1$. However, permutation polynomials which are not monomials are not easy to characterize.

In this paper we consider permutation binomials. These permutation polynomials are well studied, see [1,2,9,11,12,15]. In [1], Carlitz proved that for a sufficiently large field $\mathbb{F}_q$, there exists an element $a \in \mathbb{F}_q$ such that $x^{\frac{q-1}{3}+1} + ax$ is a permutation polynomial. However, the proof is not entirely transparent. In this paper we consider this problem over a finite field with characteristic 2, i.e.,

F. Özbudak and F. Rodríguez-Henríquez (Eds.): WAIFI 2012, LNCS 7369, pp. 236–246, 2012.
© Springer-Verlag Berlin Heidelberg 2012

$q = 2^n$. We prove that for any even $n > 4$, one can always find an element $a \in \mathbb{F}_{2^n}$ such that

$$x^{\frac{2^n-1}{3}+1} + ax$$

is a permutation polynomial of $\mathbb{F}_{2^n}$. Note that if $k = 3$, then $\frac{2^n-1}{3}$ is an integer if and only if n is even. Our proof is based on the following criterion of permutation polynomials of the form $Q(x) = x^r P(x^{\frac{q-1}{m}})$ given by Wan and Lidl in [14].

Theorem 1 ([14]). *Let m and n be two positive integers such that m divides $q - 1$. Let α be a primitive element in $\mathbb{F}_q$ and assume P is a polynomial in $\mathbb{F}_q[x]$. Then $Q(x) = x^r P(x^{\frac{q-1}{m}})$ is a permutation polynomial of $\mathbb{F}_q$ if and only if the following conditions are satisfied*

1. *$\gcd(r, \frac{q-1}{m}) = 1$,*
2. *for all i, $0 \le i < m$, $P(\alpha^{i\frac{q-1}{m}}) \ne 0$,*
3. *for all j, $0 \le i < j < m$, $Q(\alpha^i)^{\frac{q-1}{m}} \ne Q(\alpha^j)^{\frac{q-1}{m}}$.*

We note here that results on permutation binomials of the kind considered in this article have already been proved in substantial generality in [2], [11], [9]. However, all of these directly or indirectly use heavy tools from function fields, namely Weil's proof of Riemann hypothesis for function fields. However, our proof is very simple and direct. It uses, along with some elementary techniques of finite fields, the criterion of permutation polynomials of the form $x^r P(x^{\frac{q-1}{m}})$ given in [14], whose derivation is again very elementary. As a byproduct of this proof, we have been able to obtain the exact number of $a \in \mathbb{F}_{2^t}$ for which $x^{\frac{2^n-1}{3}+1} + ax$ is a permutation polynomial of $\mathbb{F}_{2^n}$, where $n = 2^i t$ and t is odd. Our result shows that there is a large number $(\frac{2^{t+1}-1}{3})$ of such a even in the subfield $\mathbb{F}_{2^t}$ of $\mathbb{F}_{2^n}$. This is encouraging from the perspective of searching for such permutation polynomials for the purposes of coding theory and cryptography. These results are presented in Section 2.

In Section 3, we completely characterize permutation polynomials of the form $x^{2^n+2} + ax$ over $\mathbb{F}_{2^{2n}}$. This has already been done in [3], and our result is a subresult of [3] in the sense that there the authors have additionally shown the nonexistence of permutation polynomials of the form $x^{2^k+2} + ax$ over $\mathbb{F}_{2^{2n}}$ apart from the case $k = n$. However, their proof is quite involved and makes use of results related to Walsh spectrum of quadratic Boolean functions. Whereas, similar to the previous case, our proof of this result is direct and elementary, and it is based on Theorem 1 along with some elementary techniques of finite fields.

In Section 4, we discuss the cryptographic relevance of these permutation binomials.

Throughout this article, we will make use of the following functions. For a detailed account of their properties we refer to [10].

The function *Norm* is defined as

$$N_{\mathbb{F}_{q^m}/\mathbb{F}_q}(\gamma) = \gamma^{\frac{q^m-1}{q-1}},$$

for $\gamma \in F_{q^m}$. The function *Trace* is defined as

$$Tr_{\mathbb{F}_{q^m}/\mathbb{F}_q}(\gamma) = \gamma + \gamma^q + \ldots + \gamma^{q^{m-1}},$$

for $\gamma \in F_{q^m}$.

Also, throughout this article ω denotes a primitive element of $\mathbb{F}_{2^2}$, i.e., $\omega^2 + \omega + 1 = 0$.

2   Existence of Permutation Binomial $x^{\frac{2^n-1}{3}+1} + ax$

We derive a criterion for permutation polynomials of the form $x^{\frac{2^n-1}{3}+1}+ax$. The generalization of this characterization can be found in [6, Theorem 3.7] and [13, Lemma 1]. Since we consider fields with characteristic 2, the characterization looks simpler.

Proposition 1. *Let $n = 2k$, $k > 2$ be any integer. Then*

$$W(x) = x(x^{\frac{2^n-1}{3}} + a)$$

is a permutation polynomial over $\mathbb{F}_{2^n}$ if and only if the elements

$$(1 + a)^{\frac{2^n-1}{3}}, \omega(\omega + a)^{\frac{2^n-1}{3}}, \omega^2(\omega^2 + a)^{\frac{2^n-1}{3}}$$

are all distinct.

Proof: Let α be a primitive element of $\mathbb{F}_{2^n}$, $P(x) = x + a, a \in \mathbb{F}_{2^n} \setminus \mathbb{F}_{2^2}$ be a polynomial in $\mathbb{F}_{2^n}[x]$ and $\omega = \alpha^{\frac{2^n-1}{3}}$. Since $\langle \omega \rangle = \mathbb{F}_{2^2}^*$ and $a \notin \mathbb{F}_{2^2}$, the first two conditions of Theorem 1 are satisfied.

Hence, $W(x)$ is a permutation polynomial if and only if the last condition of Theorem 1 is satisfied, which, by the fact that $W(\alpha^i)^{\frac{2^n-1}{3}} = \omega^i(\omega^i + a)^{\frac{2^n-1}{3}}, 0 \le i < 3$, is equivalent to the condition that the elements $(1 + a)^{\frac{2^n-1}{3}}, \omega(\omega + a)^{\frac{2^n-1}{3}}, \omega^2(\omega^2 + a)^{\frac{2^n-1}{3}}$ are all distinct.   □

Lemma 1. *Let $t \ge 3$ be an odd integer. Then for each $\delta \in \{0, 1\}$ there exists an element $a \in \mathbb{F}_{2^t}$ such that $N_{\mathbb{F}_{2^{2t}}/\mathbb{F}_{2^2}}(a + \omega) = \omega^{2^\delta}$.*

Proof: We show that on the coset $\mathbb{F}_{2^t} + \omega$ of the finite field $\mathbb{F}_{2^{2t}}$, the norm $N_{\mathbb{F}_{2^{2t}}/\mathbb{F}_{2^2}}$ takes the values ω, ω^2.

For each element $\alpha \in \mathbb{F}_{2^t}^*$,

$$N_{\mathbb{F}_{2^{2t}}/\mathbb{F}_{2^2}}(\alpha) = \prod_{j=0}^{t-1} \alpha^{4^j} = \alpha^{(4^t-1)/3} = \alpha^{(2^t-1)(2^t+1)/3} = 1,$$

since t is odd.

For $\beta \in \mathbb{F}_{2^t}, \alpha \in \mathbb{F}_{2^t}^*$, we have

$$N_{\mathbb{F}_{2^{2t}}/\mathbb{F}_{2^2}}(\beta + \alpha\omega) = N_{\mathbb{F}_{2^{2t}}/\mathbb{F}_{2^2}}(\alpha)N_{\mathbb{F}_{2^{2t}}/\mathbb{F}_{2^2}}(\alpha^{-1}\beta + \omega) = N_{\mathbb{F}_{2^{2t}}/\mathbb{F}_{2^2}}(\alpha^{-1}\beta + \omega),$$

since $N_{\mathbb{F}_{2^{2t}}/\mathbb{F}_{2^2}}(\alpha) = 1$. This means that the norm $N_{\mathbb{F}_{2^{2t}}/\mathbb{F}_{2^2}}$ restricted to any of the cosets $\mathbb{F}_{2^t} + \alpha\omega, \alpha \in \mathbb{F}_{2^t}^*$, takes the same values. $N_{\mathbb{F}_{2^{2t}}/\mathbb{F}_{2^2}}$ is onto $\mathbb{F}_{2^2}$ and takes the values $0, 1$ on $\mathbb{F}_{2^t}$. Therefore on each of the cosets $\mathbb{F}_{2^t} + \alpha\omega, \alpha \in \mathbb{F}_{2^t}^*$, $N_{\mathbb{F}_{2^{2t}}/\mathbb{F}_{2^2}}$ takes the values ω, ω^2.   □

We now show the existence of such permutation binomials for any even $n = 2^i t$, where t is odd. Before stating our result we note the following property of the *Norm* function.

Fact 1 Let $n = 2^i t$, where t is odd. Let $M_j = \{\alpha \in \mathbb{F}_{2^n}^* : N_{\mathbb{F}_{2^n}/\mathbb{F}_{2^2}}(\alpha) = \omega^j\}$, where $j = 0, 1$ and 2. Then $|M_0| = |M_1| = |M_2| = (2^n - 1)/3$.

Theorem 2. *For every even integer $n = 2^i t$, where $t > 2$ is odd, a permutation binomial of the form $x(x^{\frac{2^n-1}{3}} + a) \in \mathbb{F}_{2^n}[x]$ always exists.*

Proof: Let $q = 2^t$ and $m = 2^i$. We show that there exists $a \in \mathbb{F}_q$ such that the elements $(1 + a)^{\frac{q^m-1}{3}}, \omega(\omega + a)^{\frac{q^m-1}{3}}, \omega^2(\omega^2 + a)^{\frac{q^m-1}{3}} \in \mathbb{F}_{2^2}^*$ are all distinct.

By Lemma 1 we know that for each $\delta \in \{0, 1\}$ there exists $a \in \mathbb{F}_q$ such that

$$(a + \omega)^{\frac{q^2-1}{3}} = \omega^{2^\delta}. \tag{1}$$

From Fact 1, there are $(2^t - 1)(2^t + 1)/3$ elements $\beta + \alpha\omega \in \mathbb{F}_{2^{2t}}, \alpha \in \mathbb{F}_{2^t}^*, \beta \in \mathbb{F}_{2^t}$ such that $N_{\mathbb{F}_{2^{2t}}/\mathbb{F}_{2^2}}(\beta + \alpha\omega) = \omega$. Note that if $N_{\mathbb{F}_{2^{2t}}/\mathbb{F}_{2^2}}(1 + \omega) = \omega$, then $N_{\mathbb{F}_{2^{2t}}/\mathbb{F}_{2^2}}(\alpha + \alpha\omega) = \omega$ for $\alpha \in \mathbb{F}_{2^t}^*$. However, one can get only $2^t - 1$ such elements of $\mathbb{F}_{2^{2t}}$ for which the norm $N_{\mathbb{F}_{2^{2t}}/\mathbb{F}_{2^2}}(\cdot)$ is ω. Now, since $t \geq 3$, it follows that there must be an element $\beta + \alpha\omega, \beta \neq \alpha$ such that $N_{\mathbb{F}_{2^{2t}}/\mathbb{F}_{2^2}}(\beta + \alpha\omega) = \omega$. The same is true for the case $N_{\mathbb{F}_{2^{2t}}/\mathbb{F}_{2^2}}(\beta + \alpha\omega) = \omega^2$.

Therefore, for each $\delta \in \{0, 1\}$ there exists an element $a \neq 1$ in $\mathbb{F}_{2^t}$ such that $(a + \omega)^{\frac{q^2-1}{3}} = \omega^{2^\delta}$. For such a we have

$$\left((a + \omega)^{\frac{q^2-1}{3}}\right)^q = \left(\omega^{2^\delta}\right)^q$$
$$(a^q + \omega^q)^{\frac{q^2-1}{3}} = w^{2^{\delta+t}}$$
$$(a + \omega^2)^{\frac{q^2-1}{3}} = w^{2^{\delta+t}}, \tag{2}$$

since $a \in \mathbb{F}_q$ and t is odd.
And we also have

$$(1 + a)^{\frac{q^m-1}{3}} = N_{\mathbb{F}_{q^m}/\mathbb{F}_4}(1 + a)$$
$$= N_{\mathbb{F}_{q^2}/\mathbb{F}_4}(N_{\mathbb{F}_{q^m}/\mathbb{F}_{q^2}}(1 + a))$$
$$= \left(N_{\mathbb{F}_{q^2}/\mathbb{F}_4}(1 + a)\right)^{2^{i-1}}$$
$$= 1, \text{ since } (1 + a) \in \mathbb{F}_q \text{ and } a \neq 1.$$

By Equation (1) and (2) we have

$$\omega N_{\mathbb{F}_{q^m}/\mathbb{F}_4}(a+\omega) = \omega N_{\mathbb{F}_{q^2}/\mathbb{F}_4}(N_{\mathbb{F}_{q^m}/\mathbb{F}_{q^2}}(a+\omega))$$

$$= \omega \left(N_{\mathbb{F}_{q^2}/\mathbb{F}_4}(a+\omega)\right)^{2^{i-1}}$$

$$= \omega((a+\omega)^{\frac{q^2-1}{3}})^{2^{i-1}}$$

$$= \omega^{2^{\delta+i-1}+1},$$

$$\omega^2 N_{\mathbb{F}_{q^m}/\mathbb{F}_4}(a+\omega^2) = \omega^2 N_{\mathbb{F}_{q^2}/\mathbb{F}_4}(N_{\mathbb{F}_{q^m}/\mathbb{F}_{q^2}}(a+\omega^2))$$

$$= \omega^2 \left(N_{\mathbb{F}_{q^2}/\mathbb{F}_4}(a+\omega^2)\right)^{2^{i-1}}$$

$$= \omega^2((a+\omega^2)^{\frac{q^2-1}{3}})^{2^{i-1}}$$

$$= (\omega^{2^{\delta+i-1}+1})^2.$$

Now, we need $\omega^{2^{i+\delta-1}+1} \neq 1$, since it would automatically imply $\omega^{2^{i+\delta-1}+1} = \omega$ or ω^2 and we will be done. Note that $\omega^{2^{i+\delta-1}+1} \neq 1$ if and only if $i + \delta$ is odd. So, depending on i, we choose a in (1) to make $i + \delta$ odd, i.e., we choose a such that:

$$\delta := \begin{cases} 0 & \text{if } i \text{ is odd} \\ 1 & \text{if } i \text{ is even.} \end{cases}$$

$\square$

Next we prove the existence for $n = 2^i$, where $i \geq 3$.

Theorem 3. *For every integer $n = 2^i$, where $i > 2$, a permutation binomial of the form $x(x^{\frac{2^n-1}{3}} + a) \in \mathbb{F}_{2^n}[x]$ always exists.*

Proof: It is enough to show that for $n = 2^i$, $i > 2$ there always exists $a \in \mathbb{F}_{2^n} \backslash \mathbb{F}_{2^2}$ such that

$$\{(1+a)^{\frac{2^n-1}{3}}, \omega(\omega+a)^{\frac{2^n-1}{3}}, \omega^2(\omega^2+a)^{\frac{2^n-1}{3}}\}$$

are all distinct. We do that in the following two cases.

Case 1: $i = 3$, i.e., $n = 8$ and the field is $\mathbb{F}_{2^8}$.

By direct computation it can be shown that there exists an $a \in \mathbb{F}_{2^8} \backslash \mathbb{F}_{2^2}$ such that $N_{\mathbb{F}_{2^8}/\mathbb{F}_{2^2}}(1+a) = 1$, $N_{\mathbb{F}_{2^8}/\mathbb{F}_{2^2}}(\omega+a) = 1$ and $N_{\mathbb{F}_{2^8}/\mathbb{F}_{2^2}}(\omega^2+a) = 1$. Therefore, $\omega N_{\mathbb{F}_{2^8}/\mathbb{F}_{2^2}}(\omega+a) = \omega$ and $\omega^2 N_{\mathbb{F}_{2^8}/\mathbb{F}_{2^2}}(\omega^2+a) = \omega^2$. Thus

$$\{(1+a)^{\frac{2^n-1}{3}}, \omega(\omega+a)^{\frac{2^n-1}{3}}, \omega^2(\omega^2+a)^{\frac{2^n-1}{3}}\} = \{1, \omega, \omega^2\}.$$

Case 2: $i \geq 4$.

Note that if $\beta \in \mathbb{F}_{2^8}$ then

$$N_{\mathbb{F}_{2^{2i}}/\mathbb{F}_{2^2}}(\beta) = N_{\mathbb{F}_{2^8}/\mathbb{F}_{2^2}}(N_{\mathbb{F}_{2^{2i}}/\mathbb{F}_{2^8}}(\beta))$$

$$= N_{\mathbb{F}_{2^8}/\mathbb{F}_{2^2}}(\beta^{2^{i-3}})$$

$$= (N_{\mathbb{F}_{2^8}/\mathbb{F}_{2^2}}(\beta))^{2^{i-3}}.$$

This implies that for a obtained in Case 1 we have

$$N_{\mathbb{F}_{2^{2i}}/\mathbb{F}_{2^2}}(1+a) = 1, N_{\mathbb{F}_{2^{2i}}/\mathbb{F}_{2^2}}(\omega+a) = 1, N_{\mathbb{F}_{2^{2i}}/\mathbb{F}_{2^2}}(\omega^2+a) = 1.$$

Consequently,

$$\{(1+a)^{\frac{2^n-1}{3}}, \omega(\omega+a)^{\frac{2^n-1}{3}}, \omega^2(\omega^2+a)^{\frac{2^n-1}{3}}\} = \{1, \omega, \omega^2\}.$$

Hence the result follows. □

The proof of Theorem 3 shows a way to construct exceptional polynomials.

Definition 1. *A permutation polynomial $F(x)$ defined over $\mathbb{F}_{2^t}$ that also permutes infinitely many extensions of $\mathbb{F}_{2^t}$ is called exceptional polynomial.*

Corollary 1. *The permutation polynomial $x(x^{\frac{2^n-1}{3}}+a) \in \mathbb{F}_{2^{2^3}}[x]$ is an exceptional polynomial.*

Proof: It follows from the proof of Theorem 3 that permutation polynomial $x(x^{\frac{2^n-1}{3}}+a) \in \mathbb{F}_{2^{2^3}}[x]$ is also a permutation polynomial over the extension $\mathbb{F}_{2^{2^i}}$ for all $i \geq 4$. □

We now obtain the number of $a \in \mathbb{F}_{2^t}$ such that $x(x^{\frac{2^n-1}{3}}+a) \in \mathbb{F}_{2^t}[x]$ permutes the field $\mathbb{F}_{2^{2^i}t}$, where t is odd.

Proposition 2. *The number of permutation binomials of the form $x(x^{\frac{2^n-1}{3}}+a) \in \mathbb{F}_{2^{2^i}t}[x]$, where $a \in \mathbb{F}_{2^t}$ and t odd is*

$$(2^{t+1}-1)/3.$$

Proof: Let $n = 2^i t$. We count the number of $a \in \mathbb{F}_{2^t}$ such that

$$\{N_{\mathbb{F}_{2^n}/\mathbb{F}_{2^2}}(1+a), \omega N_{\mathbb{F}_{2^n}/\mathbb{F}_{2^2}}(\omega+a), \omega^2 N_{\mathbb{F}_{2^n}/\mathbb{F}_{2^2}}(\omega^2+a)\}$$

are all distinct. This is possible only in the following two ways.

1. $N_{\mathbb{F}_{2^n}/\mathbb{F}_{2^2}}(a+1) = 1, N_{\mathbb{F}_{2^n}/\mathbb{F}_{2^2}}(a+\omega) = 1$ and $N_{\mathbb{F}_{2^n}/\mathbb{F}_{2^2}}(a+\omega^2) = 1$,
2. $N_{\mathbb{F}_{2^n}/\mathbb{F}_{2^2}}(a+1) = 1, N_{\mathbb{F}_{2^n}/\mathbb{F}_{2^2}}(a+\omega) = \omega$ and $N_{\mathbb{F}_{2^n}/\mathbb{F}_{2^2}}(a+\omega^2) = \omega^2$,

Note that the existence of such an a has been shown in Theorem 2.

We observe that $N_{\mathbb{F}_{2^n}/\mathbb{F}_{2^2}}(a+\omega^2) = N_{\mathbb{F}_{2^n}/\mathbb{F}_{2^2}}((a+\omega)^{2^t})$ for odd t. Then $N_{\mathbb{F}_{2^n}/\mathbb{F}_{2^2}}(a+\omega^2) = (N_{\mathbb{F}_{2^n}/\mathbb{F}_{2^2}}(a+\omega))^{2^t}$. So for the first case it is enough to count the number of $a \in \mathbb{F}_{2^t}$ such that $N_{\mathbb{F}_{2^n}/\mathbb{F}_{2^2}}(a+\omega) = 1$. For any $a \in \mathbb{F}_{2^t}^*$, we have $N_{\mathbb{F}_{2^n}/\mathbb{F}_{2^2}}(a) = 1$ and $N_{\mathbb{F}_{2^n}/\mathbb{F}_{2^2}}(a) = N_{\mathbb{F}_{2^{2t}}/\mathbb{F}_{2^2}}(N_{\mathbb{F}_{2^{2^i}t}/\mathbb{F}_{2^{2t}}}(a)) = N_{\mathbb{F}_{2^{2t}}/\mathbb{F}_{2^2}}(a^{2^{i-1}}) = N_{\mathbb{F}_{2^{2t}}/\mathbb{F}_{2^2}}(a)^{2^{i-1}} = 1$. So $|\{a \in \mathbb{F}_{2^t} : N_{\mathbb{F}_{2^n}/\mathbb{F}_{2^2}}(a+1) = 1\}| = 2^t - 1$.

$|\{b \in \mathbb{F}_{2^{2t}} : N_{\mathbb{F}_{2^{2t}}/\mathbb{F}_{2^2}}(b) = 1\}| = (2^{2t}-1)/3$. Therefore, $|\{b \in \mathbb{F}_{2^{2t}} \setminus \mathbb{F}_{2^t} : N_{\mathbb{F}_{2^{2t}}/\mathbb{F}_{2^2}}(b) = 1\}| = (2^{2t}-1)/3 - (2^t-1) = (2^t-1)(2^t-2)/3$. It can be shown that $N_{\mathbb{F}_{2^{2t}}/\mathbb{F}_{2^2}}$ maps equal number of elements in every coset $\mathbb{F}_{2^t} + \alpha\omega$, $\alpha \in \mathbb{F}_{2^t}^*$ to a fixed element of $\mathbb{F}_{2^2}$. The number of such cosets is $2^t - 1$. Considering the coset $\mathbb{F}_{2^t} + \omega$, we have

$$|a \in \mathbb{F}_{2^t} : N_{\mathbb{F}_{2^{2t}}/\mathbb{F}_{2^2}}(a + \omega) = 1| = (2^t - 2)/3.$$

On the other hand, we also have

$$|a \in \mathbb{F}_{2^t} : N_{\mathbb{F}_{2^{2t}}/\mathbb{F}_{2^2}}(a + \omega) = \omega| = (2^{2t} - 1)/3 \cdot \frac{1}{2^t - 1} = (2^t + 1)/3.$$

Therefore, the total number of permutations with $a \in \mathbb{F}_{2^t}$ is $(2^t - 2)/3 + (2^t + 1)/3 = (2^{t+1} - 1)/3$.   □

3 Permutation Polynomials of the Form $x^{2^n+2} + ax$ over $\mathbb{F}_{2^{2n}}$

Our next result, pertaining to permutation polynomials of the form $x^{2^n+2} + ax \in \mathbb{F}_{2^{2n}}[x]$, has already been proven in [3]. There, classification of this type of permutation polynomials was used to prove results on cubic monomial bent functions of Maiorana-McFarland class. However, the method used there is rather complex and is based on the results related to Walsh spectrum of quadratic boolean functions. We derive the same result here using a direct application of Theorem 1 along with some very elementary techniques. Our proof is rather short and simple and once again demonstrates the effectiveness of Theorem 1 in obtaining exact results of this type.

Theorem 4. *Let $n \geq 3$ and $a \in \mathbb{F}_{2^{2n}}^*$. Then for odd n, $x^{2^n+2} + ax \in \mathbb{F}_{2^{2n}}[x]$ is a permutation polynomial if and only if $a^{2^n-1} \in \{\omega, \omega^2\}$. For n even there is no such permutation polynomial of this form.*

Proof: First, note that for $f(x) = x^{2^n+2} + ax$ to be a permutation polynomial it is necessary that $a \in \mathbb{F}_{2^{2n}}^* \setminus \mathbb{F}_{2^n}^*$. For, otherwise $f(x) = 0$ has two distinct solutions in $\mathbb{F}_{2^n}$.

We observe that $x^{2^n+2} + ax = x(x^{\frac{2^{2n}-1}{2^n-1}} + a)$. So we check conditions of Theorem 1 to find if this polynomial is a permutation polynomial. The first condition is trivially satisfied. Let α be a primitive element of $\mathbb{F}_{2^{2n}}$. Then for $0 \leq i < 2^n - 1$, $\alpha^{(2^n+1)i} \in \mathbb{F}_{2^n}^*$. So by the observation of the above paragraph, it follows that the second condition is also satisfied.

So $f(x)$ is a permutation polynomial if and only if the last condition is satisfied, where we need to show that $f(\alpha^i)^{2^n+1}$ is $1 - 1$ for a primitive element α of $\mathbb{F}_{2^{2n}}$, and for $0 \leq i < 2^n - 1$. Here we note that for $0 \leq i < 2^n - 1$, $(\alpha^{2^n+1})^i$ runs through all the elements of $\mathbb{F}_{2^n}^*$. Hence, it remains to show that $g(v) = v(v + a)^{2^n+1}$ induces a $1 - 1$ mapping for $v \in \mathbb{F}_{2^n}^*$. Considering $v \in \mathbb{F}_{2^n}^*$, the last condition (after some simplification) is equivalent to showing that $g(v) = v^3 + (a^{2^n} + a)v^2 + a^{2^n+1}v$ induces $1 - 1$ mapping for $v \in \mathbb{F}_{2^n}^*$. Note that

$g(v) \in \mathbb{F}_{2^n}[v]$ as $(a^{2^n}+a)^{2^n} = a^{2^n}+a$ and $(a^{2^n+1})^{2^n-1} = 1$ are both in $\mathbb{F}_{2^n}$. We simplify further by making the substitution $v = u + a^{2^n} + a$ in $g(v)$. Hence the final condition is that $f(x)$ is a permutation polynomial if and only if the mapping $h(u) = u^3 + (a^{2^{n+1}} + a^2 + a^{2^n+1})u$ is $1-1$ for $u \in \mathbb{F}_{2^n} \setminus \{a^{2^n}+a\}$. Next, we consider the following two cases.

1. For n odd, we claim that $h(u)$ is $1-1$ if and only if $a^{2^{n+1}}+a^2+a^{2^n+1}=0$. For if the condition holds then $h(u) = u^3$ is $1-1$ because $\gcd(3, 2^n-1) = 1$ for odd n. If $a^{2^{n+1}}+a^2+a^{2^n+1} \neq 0$, then $h(u) = 0$ for $u = 0$ and $u = a^{2^n} + a + a^{2^{n-1}+2^{n-1}}$. Which means $g(v_1) = g(v_2)$ for $v_1 = a^{2^n} + a \in \mathbb{F}_{2^n}^*$ and $v_2 = a^{2^{2n-1}+2^{n-1}} = (a^{2^n+1})^{2^{n-1}} \in \mathbb{F}_{2^n}^*$.
   Note that $a^{2^{n+1}}+a^2+a^{2^n+1} = a^2((a^{2^n-1})^2 + a^{2^n-1} + 1)$. Therefore, $a^{2^{n+1}}+a^2+a^{2^n+1} = 0$ if and only if a^{2^n-1} is a root of the equation $y^2 + y + 1 = 0$ in $\mathbb{F}_{2^{2n}}$, i.e., $a^{2^n-1} \in \{\omega, \omega^2\}$.
2. For n even, if $a^{2^{n+1}}+a^2+a^{2^n+1} \neq 0$ then by the same argument as above it follows that $h(u)$ is not $1-1$ for $u \in \mathbb{F}_{2^n} \setminus \{a^{2^n}+a\}$. If $a^{2^{n+1}}+a^2+a^{2^n+1} = 0$, then also $h(u) = u^3$ is not $1-1$, because for even n, $3 \mid 2^n-1$. $\qquad\square$

Next corollary follows from the above theorem by noting that $a^{2^n-1} = \omega$ (ω^2) if and only $a \in \omega^2\mathbb{F}_{2^n}^*$ $(\omega\mathbb{F}_{2^n}^*)$.

Corollary 2. *Let n be odd, then $x^{2^n+2} + ax \in \mathbb{F}_{2^{2n}}[x]$ is a permutation polynomial if and only if $a \in \omega\mathbb{F}_{2^n}^* \bigcup \omega^2\mathbb{F}_{2^n}^*$, where $\omega^2 + \omega + 1 = 0$ in $\mathbb{F}_{2^{2n}}$. Hence the number of such a is exactly $2(2^n-1)$.*

4 Cryptographic Relevance of Permutation Polynomials

Permutation polynomials have substantial applications in cryptography. For example, in block ciphers, permutation polynomials are used as S-boxes which are important building blocks. In this section, we emphasize the influence of the permutation polynomial $x(x^{\frac{2^n-1}{3}} + a)$ in cryptography.

A block cipher can be viewed as a mapping $E_k : \mathbb{F}_{2^n} \to \mathbb{F}_{2^m}$, where k is chosen from the key-space.

Definition 2. *Let $F : \mathbb{F}_{2^n} \to \mathbb{F}_{2^m}$. For $a \in \mathbb{F}_{2^n}$, the function $D_a F$ given by*

$$D_a F(x) = F(x) + F(x+a), \text{ for all } x \in \mathbb{F}_{2^n}$$

is called the derivative *of F in the direction of a. Further, $a \in \mathbb{F}_{2^n}^*$ is said to be a* linear structure *of F if the function $D_a F$ is constant.*

In [7], Evertse showed an attack on block ciphers by exploiting their linear structures. This brings the motivation to study the linear structures of a polynomial and has been studied in [5]. Later on in [4], Charpin and Sarkar studied permutation polynomials with linear structures. A polynomial of the form

$$L(x) = \sum_{k=1}^{n-1} c_k x^{2^k}, \ c_k \in \mathbb{F}_{2^n} \text{ not all zero, is called a linearized polynomial and for}$$

any constant $A \in \mathbb{F}_{2^n}$, the polynomial $L(x) + A$ is called affine. It is clear that any $a \in \mathbb{F}_{2^n}^*$ is a linear structure of affine and linearized polynomials. Note that a polynomial $F(x)$ has linear structure if and only if $F(x) + L(x)$ has a linear structure. We state a result from [4] on polynomials which cannot have a linear structure.

Theorem 5. *[4, Theorem 4] Let $\alpha \in \mathbb{F}_{2^n}$. Let r and s be integers such that*

$$1 \leq r, s \leq 2^n - 2.$$

Define the functions over $\mathbb{F}_{2^n}$

$$F(x) = \lambda(x^r + \alpha x^s) + L(x), \ \lambda \in \mathbb{F}_{2^n}^*, \tag{3}$$

where L is any affine function. Then such a function F has no linear structure unless it is affine.

Corollary 3. *The permutation polynomials of the form $x^{\frac{2^n-1}{3}+1} + ax$ do not have any linear structure.*

4.1   Boolean Functions from Permutation Polynomials

Any polynomial $F(x) \in \mathbb{F}_{2^n}[x]$ gives rise to the Boolean function $f : \mathbb{F}_{2^n} \rightarrow \mathbb{F}_2$ defined by $f(x) = Tr_{\mathbb{F}_{2^n}/\mathbb{F}_2}(F(x))$. The so-called *component* functions $f_\lambda(x)$ of a polynomial $F(x)$ are defined by $f_\lambda(x) = Tr_{\mathbb{F}_{2^n}/\mathbb{F}_2}(\lambda F(x))$ for all $\lambda \in \mathbb{F}_{2^n}^*$. Boolean functions which have good cryptographic properties are used in LFSR based stream ciphers. To maintain the pseudorandomness of the cipher sequence, the Boolean function is required to be balanced.

Definition 3. *A Boolean function $f : \mathbb{F}_{2^n} \rightarrow \mathbb{F}_2$ is said to be balanced if $|\{x \in \mathbb{F}_{2^n} : f(x) = 1\}| = 2^{n-1}$.*

The relation between a permutation polynomial and a balanced Boolean function is exhibited in the following result.

Theorem 6. *[10, Theorem 7.7] The polynomial $F(x)$ over $\mathbb{F}_{2^n}$ is a permutation if and only all its component functions*

$$f_\lambda(x) = Tr_{\mathbb{F}_{2^n}/\mathbb{F}_2}(\lambda F(x)), \quad \lambda \in \mathbb{F}_{2^n}^*,$$

are balanced.

Therefore we have the following result.

Corollary 4. *For even $n > 4$, there exists an $a \in \mathbb{F}_{2^n}^*$ such that the Boolean function $Tr_{\mathbb{F}_{2^n}/\mathbb{F}_2}(x^{\frac{2^n-1}{3}+1} + ax)$ is balanced.*

For some n, we have $\gcd(\frac{2^n-1}{3} + 1, 2^n - 1) = 1$. Then the above result partially supports the conjecture proposed by Helleseth [8].

Conjecture 1. [8] For all integer k coprime with $2^n - 1$, there exists $a \in \mathbb{F}_{2^n}^*$ such that $Tr_{\mathbb{F}_{2^n}/\mathbb{F}_2}(x^k + ax)$ is a balanced Boolean function.

Nonlinearity is another important cryptographic property of Boolean functions. A linear Boolean function is of the form $Tr_{\mathbb{F}_{2^n}/\mathbb{F}_2}(\lambda x)$. The minimum among all the distances from a Boolean function f to all the linear functions is the measure of nonlinearity of f. To resist affine approximation attack against a stream cipher better, the Boolean function should have high nonlinearity. For even n, the maximum nonlinearity is $2^{n-1} - 2^{n/2-1}$ and the functions that achieve this nonlinearity are called *bent functions*. Maiorana-McFarland function is an important class of bent functions.

Let $n = 2t$. Let us consider a Boolean function f defined by

$$f : (x, y) \in \mathbb{F}_{2^t} \times \mathbb{F}_{2^t} \mapsto Tr_{\mathbb{F}_{2^t}/\mathbb{F}_2}(x\pi(y) + h(y)) \tag{4}$$

where both π and h are functions on $\mathbb{F}_{2^t}$. Then, f is a bent function if and only if π is a bijection. In this case, f is said to belong to the class of Maiorana-McFarland bent functions.

In [4, Theorem 7] it was proved that a Maiorana-McFarland bent function of the form (4) has affine derivative if and only if the corresponding permutation polynomial $\pi(y)$ does not have any linear structure. As we have seen in Corollary 3, $\pi(y) = y^{\frac{2^n-1}{3}+1} + ay$ does not have any linear structure, therefore, the function given by (4) does not have any affine derivative. We state this result in the following.

Theorem 7. *Let $t > 4$ be even and $a \in \mathbb{F}_{2^t}^*$ such that $\pi(y) = y^{\frac{2^t-1}{3}+1} + ay$ is a permutation polynomial over $\mathbb{F}_{2^t}$. Then the Maiorana-McFarland bent function given by*

$$f : (x, y) \in \mathbb{F}_{2^t} \times \mathbb{F}_{2^t} \mapsto Tr_{\mathbb{F}_{2^t}/\mathbb{F}_2}(x\pi(y) + h(y)),$$

where h is any function on $\mathbb{F}_{2^t}$, does not have any affine derivative.

5   Conclusions

Our main contribution in this article is the derivation of existential and counting results of certain classes of permutation binomials using elementary techniques. Our methods are much simpler than the ones given in [3, 6, 11, 13]. It would be interesting as well as challenging to explore the effectiveness of these techniques for other classes of permutation polynomials.

Acknowledgment. The authors would like to thank the anonymous reviewers for pointing out the references [6, 12, 13, 15].

References

1. Carlitz, L.: Some theorems on permutation polynomials. Bull. Amer. Math. Soc. 68(2), 120–122 (1962)

2. Carlitz, L., Wells, C.: The number of solutions of a special system of equations in a finite field. Acta Arithmetica 12, 77–84 (1966)
3. Charpin, P., Kyureghyan, G.M.: Cubic monomial bent functions: a subclass of $\mathcal{M}^*$. Siam J. Disc. Math. 22(2), 650–665 (2008)
4. Charpin, P., Sarkar, S.: Polynomials with linear structure and Maiorana-McFarland construction. IEEE Trans. on Inf. Theory 57(6), 3796–3804 (2011)
5. Dubuc, S.: Characterization of linear structures. Des. Codes Cryptography 22, 33–45 (2001)
6. Evans, A.B.: Orthomorphism graphs of groups. Lecture Notes in Mathematics, vol. 1535. Springer, Berlin (1992)
7. Evertse, J.-H.: Linear Structures in Block Ciphers. In: Price, W.L., Chaum, D. (eds.) EUROCRYPT 1987. LNCS, vol. 304, pp. 249–266. Springer, Heidelberg (1988)
8. Helleseth, T.: Some results about the cross-correlation function between two maximal linear sequences. Discrete Math. 16(3), 209–232 (1976)
9. Laigle-Chapuy, Y.: Permutation polynomials and applications to coding theory. Finite Fields and Their Applications 13, 58–70 (2007)
10. Lidl, R., Niederreiter, H.: Finite Fields, 2nd edn. Encyclopedia of Mathematics and its Applications, vol. 20. Cambridge University Press (1997)
11. Masuda, A.M., Zieve, M.E.: Permutation binomials over finite fields. Trans. of the American Mathematical Society 361(8), 4169–4180 (2009)
12. Niederreiter, H., Robinson, K.H.: Complete mappings of finite fields. J. Australian Math. Soc. Ser. A 33, 197–212 (1982)
13. Niederreiter, H., Winterhof, A.: Cyclotomic $\mathscr{R}$-orthomorphisms of finite fields. Discrete Math. 295, 161–171 (2005)
14. Wan, D.Q., Lidl, R.: Permutation polynomials of the form $x^r f(x^{\frac{q-1}{m}})$ and their group structure. Monatsh. Math. 112(2), 149–163 (1991)
15. Turnwald, G.: Permutation polynomials of binomial type. In: Contributions to General Algebra, Teubner, Stuttgart, vol. 6, pp. 281–286 (1988)

Author Index